*March, Peyton C.; Harrison, P.C.*

# The Airplane Propeller

*Inktank publishing*

*March, Peyton C.; Harrison, P.C.*

**The Airplane Propeller**

*Inktank publishing, 2018*

*www.inktank-publishing.com*

*ISBN/EAN: 9783747756430*

 

# THE

# AIRPLANE PROPELLER

PREPARED BY THE PROPELLER SECTION
ENGINEERING DIVISION, McCOOK FIELD
DAYTON, OHIO

January, 1920

WASHINGTON
GOVERNMENT PRINTING OFFICE
1921

WAR DEPARTMENT,
WASHINGTON, *January 24, 1920.*

The following publication, entitled "The Airplane Propeller," is published for the information and guidance of all concerned.
[062.11, A. G. O.]

BY ORDER OF THE SECRETARY OF WAR:

PEYTON C. MARCH,
*General, Chief of Staff.*

OFFICIAL:
P. C. HARRIS,
*The Adjutant General.*

3

# TABLE OF CONTENTS.

# BUREAU OF AIRCRAFT PRODUCTION.

## SPECIFICATIONS PERTAINING TO PROPELLERS.

| Specification No. | Subject. |
|---|---|
| 11040 | Copper sheet. |
| 11043A | Soft brass sheet. |
| 11045 | Monel metal sheet or strip. |
| 14000B | Hide glue certified for use in airplane construction. |
| 14001C | Spar varnish for use in airplane construction. |
| 14008A | Liquid filler and priming varnish for finishing airplane propellers. |
| 14012 | Orange shellac varnish. |
| 14019 | Paste wood filler. |
| 14020A | Casein glue certified for use in airplane construction. |
| 14024 | Application of certified casein joint glue. |
| 15002C | Specific gravities of woods used in aircraft construction. |
| 15020B | Properties of woods. |
| 15022C | Woods used in propeller construction. |
| 15026B | Birch lumber for use in propeller construction. |
| 15028A | Mahogany lumber for use in propeller construction. |
| 15029B | Oak lumber for use in propeller construction. |
| 15030B | Walnut lumber for use in propeller construction. |
| 15031B | Cherry lumber for use in propeller construction. |
| 15032A | Maple lumber for use in propeller construction. |
| 15037B | Kind and character of wood for various aircraft parts. |
| 15039A | Birch lumber for use in spliced and jointed propellers. |
| 15040A | Cherry lumber for use in spliced and jointed propellers. |
| 15041A | Oak lumber for use in spliced and jointed propellers. |
| 15042A | Mahogany lumber for use in spliced and jointed propellers. |
| 15043A | Walnut lumber for use in spliced and jointed propellers. |
| 15050 | Poplar lumber for use in propeller construction. |
| 16003B | Linen for airplane covering. |
| 16004A | Mercerized cotton airplane fabric (grade A). |
| 16005A | Mercerized cotton airplane fabric (grade B). |
| 20500A | General kiln-drying process for airplane stock. |
| 20504B | The determination of moisture content in wood. |
| 20505A | The determination of specific gravity of wood. |
| 20506 | The determination of specific gravity of wood. |
| 21001C | Handling and testing of hide glue. |
| 24109A | Pigskin for use in tipping propellers. |
| 29500D | Construction of airplane propellers. |
| 29511 | Application of aluminum leaf finish to airplane propellers. |
| 29512A | Propeller for wind-driven airplane generators. |
| 40538B | Packing airplane propellers for export shipment. |
| 40545 | Boxing wood propellers for generators for export shipment. |

# THE AIRPLANE PROPELLER.

## CHAPTER I.

### PROPELLER DESIGN AND LAYOUT.

1. *General.*—There are three methods of designing propellers: First, by building up several true screws and testing these one after another on the plane in flight; second, by making a small model, running a wind tunnel test on the same, and later scaling this model up until it gives the desired results when built as a propeller for an airplane; third, by a purely theoretical method guided by empirical data obtained from flight tests. This latter method is the one dealt with in the following. In this method, the propeller blade is considered as a series of wing sections similar to those of an airplane, and having the same lift, drift, and angle of attack characteristics, etc., but with the difference that the direction of movement of the propeller blade from tip to root is continually varying as is also the shape of the blade sections.

The blade, therefore, is considered as divided in units, usually 6 inches in length, and each of these units treated as a separate air foil having known values for lift and $L/D$, a known direction of motion and a known angle of attack. By assigning a unit length to such an air foil with air forces acting upon it, it is possible to compute the components of thrust and torque per unit of length for each of the propeller-blade sections or separate air foils under consideration. From this data curves can be constructed, one showing the thrust as ordinates and the radius as abscissae and a second giving the torque or horsepower absorbed as ordinates and the radius as abscissae. Then, by integrating these two curves, the total power absorbed and the total thrust delivered by the propeller can be obtained. Two such curves are shown in figure 1.

After making the above analysis, the horsepower delivered by the propeller can be obtained by multiplying its total thrust delivered in pounds by the airplane speed in feet per second and dividing by 550. The horsepower delivered by the propeller divided by the horsepower absorbed gives the efficiency of the propeller. It is found in practice, however, that the results obtained by this method, which is sometimes known as the Drzwiecki method, always gives efficiencies by far too high. The reason for this is that the effect of the slip stream or cylinder column of air acted upon by the propeller and parallel to its axis is not taken into account directly. As the velocity of the slip stream is higher than the forward speed of the plane, this results in a blast of air being driven in a backward direction as the plane moves forward. It follows, therefore, that the actual forward speed of the plane relative to still air is not as great as the speed of the airplane when referred to the slip stream.

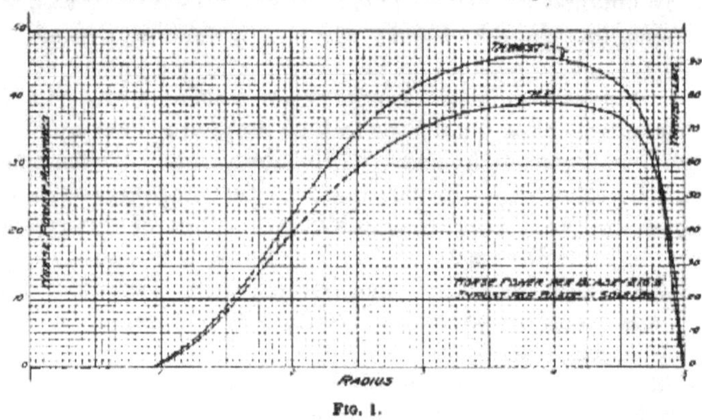


FIG. 1.

In general, the component of the velocity of the slip stream parallel to the direction of flight can be obtained from the following approximate formula:

$$T=\frac{\rho}{g} A\ V_1\ v \quad \left(\text{strictly speaking, } T=\frac{\rho}{g}A(V_1+v)V_1\right). \qquad \text{I}$$

Where—

$T$ = thrust of propeller in pounds.

$\frac{\rho}{g}$ = weight of one cubic foot of air divided by the acceleration due to gravity.

$A$ = area of slip stream in square feet as it leaves the propeller.

$V_1$ = velocity of the airplane in feet per second.

$v$ = component of the velocity of the slip stream parallel to the direction of motion and relative to the air.

The thrust of the propeller which is used in the preceding formula is equal to the product of the horsepower, the efficiency of the propeller and 550 divided by the velocity of the plane in feet per second.

There is a good deal of doubt as to the true area of the slip stream immediately behind the propeller and also as to what per cent of the ultimate velocity of the slip stream is acquired before the air reaches the disk of the propeller. Good results are obtained by considering that two-thirds of the velocity of the slip stream—i. e., $v$ in the above formula—occurs before the disk of the propeller, and that the area $A$ of the slip stream is equal to the propeller disk area less the area of a disk at the center having a diameter equal to one-fifth of the propeller diameter. That is—

$$A\ \frac{\pi}{4}\left[D^2-\left(\frac{D}{5}\right)^2\right] \qquad \text{II}$$

2. *Determining the propeller efficiency.*—Given the airplane speed at its highest velocity, the rotational speed of the propeller, the maximum horsepower of the engine at the corresponding rotational

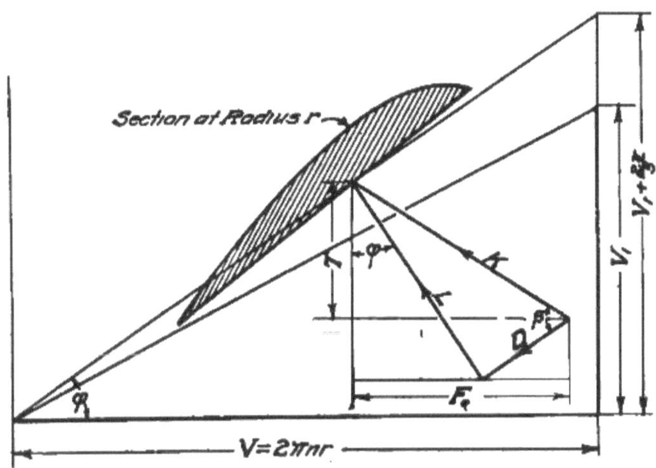


FIG. 2.

speed, and the diameter of the propeller which it is proposed to use, the first step in the propeller design is to determine the best propeller efficiency obtainable under these conditions. It is necessary to proceed by a process of trial and error in determining this efficiency.

First, the efficiency is assumed. Using this value and the data given above, the slip stream velocity can be computed by the formula—

$$v=\frac{T}{\frac{\rho}{g}AV_1} \qquad \text{III}$$

Two-thirds of this velocity is then added to the given airplane speed and from this the true effective pitch angle of the propeller at any station can be found by the formula—

$$\operatorname{Tan}\phi=\frac{V_1+\frac{2v}{3}}{V} \qquad \text{IV}$$

Where, as shown in figure 2—

$\phi$ = the true effective pitch angle of the propeller at any given station.

$V=2\pi\ r\ n$ = velocity of airfoil at the given section in feet per second.

$V_1$ = velocity of the airplane in feet per second.

$v$ = component of the velocity of the slip stream parallel to the direction of motion of the airplane and relative to the air.

$r$ = radius of the propeller blade at the given section.

$n$ = propeller speed in revolutions per second.

$L$ = lift force in pounds normal to direction of motion.

$D$ = drag force in pounds parallel to direction of motion.

$K$ = resultant of $L$ and $D$.

$F_q$ = torque force in pounds.

$T$ = thrust force in pounds.

$e$ = efficiency of given section.

After obtaining the true effective pitch angles at the different stations, the true angle of attack at each station can be selected and the corresponding values of lift-drift ratio may be picked from the $L/D$ curve for the section. In general, it is not necessary to compute the efficiency separately for each section of the propeller blade, but only at a representative point. In the case of the three propeller blade shapes which are later shown, this representative point may be taken as 75 per cent of the radius. The efficiency at this radius may be considered as the efficiency of the propeller as a whole. At this point the slip stream efficiency is equal to—

$$\frac{V_1}{V_1+\frac{2v}{3}}$$

and the airfoil efficiency is equal to—

Tan $(\beta-\phi)$ tan $\phi$, where, as shown in figure 2—

$\beta=\tan^{-1}\left(\frac{L}{D}\right)$, and

$$\phi=\tan^{-1}\left(\frac{V_1+\frac{2v}{3}}{V}\right)$$

The true efficiency will then be the product of these two, or—

$$\text{Tan } (\beta-\phi) \tan \phi \left( \frac{V_1}{V_1+\frac{2v}{3}} \right) \qquad \text{V}$$

This statement will probably require further explanation. In the first place, efficiency must be defined as the ratio between the horsepower actually required to drive the plane through the air at a given speed and the horsepower delivered to the propeller at the same speed.

The horsepower used in driving the plane equals $\frac{TV_1}{550}$, and the horsepower delivered to the propeller equals $\frac{F_q V}{550}$; therefore the efficiency is expressed in general terms by—

$$\frac{TV_1}{550} \div \frac{F_q V}{550} = \frac{TV_1}{F_q V}$$

Then the expression for airfoil efficiency is deduced as follows (see fig. 2):

$$K = L \sin \beta + D \cos \beta \qquad (1)$$

$$T = K \cos (\phi + 90^\circ - \beta) \qquad (2)$$

$$TV_1 = K \cos (\phi + 90^\circ - \beta) V_1 \qquad (3)$$

But $V_1$ can be written $\left( V_1 + \frac{2v}{3} \right) \left( \frac{V_1}{V_1+\frac{2v}{3}} \right)$

Substituting this value in the left-hand member of 3, we have—

$$TV_1 = K \cos (\phi = 90^\circ - \beta) \left( V_1 + \frac{2v}{3} \right) \left( \frac{V_1}{V_1+\frac{2v}{3}} \right) \qquad (4)$$

$$F_q = K \sin (\phi + 90^\circ - \beta) \quad \text{(From figure 2)} \qquad (5)$$

$$F_q V = K \sin (\phi + 90^\circ - \beta) V \qquad (6)$$

From (4), (6)—

$$e = \frac{TV_1}{F_q V} = \frac{K \cos (\phi = 90^\circ - \beta) \left( V_1 + \frac{2v}{3} \right) \left( \frac{V_1}{V_1+\frac{2v}{3}} \right)}{K \sin (\phi + 90^\circ - \beta) V} \qquad (7)$$

From IV and (7)—

$$= \frac{K \cos (\phi + 90^\circ - \beta)}{K \sin (\phi + 90^\circ - \beta)} \tan \phi \left( \frac{V_1}{V_1+\frac{2v}{3}} \right) \qquad (8)$$

$$= \sec (90^\circ - (\beta - \phi)) \tan \phi \left( \frac{V_1}{V_1+\frac{2v}{3}} \right) \qquad (9)$$

$$= \tan (\beta - \phi) \tan \phi \left( \frac{V_1}{V_1+\frac{2v}{3}} \right);$$

which is formula *V* given above.

In figure 3 the values of tangent $(\beta-\phi)\tan\phi$ are plotted against $\phi$ for different values of $\beta$, or, in other words, against different values of $L/D$. $\phi$ is plotted as the cotangent.

After having determined that the diameter, slip, and efficiency are satisfactory, the design of the propeller may be proceeded with. For

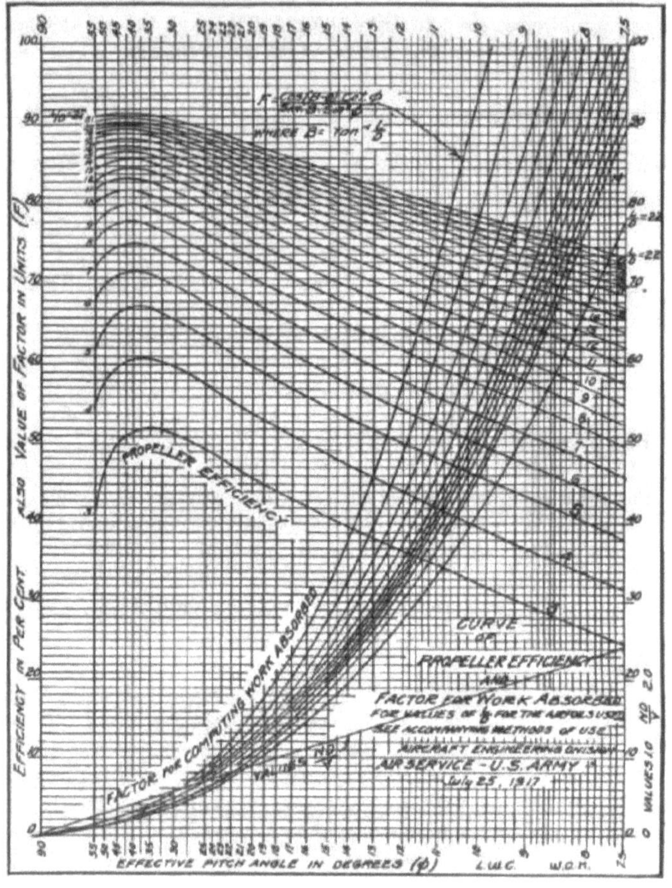


FIG. 3.

this discussion, however, a power analysis will be worked through for a propeller which has already been built and tested, and from this analysis an empirical formula will be derived.

Before proceeding, a word of explanation should be given concerning the minimum camber curve shown in figure 4. The thickness of the propeller blade depends upon stress considerations, but for routine design work this curve of minimum camber ratio plotted against per cent radius is used. In connection with the plan forms given, and when used for propellers having an aspect ratio not greater than 6, and a given speed not greater than 1,800 r. p. m., the curve of minimum camber ratio can be used with safety.

Higher rotative speeds or greater aspect ratios would require thicker blade sections. For convenience, the ordinates for certain per cent increases of the minimum camber curve have been tabulated below the curve.

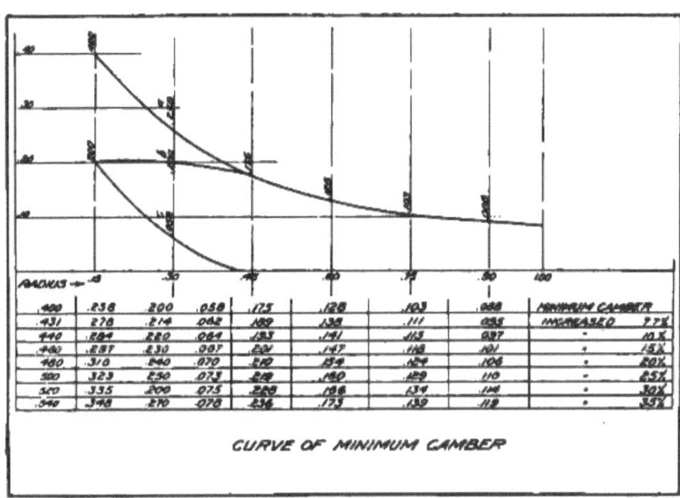


| | | | | | | | | |
|---|---|---|---|---|---|---|---|---|
| .400 | .238 | .200 | .058 | .175 | .128 | .103 | .088 | MINIMUM CAMBER |
| .431 | .278 | .214 | .062 | .189 | .138 | .111 | .095 | INCREASED 7.7% |
| .440 | .284 | .220 | .064 | .193 | .141 | .113 | .097 | " 10% |
| .460 | .297 | .230 | .067 | .201 | .147 | .118 | .101 | " 15% |
| .480 | .310 | .240 | .070 | .210 | .154 | .124 | .106 | " 20% |
| .500 | .323 | .250 | .073 | .219 | .160 | .129 | .110 | " 25% |
| .520 | .335 | .200 | .075 | .228 | .166 | .134 | .114 | " 30% |
| .540 | .348 | .270 | .078 | .236 | .173 | .139 | .118 | " 35% |

FIG. 4.

The Drzwiecki method of calculating the power absorbed, thrust, and efficiency is briefly as follows:

In figure 5, let—

$r$ = radius of section in feet.

$V = 2\pi rn$, where $n$ = revolutions per second.

$V_1$ = airplane velocity in feet per second.

$V_a$ = airfoil velocity in feet per second.

$L$ = lift in pounds.

$D$ = drift in pounds.

$\phi$ = effective pitch angle.

$c$ = chord of blade section in feet.

Also—

$\Delta r$ = elementary length of blade section in feet (in practice usually taken as 0.5 feet).

$\Delta s = s\,\Delta r$ = area of surface in square feet.

$Q$ = torque in foot pounds.
$T$ = thrust in pounds.
$e$ = efficiency.
$Ky$ = lift coefficient in absolute units.

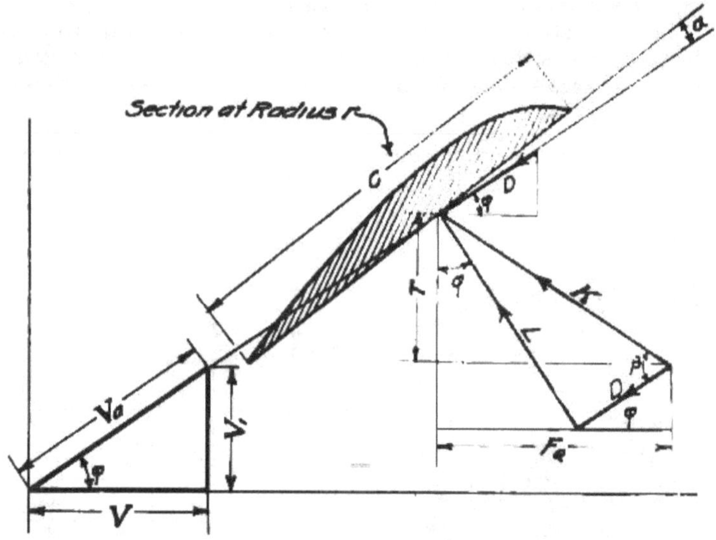


DIAGRAM ILLUSTRATING DRZWIECKI METHOD OF PROPELLER POWER ANALYSIS

FIG. 5.

$Kx$ = drift coefficient in absolute units.

$\gamma = \frac{L}{D} = \frac{Ky}{Kx}$

$\rho$ = density of air.

$g$ = acceleration due to gravity.

$\frac{\rho}{g} = \frac{0.07608}{32.1} = 0.00237$ (approximately) for standard air.

Then—

$$L = Ky \frac{\rho}{g} S V_a^2 \tag{1}$$

$$= Ky \frac{\rho}{g} S \frac{V_1^2}{\sin^2 \phi} \tag{2}$$

$$D = Kx\frac{\rho}{g}S V_a^2 = Kx\frac{\rho}{g}S\frac{V_1^2}{\sin^2\phi} = \frac{Ky\frac{\rho}{g}S V_1^2}{\gamma \sin^2\phi} \qquad (3)$$

$$D \cos\phi = \frac{Ky\frac{\rho}{g}S V_1^2(\cos\phi)}{\gamma \sin\phi\ (\sin\phi)} = \frac{Ky\frac{\rho}{g}S V_1^2}{\gamma \sin\phi \tan\phi} \qquad (4)$$

$$\text{Tan}\ \phi = \frac{V_1}{V} \qquad (5)$$

$$D \cos\phi = \frac{Ky\frac{\rho}{g}S V_1^2 V}{\gamma \sin\phi\ V_1} = \frac{Ky\frac{\rho}{g}S V_1 V}{\gamma \sin\phi} \qquad (6)$$

From equation (2) it will be seen that—

$$L \sin\phi = \frac{Ky\frac{\rho}{g}S V_1^2}{\sin\phi} \qquad (7)$$

In figure 5, $F_q$ represents the component torque force opposing the rotation of the propeller.

$$F_q = (D\cos\phi + L\sin\phi) \qquad (8)$$

Substituting (6) and (7) in (8)—

$$F_q = \left(\frac{Ky\frac{\rho}{g}S V_1 V}{\gamma \sin\phi} + \frac{Ky\frac{\rho}{g}S V_1^2}{\sin\phi}\right) \qquad (9)$$

In feet, pounds, seconds, units, this equation becomes, for standard air—

$$F_q = .00237\left(\frac{Ky\ S\ V_1\ V}{\gamma \sin\phi} + \frac{Ky\ S\ V_1^2}{\sin\phi}\right) \qquad (10)$$

$$= .00237\ S\ Ky\left(\frac{V_1\ V}{\gamma \sin\phi} + \frac{V_1^2}{\sin\phi}\right) \qquad (11)$$

$$= \frac{.00237\ S\ Ky\ V_1\ (V_1\gamma + V)}{\gamma \sin\phi} \qquad (12)\quad \text{VI}$$

$$\text{Power absorbed} = \frac{.00237\ S\ Ky\ V_1\ (V_1\gamma + V)\ V}{\gamma \sin\phi\ 550} \qquad (13)\quad \text{VII}$$

In practice, the propeller blade is divided into sections of short length and this equation applied to each section separately, as will be

161988—21——2

explained later. A summation of the results thus obtained gives the power absorbed for the whole blade.

Thrust, $T = .00237\ S\ (L \cos\phi - D \sin\phi)$ (14)

$$= .00237\ S\left(\frac{Ky\ V_1^2 \cos\phi}{\sin^2\phi} - \frac{Ky\ V_1^2 \sin\phi}{\gamma \sin^2\phi}\right) \qquad (15)$$

$$= .00237\ S\left(\frac{Ky\ V_1^2}{\sin\phi}\frac{V}{V_1} - \frac{Ky\ V_1^2}{\gamma \sin\phi}\right) \qquad (16)$$

$$= .00237\ S\frac{Ky\ V_1}{\gamma \sin\phi}(V\gamma - V_1) \qquad (17)\quad \text{VIII}$$

Efficiency, $e = \frac{\text{work delivered}}{\text{work absorbed}}$ (general equation for efficiency). (18)

$$= \frac{T\ V_1}{F_q\ V} \qquad (19)$$

Substituting from (12) and (17)—

$$= \frac{\frac{Ky\ V_1}{\gamma \sin\phi}(V_1\gamma - V)\ .00237\ S\ V_1}{\frac{.00237\ S\ Ky\ V_1\ (V\gamma + V_1)\ V}{\gamma \sin\phi}} \qquad (20)$$

$$= \frac{(V\gamma - V_1)\ V_1}{(V_1\gamma + V)\ V} = \frac{\gamma V^2 \tan\phi - V^2 \tan\phi}{\gamma V^2 \tan\phi + V^2} \qquad (21)$$

From the figure it is seen that $\beta = \tan^{-1}\frac{L}{D} = \tan^{-1}\gamma$; therefore (21) can be written in the form—

$$e = \frac{(\tan\beta - \tan\phi)\ \tan\phi}{\tan\phi\ \tan\beta + 1} \qquad (22)$$

$$= \tan\ (\beta - \phi)\ \tan\phi. \qquad (23)\quad \text{IX}$$

This expression for efficiency does not take into account the slip stream effect.

(See fig. 3 for plotted values.)

In making a power analysis by the Drzwiecki method, these formulæ—VII, VIII, and IX—are applied separately to each elementary section of the blade. As an illustration, the following analysis is made of the propeller detailed in figure 6. From this drawing the blade widths, thicknesses, camber ratios, i. e., thickness divided by width, and the blade angles are found for each 6-inch station. These are tabulated below, together with the airplane speed $V$ and the r. p. m. of the propeller.

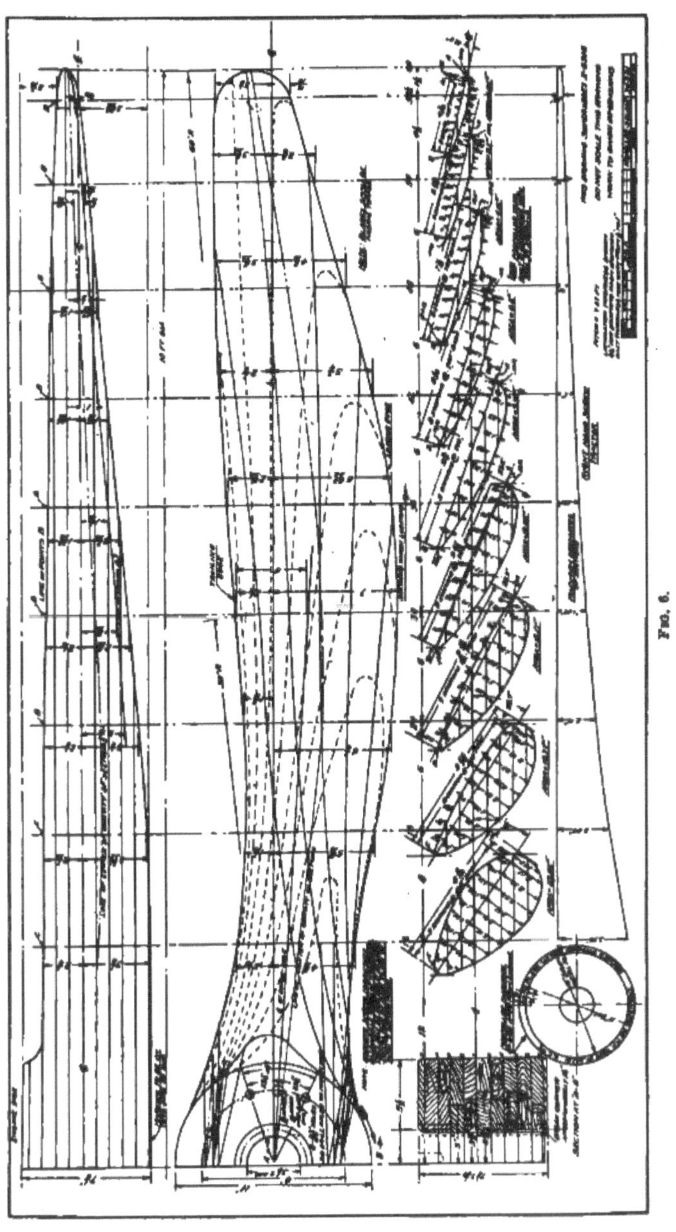

Fig. 6.

From the forward speed and r. p. m. the effective angle can be computed and the angle of attack $\alpha$ found by subtracting these angles from the given blade angles. Knowing the angle of attack

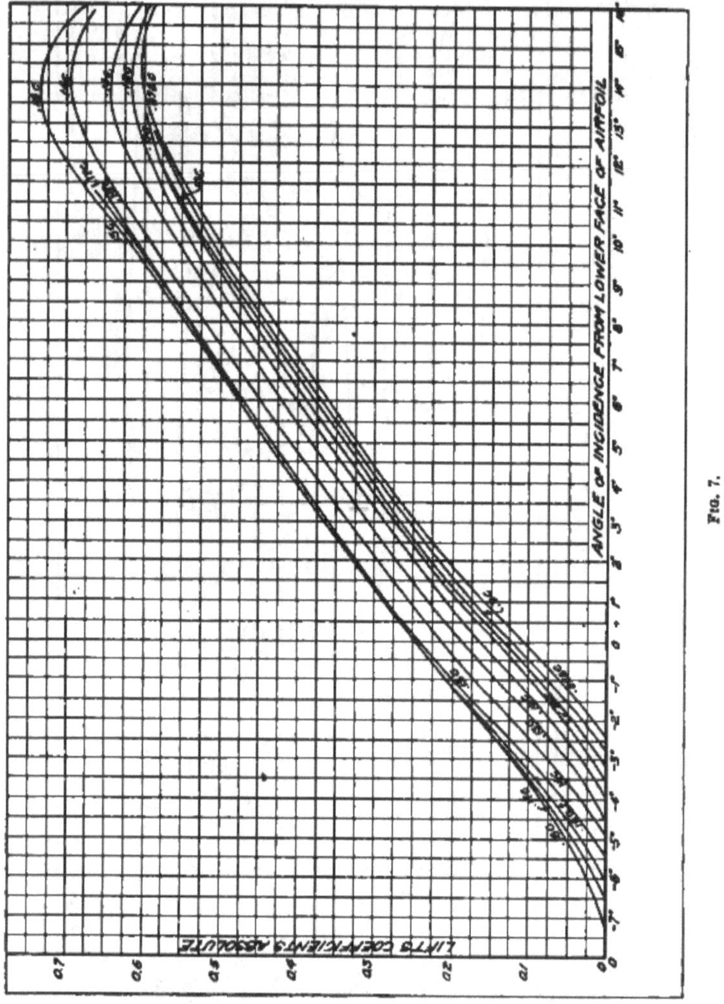


Fig. 7.

and the camber ratio, the lift coefficient and $L/D$ for each section is read directly from the curves given in figures 7 and 8, respectively. These curves are obtained from airfoils tested at 30 miles per hour and should be scaled up for airfoils moving at a velocity comparable to that of a propeller-blade section. The scaled-up coefficients are given in figures 9 and 10. However, the original coefficients are used in general propeller design because, through experience with these

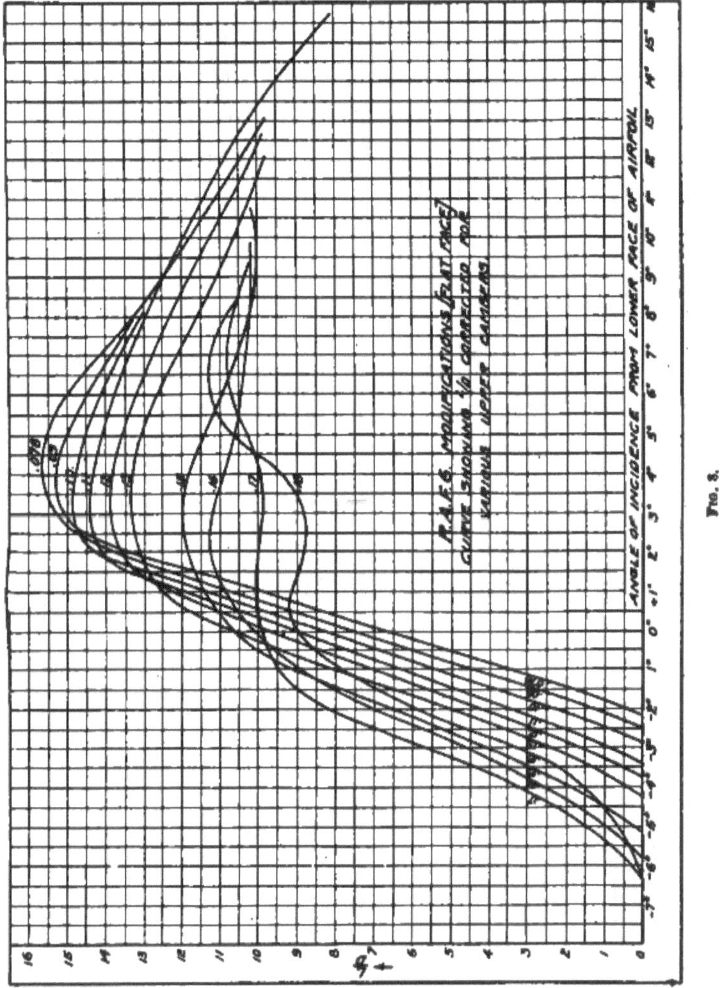


FIG. 8.

curves, blade factors for different plan forms have been carefully obtained. This will be explained later.

In the curves of figures 7, 8, 9, and 10 the assumption is made that the aerodynamic characteristics are functions of the camber ratios.

This assumption is justifiable in view of the limited range of possible airfoil shapes which could be employed in a propeller of conventional design.

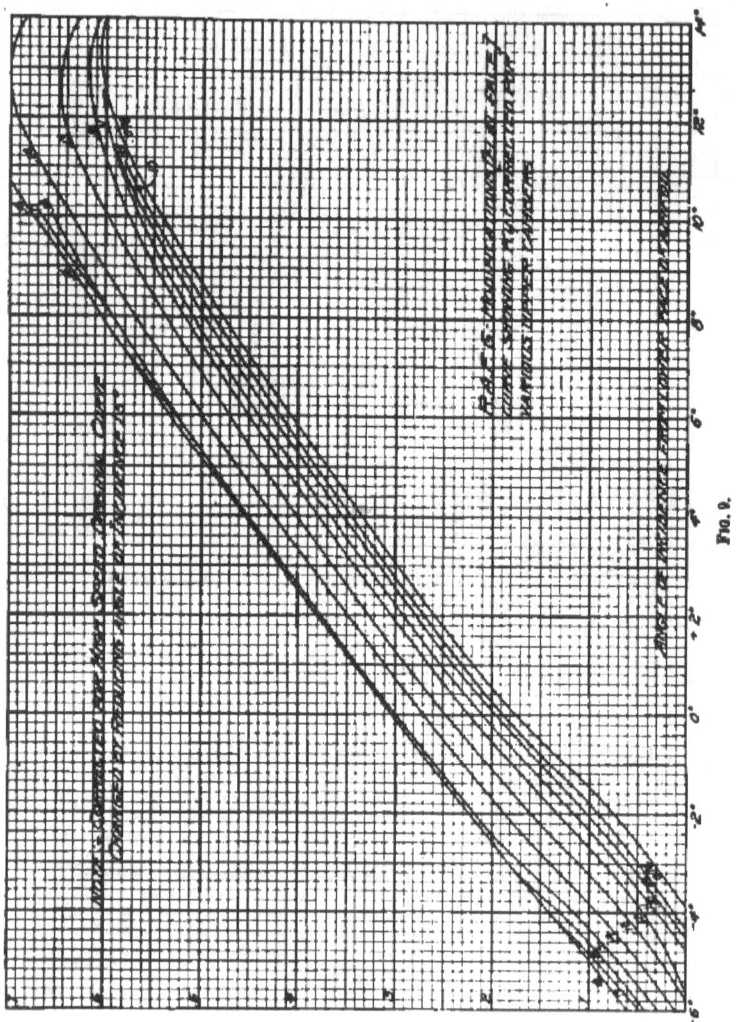

FIG. 9.

In selecting the *Ky* or *L/D* characteristics for an airfoil having a camber ratio lying between any two camber ratios for which curves are plotted, it is necessary to perform an interpolation.

The thrust at each station of the blade may be computed by Formula VIII, or by first figuring the efficiency and then multiplying the work absorbed by the efficiency and dividing the velocity of the

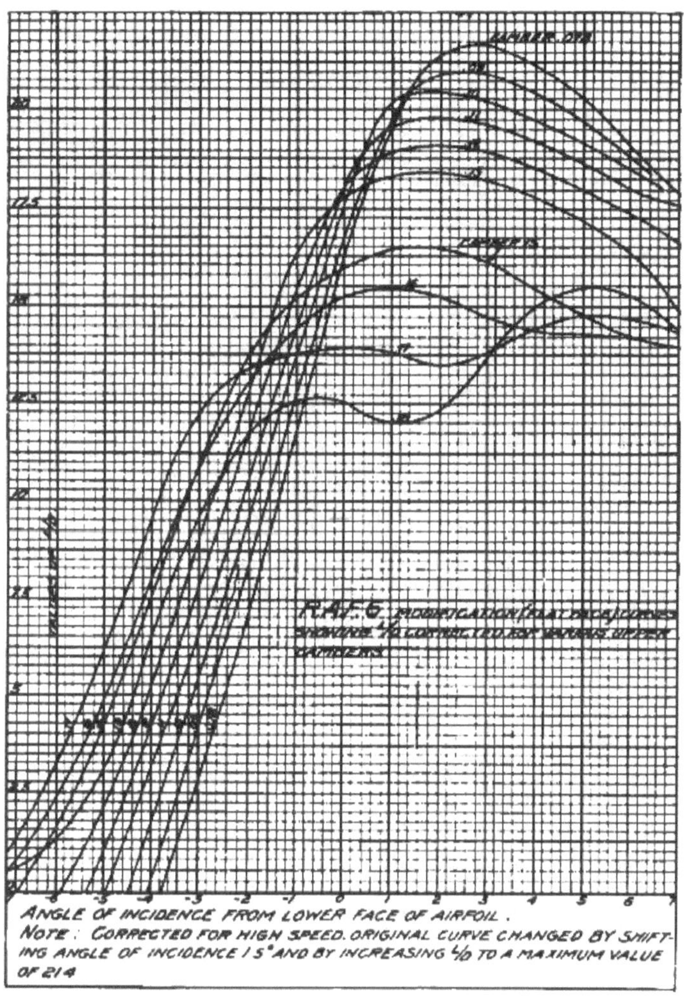


FIG. 10.

plane in feet per second. This method has been used in the following sample calculation. Also, since the effective pitch angle and $L/D$ for each section have already been found, the efficiency for each section can be read directly from the curves in figure 3.

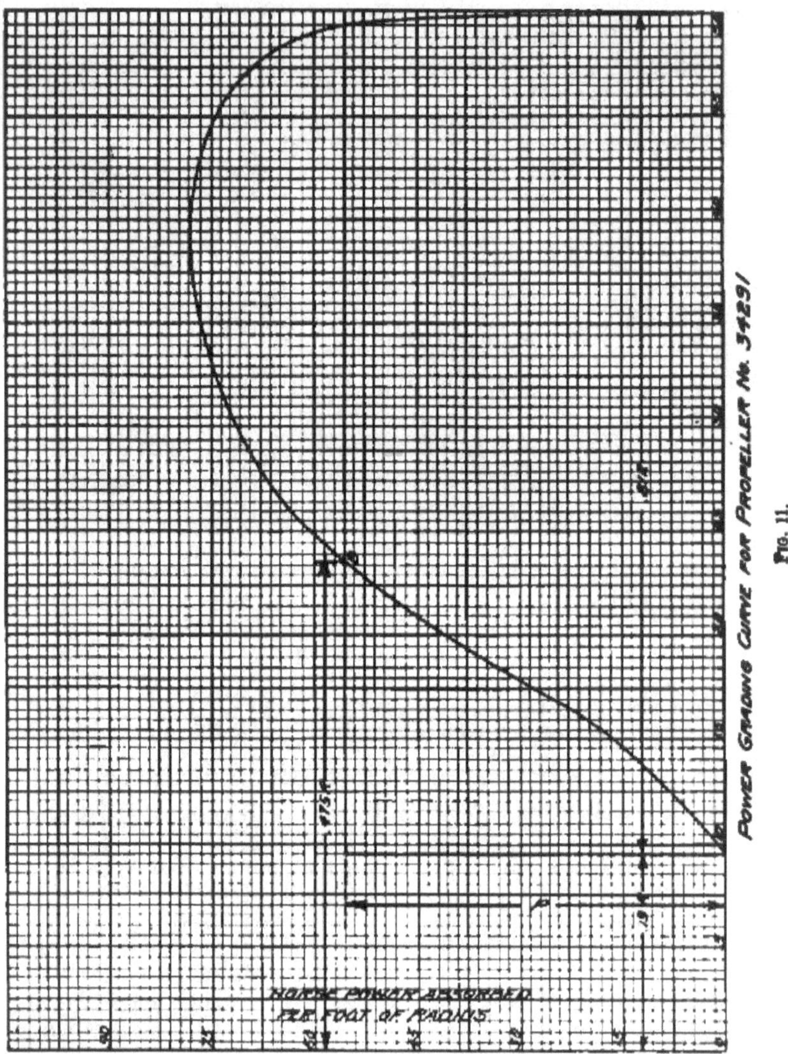


POWER GRADING CURVE FOR PROPELLER No. 34291

FIG. 11.

*Sample design by Drzwiecki method using propeller No. 34291.*

Forward speed = 123 m. p. h.
R. p. m. = 1760 (flight near ground).
Diameter = 10 ft.
Effective pitch 6 = 16 ft.

$V_1$ = forward speed in feet/sec. = 1.467×123 m. p. h. = 180.5.
$n$ = revolutions/sec. = 1760/60 = 29.3.

(See fig. 6.)

| Station radius in feet | 1.0 | 1.5 | 2.0 | 2.5 | 3.0 | 3.5 | 4.0 | 4.5 | 4.87 |
|---|---|---|---|---|---|---|---|---|---|
| Blade angle in degrees $(\phi+\alpha)=\theta$ (from fig. 6) | 35 | 32.7 | 28.3 | 23.6 | 20.2 | 17.8 | 15.8 | 14.3 | 13.6 |
| Effective pitch angle $\phi$ | 44.5 | 33.2 | 26.1 | 21.4 | 18.1 | 15.6 | 13.8 | 12.3 | 11.4 |
| Angle of attack | −9.5 | −.5 | 2.2 | 2.2 | 2.1 | 2.2 | 2.0 | 2.0 | 2.2 |
| Camber ratio, $\frac{\text{max. ordinate}}{c}$ (fig. 6) | 0.49 | 0.328 | 0.234 | 0.168 | 0.135 | 0.111 | 0.097 | 0.086 | 0.063 |
| Lift coefficient, $Ky$ (from fig. 7) | 0.06 | 0.24 | 0.34 | 0.33 | 0.285 | 0.24 | 0.22 | 0.209 | 0.213 |
| Lift drift = $L/D$ = (from fig. 8) | 3.5 | 6.5 | 8.2 | 10.3 | 12.9 | 14.3 | 14.1 | 13.8 | 14 |
| $\Delta\gamma$, length of element (feet) | 0.5 | 0.5 | 0.5 | 0.5 | 0.5 | 0.5 | 0.5 | 0.5 | 0.25 |
| $c$, mean width of element (feet) | 0.68 | 0.727 | 0.794 | 0.82 | 0.812 | 0.753 | 0.642 | 0.487 | 0.531 |
| Area of element, $S$ (square feet) | 0.340 | 0.364 | 0.397 | 0.410 | 0.406 | 0.377 | 0.321 | 0.244 | 0.088 |
| Sine $\phi$, $\frac{V_1}{V_a}$ | 0.701 | 0.548 | 0.440 | 0.355 | 0.311 | 0.269 | 0.239 | 0.213 | 0.196 |
| $V_1$ | 632 | 1,173 | 1,480 | 1,860 | 2,330 | 2,580 | 2,550 | 2,460 | 2,530 |
| $V=(2\pi r n)$ | 184 | 276 | 368 | 460 | 552 | 645 | 736 | 830 | 897 |
| $(V_1+V)$ | 816 | 1,449 | 1,848 | 2,320 | 2,882 | 3,225 | 3,286 | 3,290 | 3,427 |
| Work absorbed by element/second (from Formula VI) | 445 | 4,200 | 10,900 | 16,450 | 19,640 | 20,900 | 20,500 | 20,600 | 8,900 |
| H. P. absorbed by element work absorbed/550 | 0.81 | 7.65 | 19.8 | 29.9 | 35.7 | 38.0 | 39.4 | 37.5 | 16.2 |
| H. P. per foot of length | 1.62 | 15.3 | 39.6 | 59.8 | 71.4 | 76.0 | 78.8 | 75.0 | 64.8 |
| Efficiency (per cent) (from Formula IX) | 55.8 | 72.6 | 74.9 | 76.7 | 78.6 | 78.3 | 76.2 | 73.6 | 72.9 |
| Thrust (pounds) = $\frac{F_a V_a}{V_1}$ | 1.38 | 16.9 | 45.1 | 70.0 | 85.5 | 90.6 | 91.5 | 84.1 | 36.0 |

From curves, figure 1.—Horsepower per blade equals 216.6; horsepower absorbed equals 433; thrust per blade equals 501.8 pounds; total thrust equals 1,004 pounds.

Efficiency, $E$, for whole propeller equals $\frac{T\times V_1}{H.P.\times 550}=\frac{1{,}004\times 180.5}{433\times 550}=76$ per cent.

Figure 11 gives the power-grading curves for one blade of this propeller. By integrating this power-grading curve and dividing the resultant area by the length $AC$—that portion of the blade which was loaded—a rectangle is found which represents a uniform absorption of horsepower over the loaded portion of the blade. The height of this rectangle is equal to the ordinate of the power-grading curve at the point $B$, which occurs .475 R from the center of the hub. From this it is seen that if the loaded portion of the propeller blade—from the tip in .81 R—absorbs power uniformly per unit length and equal to the amount absorbed at $B$, the intersection of the two curves, the total horsepower absorbed will be the same as previously obtained.

By taking advantage of this fact and applying the power formula at this point for the whole blade an empirical formula may be derived as follows:

The equation for horsepower absorbed by each elementary length of blade is given in Formula VII, as follows, where the horsepower absorbed for each elementary blade length is designated by ΔP. (See fig. 12.)

$$\Delta P=\frac{\frac{\rho}{g}\Delta S\, Ky\, V_1(V_1\gamma+V)\,V}{550\,\gamma\,\sin\,\phi} \qquad (1)$$

$$= \frac{\frac{\rho}{g} c \Delta r\, K y\, V_1 (V_1 \gamma + V) V}{550\ \gamma \sin \phi} \tag{2}$$

Therefore,

$$\frac{\Delta P}{\Delta r} = \frac{\frac{\rho}{g} c\, K y\, V_1 (V_1 \gamma + V) V}{550\ \gamma \sin \phi} \tag{3}$$

In figure 11—

$R$ = tip radius.

$B$ = intersection of actual power grading curve with equivalent curve.

Abscissa of point $B = .475\ R$; let ordinate $= p$.

Loaded portion of blade extends over .81 $R$ measured from the extreme tip in toward the hub.

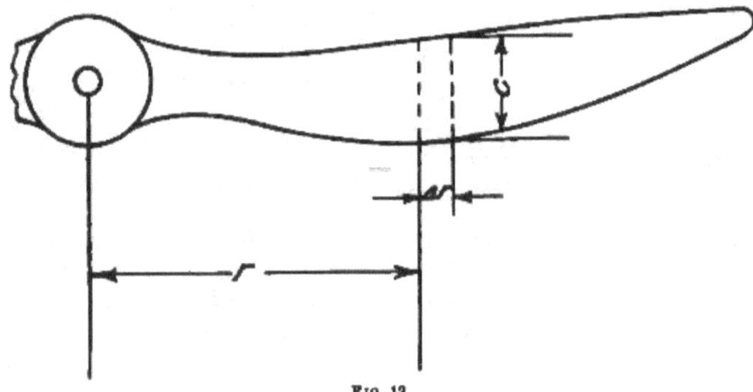


FIG. 12.

It has already been shown that the total horsepower absorbed per blade is measured by the area of the rectangle, figure 11, whose base is $A\ C$ and whose altitude is $p$, a constant.

By applying equation (3) at the point, .475 $R$, we can now write—

$$\frac{dP}{dr} = P \tag{4}$$

Therefore, $r = R$

$$P = \int_{r=.19\,R}^{r=R} p\, dr = .81\ R\ p \tag{5}$$

This may be visualized by comparing the area under the power grading curve with the area under the equivalent rectangle. (See fig. 11.)

In the following calculation, the maximum blade width $C$, is assumed to be at the station .475 $R$. This is true for practically all plan forms in use by the Air Service designers.

Assume an aspect ratio of 6, which is commonly considered good practice. Then—

$$C=\frac{R}{6}=\frac{D}{12} \tag{6}$$

From equations (3), (4), and (5) we have—

$$P=.81\ R\left[\frac{\frac{\rho}{g}C\ Ky\ V_1\ (V_1\gamma+V)}{550\ \gamma\sin\phi}\right] \tag{7}$$

$$=\frac{.81\ R\ \frac{\rho}{g}V_1\ V\left(V_1+\frac{V}{\gamma}\right)Ky\ C}{550\times\sin\phi} \tag{8}$$

Let $f$ = air-blade factor, representing the ratio of the actual power absorbed by a propeller in test, to the computed power absorbed.

Then, if $P$ = computed power absorbed, and $P_1$ = actual power absorbed—

$$P_1=Pf \tag{9}$$

Assume—

$$\gamma=10, \text{ for airfoil of camber ratio}=.165. \tag{10}$$

From figure 11—

$$\frac{V_1}{\sin\phi}=(V_1^2+V^2)^{\frac{1}{2}} \tag{11}$$

We can now write—

$$P_1=\frac{\rho}{g}\frac{.81R}{550}V\left(V_1+\frac{V}{10}\right)(V_1^2+V^2)^{\frac{1}{2}}Ky\ Cf \tag{12}$$

Substituting from (6)—

$$=\frac{\rho}{g}\frac{.81R^2}{6\times550}V\left(V_1+\frac{V}{10}\right)(V_1^2+V^2)^{\frac{1}{2}}Kyf \tag{13}$$

$$=\frac{\rho}{g}\frac{.81}{4\times6\times550}VD^2\left(V_1+\frac{V}{10}\right)(V_1^2+V^2)^{\frac{1}{2}}Kyf \tag{14}$$

$$=\frac{\rho}{g}\frac{.81}{13200}VD^2\left(V_1+\frac{V}{10}\right)(V_1^2+V^2)^{\frac{1}{2}}Kyf \tag{15}$$

$$=145\times10^{-9}D^2V\left(V_1+\frac{V}{10}\right)(V_1^2+V^2)^{\frac{1}{2}}Kyf \tag{16}$$ X

For any other aspect ratio, Formula X becomes—
H. P. absorbed per blade

$$=\frac{6}{\text{aspect ratio}}\times 145\times 10^{-9}D^2V\left(V_1+\frac{V}{10}\right)(V_1^2+V^2)^{\frac{1}{2}}\,Kyf \qquad \text{XI}$$

A sample calculation is made below, using the formula just derived. The data furnished the propeller designer by the designer of the airplane on which the propeller is to be used will be taken as identical with that in the previous case.

### APPROXIMATE METHOD.

$V_1$ = airplane speed = 123 mi./hr. = 180.5 feet/sec.
H. P. = 402 at 1,760 r. p. m. = 29.3 r. p. s.
$D$ = 10 feet.
Effective pitch, $P_e$ = 6.16 feet. Number of blades = 2.
Aspect ratio = 6.
$D^2$ = 100.
$V = 29.3\pi\times .475\ D$ = 437 feet/sec. (At station .475 $R$.)

$$\left(V_1+\frac{V}{10}\right)=180.5+43.7=224.2 \text{ feet/sec.}$$

$(V_1^2+V^2)^{\frac{1}{2}}=(32600+191000)^{\frac{1}{2}}=(223600)^{\frac{1}{2}}=473$ feet/sec.
Let $Kyf=kp$.

Substituting the above values in Formula XI, we have—

H. P. absorbed per blade

$$=\frac{6}{6}\times 145\times 10^{-9}\times 100\times 437\times 224.2\times 473\ kp=671\ kp$$

or—

$$201=671\ kp$$

Therefore—

$$kp=.299$$

Since the aspect ratio is 6, the minimum camber curve, figure 4, can be used with safety.

The camber ratio at .473 $R$ from the minimum camber curve is .165

Use plan form No. 1 with a blade factor of .90—

$$Ky=\frac{kp}{f}=\frac{.299}{.90}=.333.$$

Now find the angle of attack corresponding to this value of $Ky$ from the lift curves given in figure 7.

Follow the lift curve for camber of .165 until it crosses the horizontal line representing a $Ky$ of .333; on the vertical line through this intersection read the angle of attack. For this particular case, the angle of attack is found to be 2.2°. Thus the one unknown dimension of the blade is solved. Should this angle of attack come out too high or too low it will be necessary to change the aspect ratio of the blade or to increase the camber ratio of the sections. When the aspect ratio runs higher than 6—for instance, up to 7—the camber of the blade should be increased from 25 to 35 per cent. For a design with a high-aspect ratio, birch is the only suitable wood, and even when built of birch a new design should be given a destructive whirling test before the propeller is put into service.

It is now necessary to find the efficiency, true angle of attack, and angle from no lift. For these calculations, consider the section at the 75 per cent radius. (See art. 2.)

Assume $e=76$ per cent.

$$\text{Thrust}=\frac{\text{H. P.}\times e\times 550}{V}=\frac{402\times 0.76\times 550}{180.5}=930 \text{ pounds.}$$

$$\text{Disk area}=.7854\ \left(D^2-\left(\frac{D}{5}\right)^2\right)=75 \text{ square feet.}$$

$$v=\frac{930}{0.00237\times 180.5\times 75}=29 \text{ feet per sec.} \quad \text{(Formula III.)}$$

$$\text{Slip}=\frac{29}{180.5}=16 \text{ per cent.}$$

Take 2/3 of total slip as occurring before the disk,

$2/3\times 29=19.3$ feet per sec. (See fig. 2.)

$V$ at $0.75=0.75\ D\times\pi\times n=690$ feet/sec.

Camber at 0.75 $R$ for minimum camber $=0.103$. (From fig. 4.)

If $\phi=$ effective pitch angle of the chosen section, and $\phi_a=$ apparent effective pitch angle of this section,

Then—

$$\tan\phi=\frac{180.5+19.3}{690}=\frac{199.8}{690}=0.290$$

Therefore—

$$\phi=16.2^\circ$$

$$\tan\phi_a=\frac{180.5}{690}=0.262$$

Therefore—

$$\phi_a=14.7^\circ$$

$$\theta=\text{blade angle}=(14.7^\circ+2.2^\circ)=16.9^\circ$$

$$\begin{aligned}\text{True angle of attack}&=(14.7^\circ+2.2^\circ-16.2^\circ)\\&=(16.9-16.2)\\&=0.7^\circ\end{aligned}$$

From figure . . . (13).

Angle of no lift $=4.5^\circ$.

Angle from no lift $=(0.7^\circ+4.5^\circ)=5.2^\circ$.

Referring now to the lift drift curve for 0.103 camber, figure 8, and scaling back 5.2° from the point where it cuts the no lift line, it is found that the high point of the lift drift curve occurs here. The angle from no lift should lie between 4.5° and 8° for good design.

The $L/D$ for this section acting at high-air speeds is about 20.

From figure 3 the efficiency is read on the horizontal line passing through the intersection of the vertical line for an effective pitch angle of 16.2° and the curve for an $L/D$ of 20. In this case the efficiency is read as 84 per cent. This is not the true efficiency of the propeller, however, as it does not take the slip stream into account. The true efficiency is obtained from the values as found in figure 3 by multiplying by the slip stream efficiency $\frac{V_1}{V_1+\frac{2v}{3}}$. (See art. 1.)

Hence—

$$\text{True efficiency, } e = .84 \times \frac{180.5}{199.8} = 76 \text{ per cent.}$$

It is well to give at the end of the calculations a summary of the blade dimensions. For the preceding design these are:

Diameter, 10 feet.
Maximum blade width, 10 inches.
$Pe = 6.16$ feet.
Angle of attack, 2.2°.
Use minimum camber curve.
Drop trailing edge 1½ laminations.
Use R. A. F. 6 modified airfoil.

The diameter of the hub is taken as 1 inch for every foot of propeller diameter except in cases where the propeller hub flange is of a larger diameter. In this case the wooden hub should be increased so as to extend beyond the metal flange.

It is customary to figure the pitch of a propeller at the two-third radius. This pitch is equal to $2.1 \times D \times \tan\theta$, where $\theta$ is the blade angle at two-third radius.

In the above calculations a certain drop was given to the trailing edge to take care of short grain along the trailing edge at the tip of the propeller. A further refinement of this method is to balance the resultant moment of torque and thrust at each section by a moment due to the centrifugal force applied at the center of gravity of the section. In this latter method the thrust and torque at each section are first figured. The sections are then plotted, traced on cardboard, cut to actual size, and the center of gravity located by balancing about two axes. Each section is then plotted in the finished drawing with its center of gravity so placed with regard to the center line of the propeller that the centrifugal force acting through this point and tending to pull the center of gravity into coincidence with the center line of the propeller is exactly counterbalanced at each section by the component of thrust and torque acting on that section.

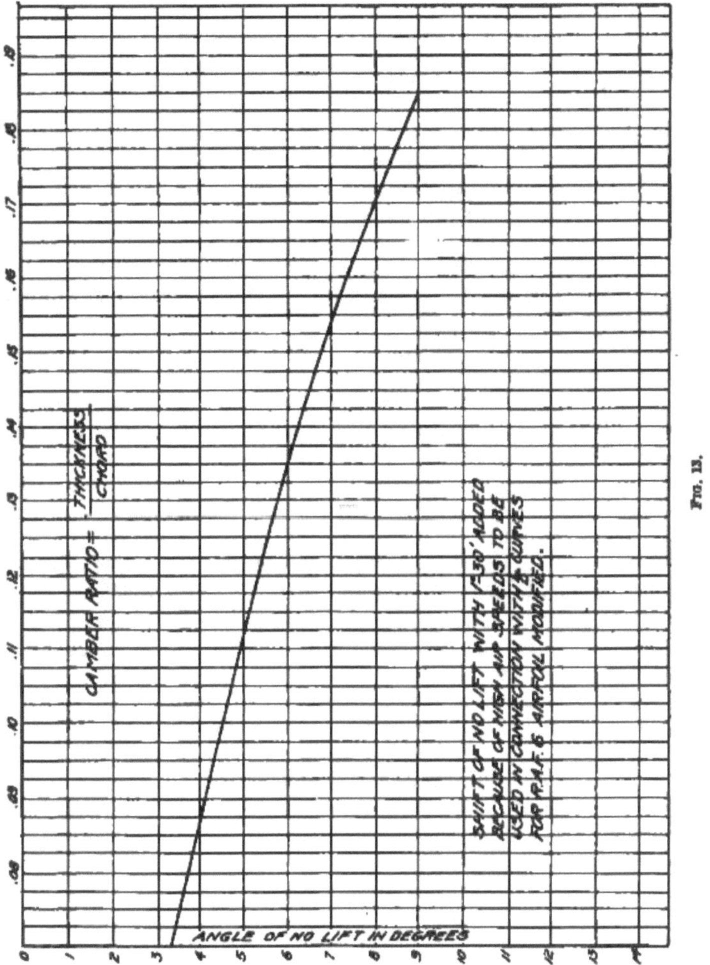


FIG. 13.

The approximate formula for horsepower absorbed given above is not in a very practical form, due to the lack of wind tunnel data on thick sections, such as occur at 0.479 $R$. The British air board, the United States Army, and the Navy Air Service use a similar formula, taking a section farther out on the blade and loading the blade for only 0.73 $R$. The point at which the section is taken is 0.595 $R$, where the camber ratio taken from the minimum camber curve is 0.13 instead of 0.165, as in the previous case. It is recommended that the formula given below be used.

$$\text{H. P. per blade} = \frac{6}{\text{aspect ratio}} \times 129 \times 10^{-8} \times D^2 X V \times \left(\frac{V_1 + V}{16}\right) (V_1^2 + V^2)^{\frac{1}{2}} Ky f. \qquad \text{XII}$$

In the above formula $V$ is the rotational velocity at 0.595 $R$. The same blade factors and unscaled lift and $L/D$ coefficients are used as in the previous formula.

In case a four-bladed propeller is being designed the horsepower per blade is taken as one-fourth the engine horsepower. In this case, however, it is necessary to multiply the blade factor by 0.95, as the blades are not as effective when there are four of them, due to interference.

A propeller designed to turn at a certain r. p. m. on a plane flying near the ground will decrease in r. p. m. as the plane climbs in altitude. This is due to the fact that the engine horsepower falls off more rapidly than the decrease in air density. The decrease in air density would tend to allow the propeller to increase in revolutions, but the decrease in engine horsepower more than offsets this. It is often desirable, therefore, to design a propeller for altitude conditions. The decrease in engine horsepower is calculated for the desired altitude from the Air Board Standard Chart for decrease of engine horsepower with altitude (fig. 14). The *ky* coefficients of the propeller blade are considered to decrease in direct proportion to the air density. Care must be taken not to run a propeller of this design with the engine wide open when flying near the ground, as the results are apt to be disastrous to the propeller or engine.

It sometimes happens that after the first propeller of a design is built it is found to turn too fast or too slow when used on a plane in the air. To correct this the blade angles on the propellers built thereafter can be changed by one minute for each r. p. m. of difference between the true and desired r. p. m. For instance, an increase of one minute in the blade angles will approximately decrease the r. p. m. of a propeller by one.

In computing airplane performance the propeller can be said to absorb the full horsepower of the engine only at such times as the engine is running with throttle wide open. A common fallacy among airplane designers, when computing the performance of a plane with the engine turning at different r. p. m.'s, and with the same propeller, is to read the engine horsepower directly from the horsepower curve for each r. p. m. This is incorrect, as the horsepower developed by an engine when throttled down to a certain

r. p. m. is not the same as the power developed by this engine running wide open but held down to this r. p. m. by a load applied either in the form of a large propeller or dynamometer.

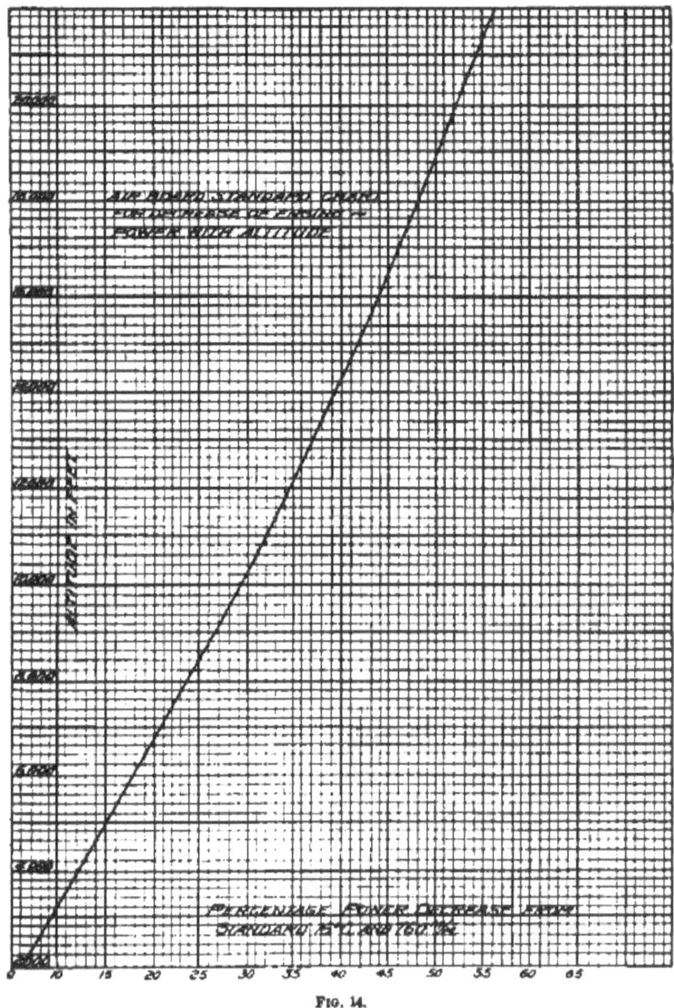


Fig. 14.

CONVERSION FACTORS FOR PRESSURE COEFFICIENTS.

To convert absolute coefficients to pounds per square foot at $V$ feet per second it is necessary to multiply by 0.00257 $V^2$.

161988—21——3

To convert absolute coefficients to pounds per square foot at $V$ miles per hour, multiply by 0.00510 $V^2$.

To convert absolute units to kilos per square meter at $V$ meters per second, multiply by 0.125 $V^2$.

To convert from kilogram meter second units to pounds per square foot at $V$ miles per hour, multiply by 0.0408 $V^2$.

To convert from kilogram meter second units to pounds per square foot at $V$ feet per second, multiply by 0.019028 $V^2$.

To convert from pounds per square foot at $V$ miles per hour to pounds per square foot at $V$ feet per second, multiply by 0.468 $V^2$.

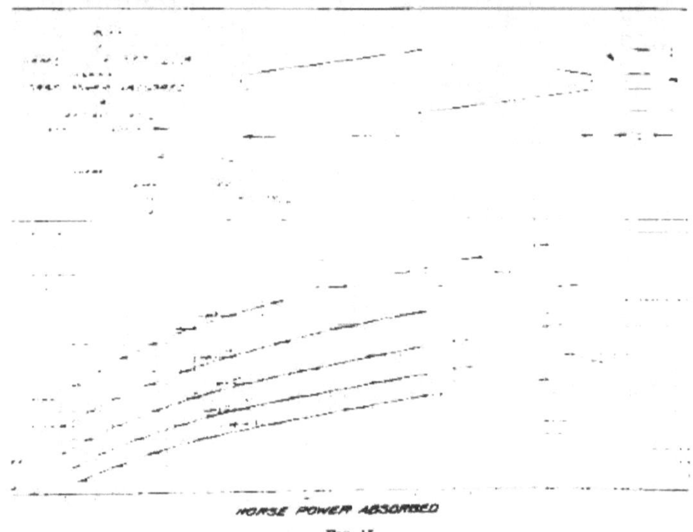


FIG. 15.

*Engine clubs.*—Clubs for holding an engine at a certain r. p. m. when running on the test stand are of two kinds—the moulinet, or club having no angle, and the propeller club, which is similar to an air propeller.

The moulinet is the simplest to make and is used when it is not desired to throw a blast of air against the engine radiator for the purpose of cooling. The necessary formula and curves for designing a moulinet club for different horsepower and r. p. m. are shown in figure 15.

The propeller club is usually made from an old airplane propeller. The propeller to be used is first run on the engine to be tested and the r. p. m. taken. The tips are then cut off by such an amount as will bring the engine up to the desired speed. The proper amount can be determined approximately by the following formula:

Let r. p. m. at which selected propeller turns on an engine $= N_1$ (determine by test).

Desired r. p. m. of club $= N_2$.

Given diameter of propeller $= D_1$.

To find $D$, desired diameter of club—

$$D = D_1 \left(\frac{N_1}{N_2}\right)^{2/3}.$$

If the computed difference in diameter is great, cut off three-quarters of the indicated amount, find the r. p. m. given by the engine using this propeller and repeat the calculation.

## FORMULÆ FOR DETERMINING THE DIAMETER OF A PROPELLER.

The following formulæ, together with a nomogram, show the approximate relation between the diameter, revolutions, horsepower absorbed by a propeller, and speed of the machine to which it is fitted.

The conditions assumed do not apply absolutely in practice; therefore, the formulæ are not suitable for the complete design of a propeller. They will, however, be very useful to the propeller designer in enabling him to settle quickly a diameter on which to base the remainder of his design, and particularly to the machine designer in giving him an exact idea of what clearance he should leave if he wishes the most efficient propeller for his conditions.

For two-bladed propellers: $\text{H. P.} = \frac{1.73}{10^9} \frac{Ky}{AR} D^4 N^3 V_1 \left(\frac{ND}{V_1}\right) 0.1.$

For four-bladed propellers: $\text{H. P.} = \frac{3.34}{10^9} \frac{Ky}{AR} D^4 N^3 V_1 \left(\frac{ND}{V_1}\right) 0.1.$

Where $AR = \frac{\text{tip radius}}{\text{max. blade width}}$, and $Ky =$ absolute lift coefficient.

In the nomogram we have taken $AR = 5.75$, and $Ky = 0.25$.

Use of nomogram, figure 16.

Example: Required diameter necessary in the following conditions:

$$\left.\begin{array}{r} \text{H. P.} = 70 \\ \text{R. P. M.} = 1000 \\ \text{Machine speed} = 100 \end{array}\right\} \text{ground level.}$$

(1) Join the two speeds 100 m. p. h. and 1000 r. p. m. by a line.

(2) Mark the point at which this line cuts the reference line.

(3) Draw a line through this point and through the point representing 70 H. P. on the H. P. scale, and project to cut the diameter scale.

(4) The point at which the line cuts the diameter scale shows the required diameter—8 feet 9 inches for a two-blader or 7 feet 6 inches for a four-blader.

In making the choice, however, between the two mentioned above it should be borne in mind that the four-blader will be about 2 per cent less efficient than the two-blader indicated by the nomogram.

## PROPELLER LAYOUT.

Propeller layout may for convenience be divided into two parts: First, the preliminary sketch made up from the original data from which is taken the dimensions of the blade, width, thickness, etc.,

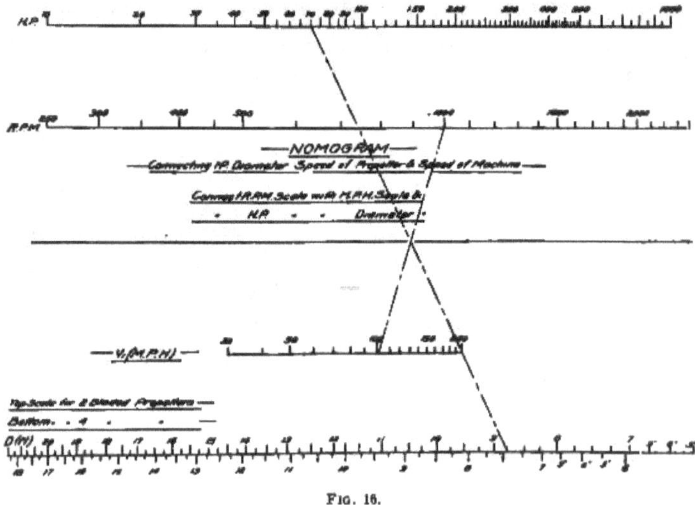


FIG. 16.

at the various station points; second, propeller layout or drawing proper.

The propeller designed in the previous section has been chosen as an example. A drawing of this propeller is shown in figure 6.

All calculations and figures given in the following explanation will apply to this particular drawing in order to show propeller designing in one single case in its entirety, enabling the reader to gain a clear idea by developing this plan step by step.

The data necessary for the layout is as follows:

- Diameter of propeller.
- Maximum blade width.
- Plan form of blade to be used.
- Effective pitch ($P_e$) in feet.
- Angle of attack.

Camber curve for determining thickness of section.
Drop of trailing edge.
Diameter and thickness of hub.
Airfoil section to be used.

For the particular example chosen this data is:
Diameter of propeller, 10 feet.
Maximum blade width, 10 inches.
Plan form to be used No. 1. (See fig. 18.)
Effective pitch ($P_e$), 6.16 feet.
Angle of attack, 2.2°.
Minimum camber curve. (See fig. 4.)
Trailing edge to be dropped, 1½ laminations.
Diameter of hub, 11 inches.
Thickness of hub, 7½ inches.
Airfoil section R. A. F. 6 modified.

*The preliminary sketch.*—This sketch is made on an 8 by 10 sheet of cross-section paper, or cloth, as shown in figure 17.

The blade radius is laid off in feet on the $X$ axis; the blade width, the thickness, and angles on the $Y$ axis. The first step is to lay off on the $X$ axis the radius and take stations at the following points: 0.15 R, 0.30 R, 0.45 R, 0.60 R, and 0.90 R. This is necessary because the propeller data sheets of blade widths and thicknesses are divided in the above per cents of radius.

*To find the blade width at each 6-inch station of the radius.*—The widths along the blade are determined by multiplying the maximum width given in the original data by the per cents of width as given at the points on the plan form curve which is to be used. As has been stated above, plan form No. 1 (fig. 18) will be used in this design. Figures 19 and 20 are given as examples of other plan forms.

In the example chosen the blade width at the 0.15 R station is found by multiplying 10 by 0.710, giving 7.10 inches. In like manner the other stations are found, viz, at 0.30 R station, 10×0.870 equals 8.70 inches. At 0.45 R station, 10×0.980 equals 9.80 inches; at 0.60 R, etc.

These points are plotted as $Y$ coordinates and a smooth curve drawn through them. The width of each successive 6-inch station is read directly from this curve. In drawing up the blade it is not customary to adhere strictly to either the width or thickness, as given by the curves in figure 18, of that portion of the blade within one-third radius of the hub. It is a better practice to plot in the hub sections by running the lamination lines into the hub in smooth curves, being careful to keep the width and thickness of the root section above what is called for on the curve.

After the blade widths are obtained the next step is to find the maximum thickness to be used in plotting the sections. This thickness of blade is determined from the minimum curve, or any per cent increase of same given in the design data, by multiplying the blade widths as just found by the per cent of thickness as given at the station points on the camber curve, figure 4.

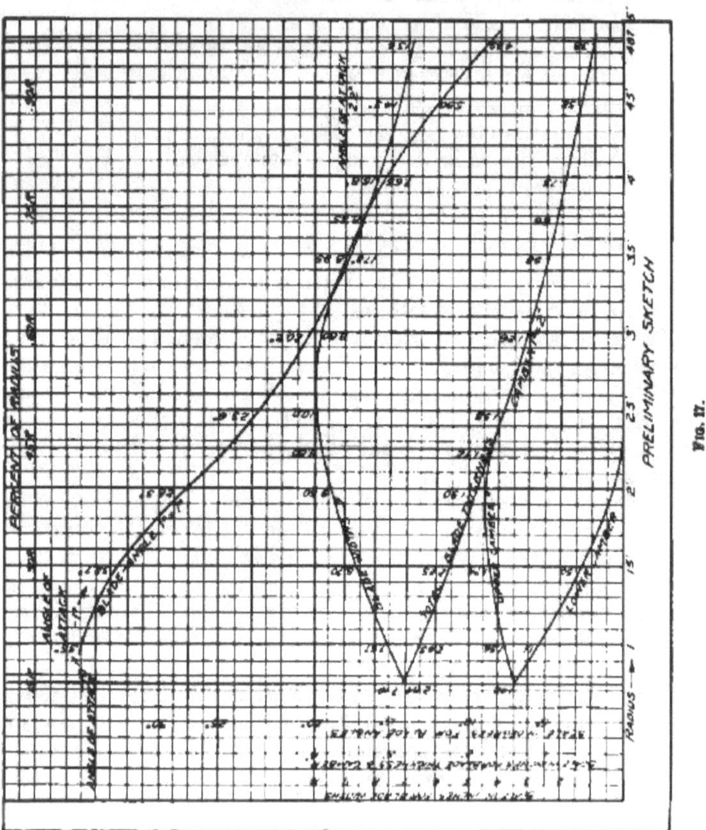


Fig. 17.

On the preliminary sketch (fig. 17) the thickness at the 0.15 R is found by multiplying the blade width just found (7.10 inches) by the ordinate, 0.4, taken from the camber curve for this station. The thickness for the 0.15 R is in this way found to be (7.10 × 0.4) = 2.84 inches.

At station 0.30 R, thickness equals 8.70 × 0.258 = 2.25 inches.

At station 0.45 R, thickness equals 9.80 × 0.175 = 1.72 inches.

At station 0.60 R, etc.

These points are then plotted as $Y$ coordinates and a smooth curve drawn through them. The thickness to be used in plotting each successive 6-inch station of the propeller blade is read directly from this curve. In case the data as given calls for a certain per cent increase in the minimum camber curve, proper ordinates may be found tabulated below the minimum camber curve.

The total camber curve, as well as the curve for the top and face camber, should always be plotted. In general, it is desired to keep the face of the blade flat as far in toward the hub as possible, cambering only after the 18-inch station is reached. Often this is not practical, as much smoother lamination lines may be secured by

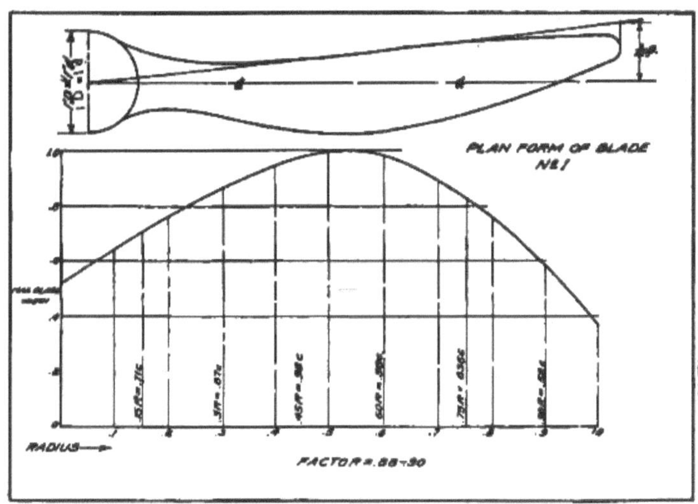


FIG. 18.

cambering the face side farther out on the blade. The total camber curve may then be used as a guide to the total thickness of the section desired.

The ordinates of the airfoils may now be figured and tabulated. These are found from the chord of the air foil at any station and the blade thickness at the corresponding station. The chord of the air foil to be used is divided into a certain number of parts, usually into 10 equal parts, with the leading edge division subdivided into halves and quarters, as shown in figure 21. The chord of each propeller section is divided into the same number of parts. To plot the section it is essential to know the ordinates at these different divisions. These ordinates are figured by taking the ratio of the maximum thickness as read from the preliminary data sheet to the maximum thickness of the air-foil section which is to be used and multiplying this ratio by the ordinate given on the air-foil section at each successive division.

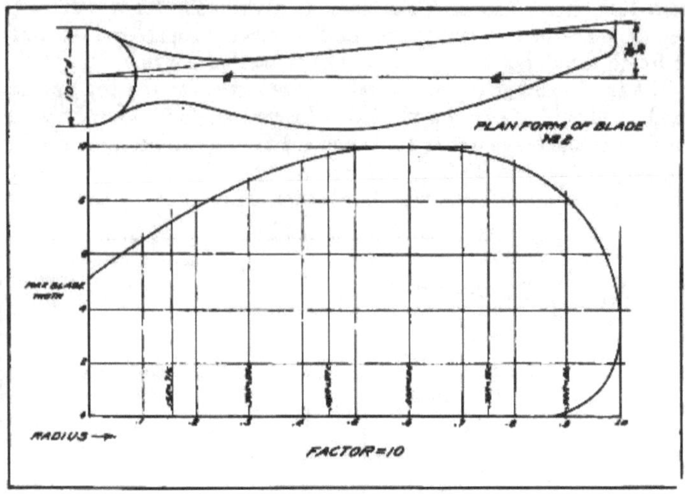


FIG. 19.

Example: Assume that the maximum thickness for some station of the propeller blade is 0.52 inch. Using the R. A. F. No. 6 modified air foil given in figure 20, the first ordinate would be $\frac{0.52}{1.00}\times 0.41=0.21$ inch; and so on for the rest of the ordinates.

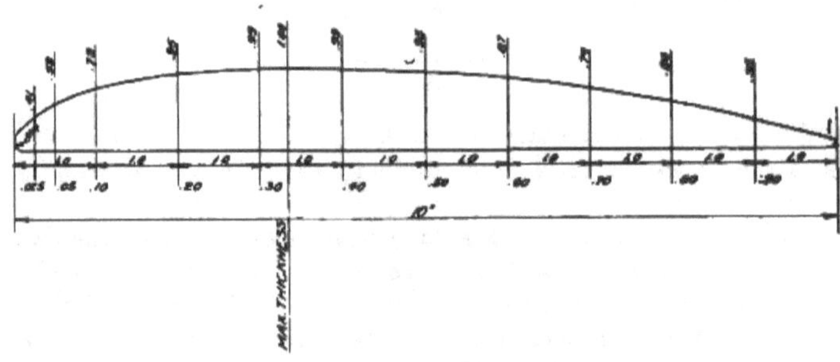


R.A.F. 6 MODIFIED

FIG. 20.

The blade angle must next be found at the different 6-inch stations. This is done by first finding the effective pitch angles of the different stations and then adding the angle of attack. The effective pitch angle at any station may be defined as the angle whose tangent is the velocity of advance divided by the peripheral velocity of the chosen station.

Let $\phi$ = apparent effective pitch angle.
$V_1$ = velocity of the plane in feet per second.
$r$ = the radius in feet of the station in question.
$n$ = revolutions per second of the propeller.

$$\text{Tangent } \phi = \frac{V_1}{2 \pi r n} \tag{1}$$

After the pitch angle for any station has been found, the angle of attack should be added to find the blade angle at that station. The angle of attack is given in the original data. It is often con-

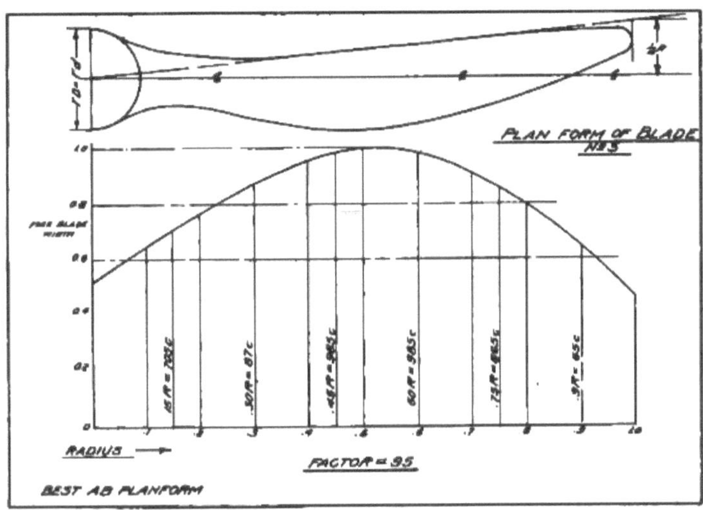


FIG. 21.

stant for all sections of the propeller blade. It may, however, be varied at the discretion of the designer to give the desired lift at each section.

A convenient method for finding the pitch angles is shown in figure 22. The inclined line $E\,B$ is drawn at any angle to the horizontal line $E\,D$. On the line $E\,B$ is laid off the effective pitch $P_e$ equal to $\frac{V_1}{n}$, where $V_1$ is the advance of the plane in feet per second and $n$ is the r. p. s. On this line $E\,B$ are also laid off the blade stations, using some convenient scale, such as 1 inch equals 1 foot.

On the line $E\,D$ is laid off the cotangents of angles starting at point $E$ with 90°. From the $P_e$ point $B$ a line is drawn to intersect the line $E\,D$ at the point representing the cotangent of 9°.

Then draw lines through each station division on the line $E\,B$ and parallel to the line $B\,D$ to intersect the line $E\,D$. The effective pitch

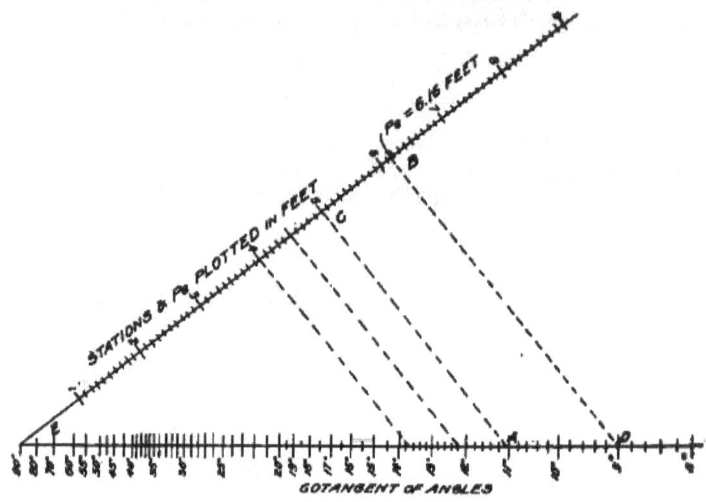


EFFECTIVE PITCH ANGLE

FIG. 22.

angles may then be read directly on the lower scale at the points of intersection. The proof for this is as follows:

Since the apparent effective pitch is equal to the advance per revolution—

$$P_e = \frac{V_1}{n} \tag{2}$$

For any station—

$$\operatorname{Tan} \phi = \frac{V_1}{2\,\pi\,r\,n}$$

$$= \frac{P_e}{2\,\pi\,r} \tag{3}$$

and

$$2\,\pi\,r \tan \phi = \frac{P_e}{r} \tag{4}$$

Therefore,

$$\frac{2\,\pi}{\cot \phi} = \frac{P_e}{r} \tag{5}$$

Now, from a table of natural cotangents, find an angle whose cotangent is equal to $2\pi$ (6.2832). This is approximately 9°.

We can now write—

$$\frac{\cot 9^\circ}{\cot \phi} = \frac{P_e}{r} \text{ (approximately).} \tag{6}$$

Substituting values given in figure 21, we have,

$ED : EA :: EB : EC$. (Accurate enough for all practical purposes.)

A convenient form for tabulating the foregoing data is shown below.

Propeller No. ................

...... blade propeller. Diameter, ...... ft. ...... in. ...... hand screw. Radius, ...... ft. ...... in.

| | |
|---|---|
| Engine.................................... | Diameter of propeller hub.................... |
| Motor H. P................................ | Thickness of propeller hub................... |
| Motor speed............................... | T. E. to advance........................lams |
| Plane speed............................... | Maximum blade width.......................... |
| Plan form of blade........................ | Calculations by.............. Date.............. |
| ........................... camber curve. | Commenced by.............. Date.............. |
| ........................... air foil. | Finished by.............. Date.............. |

$P_e =$

Stations.......................................................................

Effective pitch-angle..........................................................

Angle of attack................................................................

Blade angle....................................................................

Blade widths...................................................................

Maximum thickness..............................................................

Air foil........................................................................

...............................................................................

...............................................................................

...............................................................................

Here are tabulated the divisions into which the chord of the air foil is divided and the corresponding ordinate plotted in the proper station column.

.........................................

.........................................

.........................................

*The propeller layout.*—The propeller blade may now be laid out without further computation.

This layout should always be full size, showing half the blade only. There are usually two projections of the propeller drawn, and the cross sections of the blade at each 6-inch station of the radius commencing 12 inches from the center line of the motor shaft. (See fig. 6.)

A convenient arrangement is to locate the plan view on the center line drawn through the middle of the sheet, with the side view (or elevation) taken in the first quadrant above. The cross sections of the blade at each 6-inch station, taken looking toward the tip of the propeller, are plotted below the plan. Both the side elevation and the sections are considered in the first quadrant and might be plotted below the plan, or even one on top of the other. The maximum thickness of the blade is plotted at the bottom of the sheet below the blade sections. For the above arrangement it is almost essential to use a sheet of paper 42 inches in width. The paper should be ruled very accurately. Any inaccuracy in ruling will show up later and cause trouble. If a number of propellers are to be laid out it pays materially to have the ruling done mechanically.

The best method of procedure in laying out propellers is to draw the propeller directly on vellum tracing paper. This will net a large saving of time, as no retracing of the finished design is necessary.

When working directly on vellum tracing paper it will be well to adhere to the following suggestions:

If the lamination lines are ruled in pencil, not mechanically, but by hand, they should be drawn on the reverse side of the paper. This will keep the lines from erasing on changing the sections. Ruling should be done with a pointed pencil in order to insure a sharp, distinct line. The line should be heavy enough to work from but not heavy enough to blue print. Only such lamination lines appear on the finished drawing as actually come within the borders of the plan view, side elevation, and sectional views.

When the lamination lines are ruled mechanically this ruling should be done with a faint, red-colored water ink in the following manner:

A pencil sample of the ruling, giving the exact dimension of all lines, size of sheet, etc., should be submitted to the printer.

The ruling can be done on a large horizontal ruling machine. The printer will arrange the pens on the ruling beam according to the sample submitted. A small length of paper is then turned through the machine and this sheet is used for checking, so as to make absolutely sure that all lamination lines are very accurately spaced. Should any lamination line be found in error the printer can easily reset the pen found out of place. This sample test should be continued until the machine is perfectly set.

A roll of 42-inch vellum paper can now be fed through the machine continuously, providing the paper is started through the ruling machine straight. This will eliminate kinks in the lamination lines. Only the lamination lines can be ruled mechanically. The station lines commencing 13 inches from the center of the hub must be ruled in by hand when the sheet has been cut to the desired length.

After the paper has been ruled it should be rolled tightly and replaced in the tubes, and taken out only as needed. This is done because light will eventually bleach out the lines of water-colored ink. The horizontal necessary lines are shown in figure 23.

(*a*) Lamination lines near the top of the sheet for the side elevation. A very good width of lamination to use is $\frac{3}{4}$ inch.

(*b*) Center line for plan view of blade to be near the center of the sheet.

(*c*) Below the center line rule in another set of lamination lines for the sectional views.

(*d*) About 1 inch below the last lamination line rule in a single line used as a base for plotting maximum thicknesses.

Starting 12 inches from the center of the hub, vertical lines are drawn every 6 inches. These are station points of the blade at which the sectional views are taken.

As each section is considered to be, in general, representative of the blade for 3 inches each side of the section, the final section should be taken at a point halfway between the blade tip and the 3-inch point beyond the last even 6-inch station.

*Layout of plan view.*—The following description will be confined to the layout of propellers in which the extension of the trailing edge will pass approximately through the center of the hub in the plan

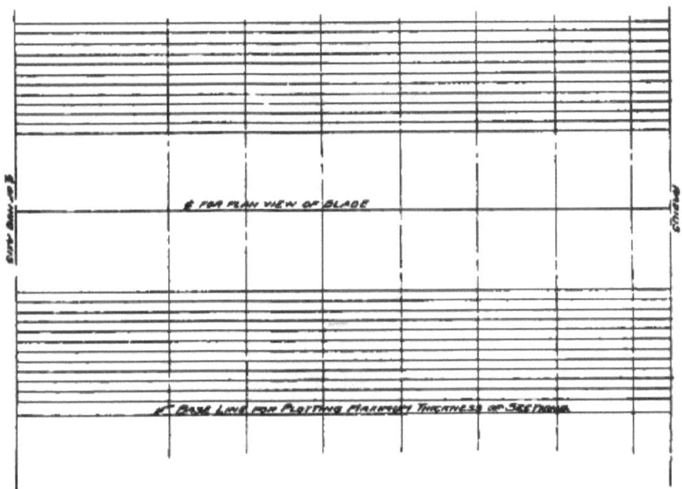


FIG. 23.

view and in the side elevation curve slightly downward toward the tip. When the propeller is so designed that the thrust and torque at each section are balanced by the displacement of the center of gravity away from the center line of the propeller, this line of center of gravities plotted in both projections is used instead of the trailing edge for laying out the propeller.

*Layout for a right-hand propeller, as shown in figure 24.*—The first step is to lay in the hub circle *A* on the plan view, and to draw a radial line *B* to a point *C*, located one-tenth or one-twelfth of the radius above the center line at a distance from the center equal to the radius of the blade. A smooth curve *D* is drawn tangent to the upper side of the hub circle and to this line at a point about one-third out. Starting at the two-thirds point, the radial line is curved down until it runs, at the end, very nearly parallel to the center line of the plan view.

These two curves, together with the remaining piece of straight line, constitute the plan view of the trailing edge of the blade.

The trailing edge $L$ in the side elevation is next plotted. Select one of the upper laminations to represent the back face of the hub. Draw a long gradual curve $L$ from here, dropping the specified number of laminations to the blade tip. This line makes the trailing edge which is thus located in two planes. The trailing edge of the blade sections may now be located in the lower set of lamination lines, as at $J$. Through these points the corresponding

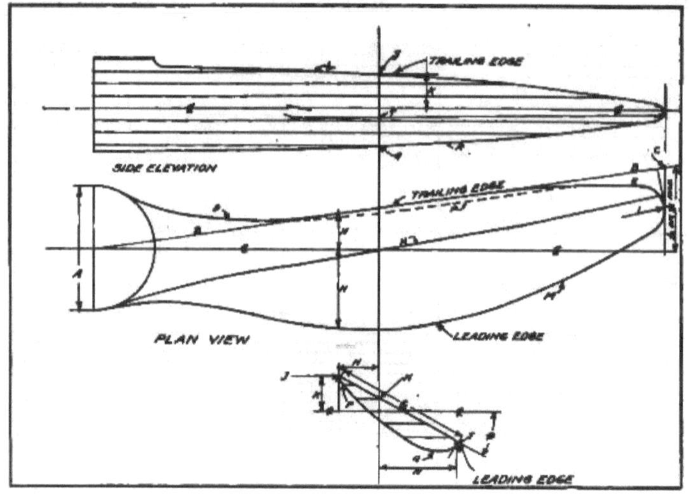


FIG. 24.

blade angles $\phi$ are drawn and the blade chord $C$ laid off along this line commencing at the trailing edge.

The leading edge of each section is then plotted in the plan view and the line $M$ connecting these points drawn in, forming a smooth curve.

The tip is formed by an arc of a circle drawn tangent to the leading and trailing edges just plotted and an arc formed by the propeller radius. In laying in the blade, the flat, or thrust face, is considered as being up and the sections taken looking toward the tip.

After the propeller outline has been plotted the face lamination lines $N$ should be plotted in the plan view. These lines must form smooth, regular curves and run tangent to the hub at the lower side.

The sections are next plotted and the corresponding lamination lines drawn in the plan view. These curves $P$ must be smooth and regular, and are shown dotted as they are on the under side of the blade. They run in tangent to the hub circle on the upper side.

The points $T$ where the angle lines intersect with the blade width lines at the leading edge of each station are then plotted.

A line through these points must form a smooth regular curve and should be drawn for about two-thirds of the way in from the tip, as shown in the side elevation by line $T$.

The point falling lowest on curve $Q$ of each sectional view is plotted in the side elevation. A curve $R$ is drawn through these points and continued into the hub. This curve must be smooth and regular, as it completes the side view of the blade and indicates any irregularities which may have occurred from the choice and plotting of the sections.

The hub in the side elevation is made of the proper thickness to meet the requirements of the motor hub. If the blade must be flattened down at the hub, this is done in such a manner that the center of gravity line of the blade runs through the center of the hub. If the blade must be made thicker at the hub, this is done by adding the necessary thickness to the engine face of the propeller.

The line of maximum thickness is plotted from the maximum thickness of each section in such a way as to give the correct thickness

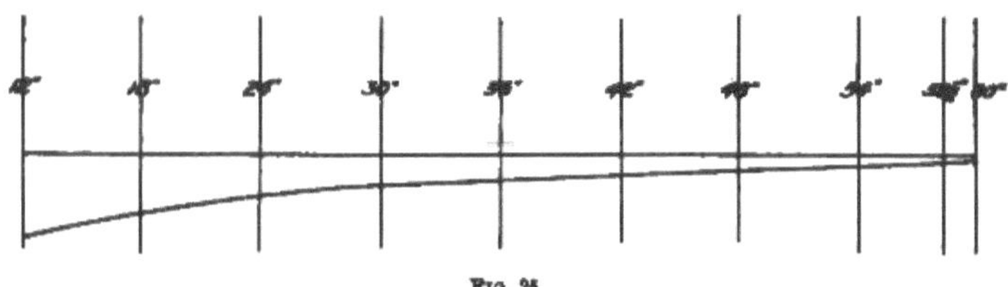

FIG. 25.

if the sections were calipered. This line of thickness must form a smooth curve, falling away regularly from the base line as it approaches the hub, as shown in figure 25.

*Dimensioning.*—The drawing should now be thoroughly dimensioned, as shown in figure 6.

*Side elevation.*—A dimension should be given from the trailing edge at the last station to the exclusion of the line representing the back face of the hub, and from the lowest point of the cambered surface to the extension of the line representing the front face of the hub.

Each station is dimensioned from the trailing edge to the center line, from the leading edge to the center line, and from the center line to the lowest point on the cambered curve.

*Plan view.*—A dimension should be given from the center line to the trailing and leading edge at each station.

The radius at the tip should be given.

The hub diameter and the drilling dimensions should appear in this view.

The method of locating the dowel should be clearly indicated, as this determines the location of the bolt holes and the position of the propeller at the time of gun fire. This location for the more common service engines is given in figure 26.

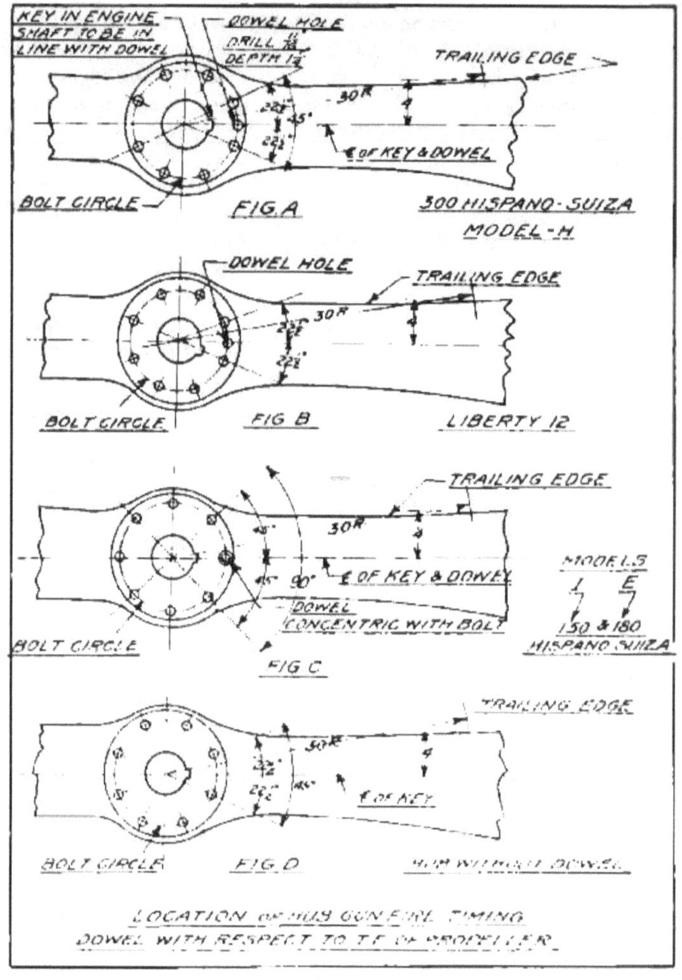


Fig. 26.

*Sectional view.*—Each section should be dimensioned as to width and thickness. The coordinates by which the section was originally laid out should be clearly indicated.

The angle between the chord of the section and the plane of rotation of propeller should be given to the nearest tenth of a degree.

The areas of the sections should be planimetered and given below the corresponding section.

The section through the hub should show the permissible wideners and the minimum distance of the widener joints from the hub hole.

The line of thickness is dimensioned at each section. This dimension will correspond in each case with the maximum ordinate of the section.

A standard hub marking, as shown in figure 27, must appear on all drawings.

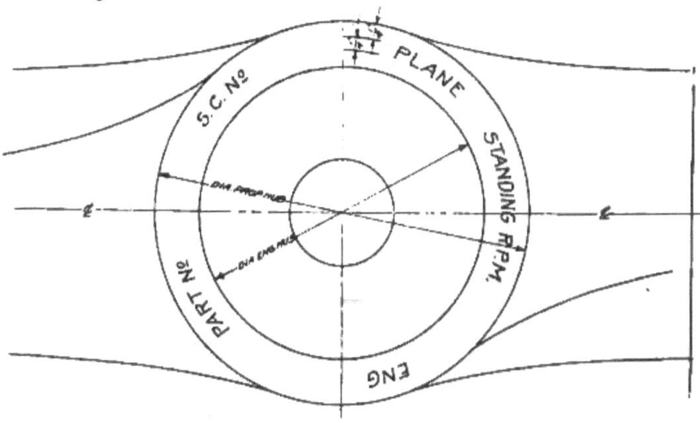


PROPELLER MARKING
ON FACE OF HUB AWAY FROM ENGINE

FIG. 27.

If the outer flange of the engine hub is of nearly the same diameter as the wood propeller hub, this marking will have to be suitably arranged on the side of the wood hub.

After the drawing has been completely dimensioned there is a considerable amount of information, either of general or special character, that must appear on the drawing. This must be supplemented from time to time by the designer, but in most cases it is covered by the following suggestions:

The leading and trailing edges of the propeller, as well as the engine face of the hub, should be plainly indicated. It is customary to take propeller sections looking toward the tip of the blade. Whether this practice or some other is followed, it should be plainly indicated on the drawing. It is well to indicate by a large subtitle,

161988—21——4

as well as by an arrow near the hub, the direction of rotation of the blade, whether right hand, left hand, pusher, or tractor.

All blades should have a fabric, hide, or other covering. The kind of covering used, together with the Government specification number for same, should appear on the drawing. Also the designer should specify of what wood or woods the propeller may be built. These woods, together with the specification number for same, should always appear upon the drawing. The specifications for all propeller materials are given in Chapter II.

The Government specification number on the construction of propellers is 29500–D. A note stating that the propeller should be

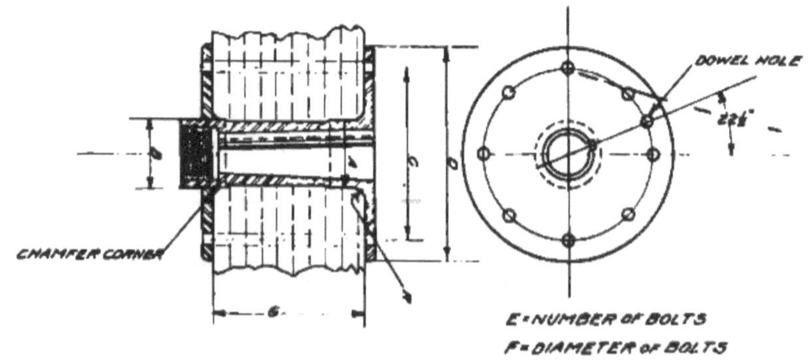


Fig. 28.

built in accordance with this specification should appear on the drawing.

The pitch figured at two-thirds radius and given in the design data should appear on the drawing.

Upon completion of the drawing the general information should be listed on a card for ready reference. On one side of the card should appear the design data and the general dimensions of the blade. On the reverse side there should be space for data obtained from at least two flight tests. The form for such a card is shown below.

Dimensions for boring propeller hubs to fit the metal hubs for a number of different aircraft engines are given in figures 28 and 29

*Propeller hub dimensions on several type aircraft engines.*

(See fig. No. 29.)

| Name of engine. | H. P. | Engine r. p. m. | Propeller r. p. m. | A. | B. | C. | D. | E. | F. | G. | Chamfer corner. | R. | Remarks. |
|---|---|---|---|---|---|---|---|---|---|---|---|---|---|
| Bugatti | 500 | 2,150 | 1,464 | 3 11/16 | 3 11/16 | 7.874 | 9 1/4 | 8 | 0.625 | 8–8 1/4 | [illegible] | [illegible] | Bore hub holes 3 1/2 |
| B. R. 2 | 230 | 1,300 | 1,300 | 2.756 | 2.756 | 6.625 | 7 7/8 | 8 | .500 | 6 | [illegible] | [illegible] | |
| Clerget | 110 | 1,200 | 1,200 | 2.756 | 2.756 | 5.905 | 7.48 | 8 | .394 | 6 | [illegible] | [illegible] | |
| Do | 130 | 1,200 | 1,200 | 2.756 | 2.756 | 5.905 | 7.48 | 8 | .394 | 6 | [illegible] | [illegible] | |
| Curtiss OX5 | 90 | 1,450 | 1,450 | 2.690 | 2.690 | 5.250 | 6 | 8 | .375 | 4 | [illegible] | [illegible] | |
| Curtiss V2 | | 1,450 | 1,450 | 3.250 | 3.250 | 8 | 10 1/4 | 8 | .760 | 5 | [illegible] | [illegible] | |
| Curtiss K12 | 375 | 2,250 | 1,350 | 2.500 | 2.500 | 8 | 10 1/4 | 8 | | | [illegible] | [illegible] | |
| Curtiss K6 | 150 | 1,700 | 1,700 | | | | | | | | | | |
| Gnome | 100 | 1,200 | 1,200 | 2.756 | 2.756 | 5.905 | 7.480 | 8 | .394 | 6 | [illegible] | [illegible] | |
| Do | 150–160 | 1,200 | 1,200 | 2.756 | 2.756 | 5.905 | 7.480 | 8 | .394 | 6 | [illegible] | [illegible] | |
| Gnat | 40 | 1,950 | 1,950 | 2.000 | 2.000 | 4.750 | 6 3/8 | 8 | 3/8 | 3 3/4 | [illegible] | [illegible] | Left hand. |
| Hall Scott A-7-A | 100 | 1,400 | 1,400 | 2.750 | 2.750 | 6 | 7 | 8 | 7/16 | 4 1/4 | [illegible] | [illegible] | |
| Hall Scott A-5-A | 125 | 1,400 | 1,400 | 2.750 | 2.750 | 6 | 7 | 8 | 7/16 | 4 1/4 | [illegible] | [illegible] | |
| Hispano Suiza-A | 150 | 1,450 | 1,450 | 3.146 | 3.146 | 6.693 | 7 1/2 | 8 | .470 | 7 | [illegible] | [illegible] | Obsolete. |
| Hispano Suiza-I | 150 | 1,600 | 1,600 | 3.146 | 3.146 | 6.693 | 7 1/2 | 8 | .470 | 6 1/2 | [illegible] | [illegible] | |
| Hispano Suiza | 180 | 1,600 | 1,600 | 2.756 | 2.756 | 5.905 | 7 1/16 | 8 | .433 | 6 | [illegible] | [illegible] | French. |
| Hispano Suiza-E | 180 | 1,750 | 1,750 | 3.146 | 3.146 | 6.693 | 7 1/2 | 8 | .470 | 6 1/4 | [illegible] | [illegible] | |
| Hispano Suiza-F | 200 | 2,100 | 1,575 | 3.146 | 3.146 | 6.693 | 7 1/2 | 8 | .470 | 6 1/4 | [illegible] | [illegible] | |
| Hispano Suiza-H | 325 | 1,800 | 1,800 | 3.146 | 3.146 | 6.693 | 7 1/2 | 8 | .560 | 6 1/4 | [illegible] | [illegible] | |
| Le Rhone | 80–90 | 1,200 | 1,200 | 2.362 | 2.362 | 4.724 | 5.905 | 8 | .394 | 4 1/4 | [illegible] | [illegible] | |
| Do | 110 | 1,200 | 1,200 | 2.756 | 2.756 | 5.905 | 7 1/2 | 8 | .433 | 5 1/4 | [illegible] | [illegible] | |
| U. S. A. 12-A | 410 | 1,700 | 1,700 | 3 1/4 | 3 1/4 | 8 | 10 | 8 | [illegible] | 7 1/4 | [illegible] | [illegible] | |
| U. S. A. 12 Epicycl | 435 | 1,900 | 1,140 | 3 11/16 | 3 11/16 | 8 | 10 | 8 | [illegible] | 8 1/4 | [illegible] | [illegible] | |
| U. S. A. 8 | 280 | 1,800 | 1,800 | 3 1/16 | 3 1/16 | 8 | 10 1/4 | 8 | [illegible] | 6 1/4 | [illegible] | [illegible] | Not standardised. |
| U. S. A. 8 | 200 | 1,800 | 1,800 | 3 1/4 | 3 1/4 | 8 | 10 | 8 | [illegible] | 6 3/4–7 1/4 | [illegible] | [illegible] | Do. |
| U. S. A. 12-C | 430 | 1,875 | 1,250 | 3 3/8 | 3 3/8 | 8 1/2 | 10 1/4 | 8 | [illegible] | 9 1/2–10 1/2 | [illegible] | [illegible] | |

FIG. 29.

[Face of card.]

Propeller No........
Superseded by......

......Dia. ......Blades. ......Screw.
Motor.................. H. P............. ............r. p. m.
Plane design............ Speed............ ............m. p. h.
Maximum blade width............ $P_e$.............
Plan form of blade............. T. E. advanced............lams.
Hub diameter............ Pitch............ Hub thickness..........
Stations..................................................................
Blade angle...............................................................
Angle of attack...........................................................
Maximum thickness.........................................................
Chord length..............................................................
Area......................................................................
Remarks (for any peculiarities of propeller)..............................

[Back of card.]

TEST DATA.

| TEST 1. | TEST 2. |
|---|---|
| Test No.............................. | .......................................... |
| Plane................................ | .......................................... |
| Engine............................... | .......................................... |
| Standing r. p. m..................... | .......................................... |
| Standing thrust...................... | .......................................... |
| Max. speed......m. p. h. Alt......ft. | ............m. p. h. Alt...........ft. |
| Max. r. p. m. level.................. | .......................................... |
| Thrust level......(from thrust meter). | .......................................... |
| Climb 10,000 ft. in...............mins. | ......................................mins. |
| Ceiling...........................ft. | ........................................ft. |

## CHAPTER II.

### MANUFACTURE OF PROPELLERS.

Although all of the various points and questions pertaining to both the materials and the construction of airplane propellers are covered by specifications gotten up and furnished by the Air Service, it is the purpose of this chapter to cover the whole story of airplane propeller manufacture, from the rough, unseasoned lumber clear through to the finished article, as we see it on the airplane ready to fly.

It is hoped that this connected story will aid the inexperienced propeller maker to a better understanding of how to apply the various specifications which are quite difficult for any but the most experienced propeller makers to follow, owing to the fact that they constitute a broken story having neither head nor tail. This is especially true for the men who have to repair propellers constantly in the flying fields and who occasionally even have to make new ones in order to keep their planes in commission. Furthermore, some of the specifications are so technical as to be difficult for the ordinary workman to follow. Although this chapter is necessarily brief, we hope to go far enough into details to make our subject clear and to help the reader to make proper use of the regular specifications, the most important of which will be referred to by number in the order in which they should be used.

These specifications must, therefore, be considered as supplementary to this chapter; hence the reader will get most of his specific details from them.

*Preliminary trip through the plant.*—Let us now turn to our subject and trace the manufacture of the propeller through all its steps in regular order from the raw material to the finished product. We will begin by getting a general glimpse of the propeller plant as a whole and then make a study of the several departments as to layout, equipment, and operation.

We will first visit the lumber sheds where the supply of green or rough lumber is stored for protection from the weather. These sheds are about the same as are usually found in ordinary lumber yards for the protection of the finer grades of lumber. Next we come to the dry kilns, which may be large or small, according to the needs of the plant. The casual observer thinks that the kilns are merely hothouses, in which the lumber is baked like bread until it is dry. While that may be true in some cases, it does not apply to the kilns in which we dry lumber for propellers, as we shall see later on.

We then come to the conditioning room where kiln-dried lumber is stored. Next is the mill where we find planing, wood inspection, band-sawing laminations, and boring and tooth-planing them. Further on we find the laminations stored in open racks in a room where men are at work weighing them and assembling them into groups before taking them to the glue room.

The hot, oppressive glue room, with its hot box, glue heater, and glue presses, is next to be visited. Here we find half-naked workmen gluing up propellers and shoving them aside in clamps to dry.

Then comes the rough shaping department, where the propellers get their first shaping by machines which are semiautomatic. Some of these are duplicating lathes, while others are special forms of multi-spindle carving machines and pattern-shaping rigs. From here we go to the seasoning room, where great numbers of half-made propellers are hanging on racks to dry and to warp, as they please.

Next we step into the carving or real propeller-making room, where each man has a propeller on a bench and is industriously shaping it to its final form as specified on the blue print.

The next operation is in the tipping room, where some sort of a sheathing or tip is applied for the purpose of making the propeller last a little longer.

When all the cutting, scraping, sanding, and tipping is finished, the propeller is taken to the finishing room, where it receives the filler, the varnish, and the finishing touches. It is now turned over to the inspection department, where it is thoroughly inspected for balance, general shape and measurements, material, and workmanship. If found to comply with the drawing and with the conditions as laid down in the several ruling specifications, it is stamped with the specified data and with the stamp of approval, and is immediately crated securely for shipment and storage.

*Detailed study of the plant (main division).*—We will now go back and study the whole course in detail.

Full-sized propeller drawings are always furnished by the Government. These drawings show only the hub and one-half the propeller, so that it is very necessary to be careful about following the center line when laying out the full-length patterns in order to have a true propeller that will line up and balance in all positions.

We can not include a full-sized drawing in this manual. Reference is made to figure 6, which is a fairly large reproduction of propeller drawing No. 34291 for explanatory purposes. Several smaller drawings (figs. 30, 31, and 32) are included merely to give the reader some idea of the variety of propellers that are in standard use. It must be borne in mind that every line on the drawing has a purpose and that the drawing should, therefore, be thoroughly studied before any attempt is made to build a propeller like it. Furthermore, when following a blue or blue-line print, it must be remembered that the print was made wet and that it has done more or less shrinking while drying. Therefore, while you are obliged to follow the print for general shape, it is necessary to work according to the given dimensions to obtain the correct shape. In other words, do not scale dimensions from the print.

The ordinary propeller is built up of laminations of various thicknesses cut from straight, flat boards. Therefore, the side elevation, as given at the top of the drawing, shows the lines between the different laminations. These are also glue lines, since thin sheets of glue extend clear through the propeller at these elevations. By tracing these lines it is also easy to see how long each lamination must be cut when sawing it from the board. Care must be exercised, however, at this point, for it is always advisable to cut the lamination from 1½ to 3 inches longer at each end and five-eighths inch wider all around, excepting near the hub, than the drawing indicates, since if the block should warp a little before being carved out some of the laminations may extend farther than is shown on the drawing. This is equally true where the surface line cuts the glue line at a very small angle on the outer half of the blade. Of course the full-length laminations will not need this extra allowance in length, since they can not change, due to the fact that the propeller will be no longer, even though it warp ever so much. Most troubles with short laminations usually occur at the last couple of feet at the end of the blade. If the rough carving is to be done by machine, we must make special allowance for this by providing extra length or extension on the long central laminations for holding the work in the machine.

The drawing shows each lamination to be three-fourths inch thick, but the manufacturer is permitted to use lumber of any thickness from 1 inch down to one-half inch. (See Specification 29500D.) If different thicknesses are used, it will be necessary for new lines to be drawn on the side elevation, spaced according to the thickness of the lumber to be used, since it is evident that the lamination lines as given on the drawing will not be right for laminations of any other thickness than three-fourths inch. Great care must be exercised in this work in order to make the laminations come out right as they approach the tips. Otherwise when it comes to carving the propeller out some of the laminations may prove to be too short and narrow and others too long and wide. If they are too long and wide they can still be cut off and worked down and will cause no disadvantage other than the waste of lumber and labor; but if they are too short and narrow they will leave holes or channels in the surface of the propeller and the whole thing is ruined and must be thrown away. The plan form or characteristic shape of the propeller is shown by the center drawing. The curved contour lines on this drawing show the edges of the laminations as they will appear on the finished propeller on the flat face. They are, therefore, also the lines where the sheets of glue between the laminations come to the surface. For this reason many people refer to them as "face glue lines." If the thickness of the laminations used differs from that shown on the drawing, they will, of course, have different widths; hence new contour lines will have to be located in order to know what shape to cut the laminations. From these the workman is able to judge the smoothness and regularity of the surface of his propeller, for if there are high or low places in the surface they will show as irregular or reverse curves in the glue lines. These same laminations must also show edges on the opposite or rounded side of the propeller. These are shown on this same drawing by means of dashed lines similar to the solid lines just discussed. In looking at this drawing the workman can imagine that he is looking at a glass propeller with the top lamination lines marked solid and the bottom or back lines dashed. In that way he can imagine thickness and the drawing will be easier to see. .

Now, by referring up to the side elevation it will be noticed that the laminations as shown on the plan or face drawing are exactly the same length as those in the side elevation.

Below the plan or surface drawing we find cross-section drawing showing what the propeller would look like if sawed off at these points. One side of each section is rather straight, while the other is quite rounded, with the thicker part near the leading edge or the edge that goes ahead in the air. A section is shown in the drawing for each 6-inch station of the outer part of the blade. Figures 33, 34, and 35 show full-sized drawings of three stations. Vertical lines from the top of the drawing at 6-inch intervals show where these different sections come in the propeller. The straight sides of these sections represent the top or face of the propeller as shown in the plan drawing, while the curved sides represent the bottom or back face where the contour lines are dashed. Draw horizontal lamination lines through these several sections, as was done in the side elevation at the top of the drawing, and you will find that the points where these lines cross the *straight sides* of the sections agree exactly with the points where the *face contour* lines cross the *vertical lines* across the plan. To check this, measure from the vertical center line (¢) which crosses the cross-section drawing and take the same measure from the center line (¢) in the plan. Proceeding in the same manner, you will find that the lamination lines in the rounded sides at points *A*, *b*, *c*, to *j*, as in figure 33, of the cross sections agree with the dashed contour lines in the plan. Likewise, if you measure from the horizontal center line to the highest points of the section, as at *A*, in figure 35, you will find that these measurements agree with the top line of the side elevation, when measured from the horizontal center line (¢) and on the same vertical line (¢) that also passes through the section. Also, the lowest points of the sections will agree with the lowest points of the side elevation. If you wish to use laminations different from the three-fourths-inch laminations, as given in the drawing, you will first draw your new thickness lines through the cross-section drawings and in the side elevation and then locate the contour lines in the plan drawing from these by means of a pair of dividers.

The last drawing at the bottom of a propeller drawing sheet is marked "Maximum thickness of sections," and is merely a diagram which has nothing to do with the shape of the propeller excepting to show how its greatest thickness gradually tapers off toward the tips. By comparison it will be found that the thickest point on each cross-section drawing agrees with this "Thickness diagram" if measured on the corresponding vertical center line. In fact, owing to the possible shrinkage of the drawing, and consequent warping, it is best

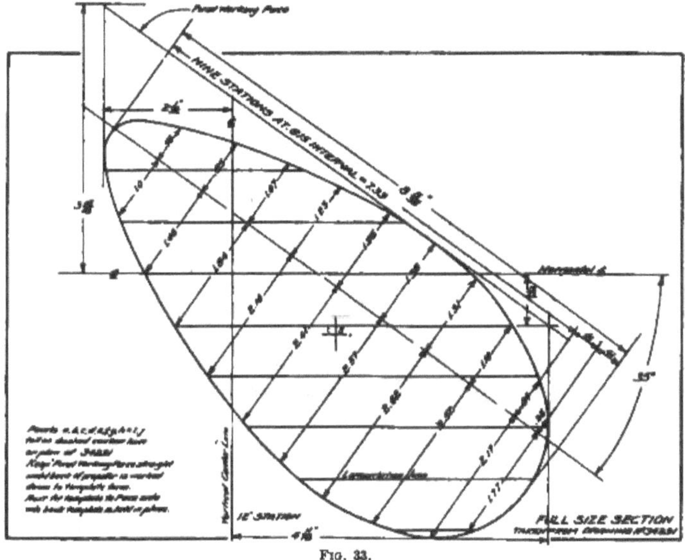


FIG. 33.

to read the dimensions on these drawings for comparison wherever dimensions are given.

Now, on many of the drawings exact dimensions are given in detail on the cross sections, as in figures 33, 34, and 35. These are to help the workman to make a metal template or sheet-metal gauge of exactly the right size and shape for each section, so that he may use them as his guides while carving and scraping the propeller to its proper size and shape at each 6-inch station. It is usually necessary to make templates for the curved side only, and then use a sheet-metal straightedge to represent the straight side. A standard template of this type is shown in figure 52. For best results, it is advisable to cut away a part of the straightedge so as to allow it to really touch the face of the propeller near the edges only, and thus avoid the danger of shaving the edges too thin. You see the flat face of the propeller is shaved down first before any attempt is made to cut the back or rounded side down to template form. If you are not prepared to make an accurate new drawing from dimensions given, the best way to find the proper shape and size for your templates when working from a shrunken blue print on which the dimensions are given is: First, carefully cut the *section drawing* from the blue print; second, cut it in two lengthwise and then crosswise by means of a knife drawn along a straight edge; third, pull these pieces apart until the section is as long and as wide as the dimensions call for. Now pin the pieces down on a soft board by means of thumb

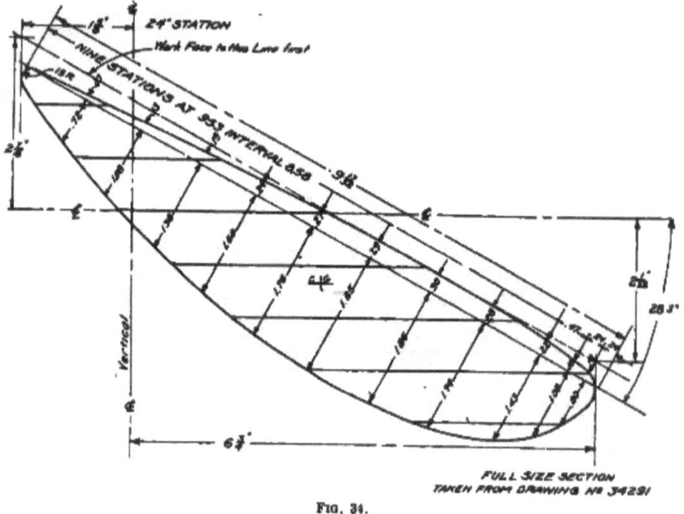


FIG. 34.

tacks and gradually work out your templates until they will exactly fit the contour of the reconstructed drawing.

In this manner the plan drawing may be enlarged to its proper size before the working patterns are made.

If the propeller is too thick or of improper form it will be loggy, and if too thin or out of balance, it will probably flutter and possibly fly to pieces and jerk the engine out of the plane.

We are fully aware that many of our readers do not need the above explanations, yet months of actual experience in both factories and flying fields has taught the writer that the great majority of the men who make, as well as those who use, airplane propellers are not quite clear on many of the above points, hence we know that those who do not need the explanations will pardon us for the sake of those who will appreciate them.

In addition to the small sample drawings contained herein, we have thought it well also to include at the conclusion of this chapter a data table showing the drawing numbers of the propellers which have been adopted. Many other propellers have been designed and built, but this list includes only those which are considered best at present. We know that this will be welcomed by a great many who have been limited to an experience with only a few different kinds of motors, planes, and propellers.

*Inspection.*—Before turning to the actual work of manufacturing we wish to call your attention to the fact that the manufacture of all propellers for the Army is governed by the specifications mentioned herein or by such others as may have been issued from time to time. Furthermore, we wish to state that the contractor is bound by his contract to comply with the above instructions, and it is up to him to do so, whether a Government inspector is on the

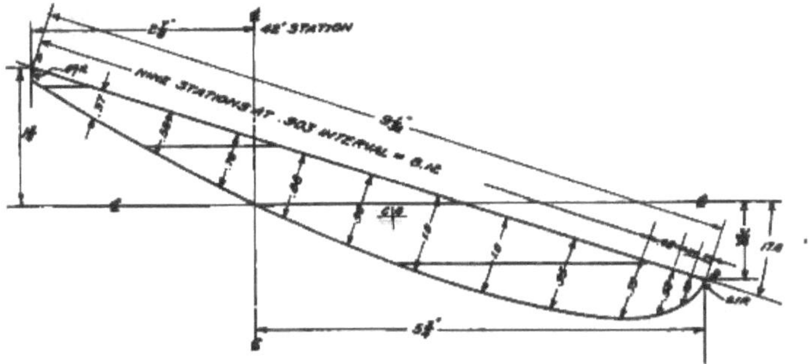

Fig. 35.—Full size section taken from drawing No. 34291.

job or not. Therefore the manufacturer must furnish such inspectors as he finds necessary to follow the work throughout from the raw material to the finished product crated for shipment. In addition to these, the Government usually furnishes inspectors who must also have free access to all parts of the plant at all times in order that they may also know that all directions are being complied with. (See chapter on inspection.) This is very essential, due to the fact that an airplane propeller may so easily be either of no value or of great danger if not made as specified. Here allow us to state that the various designs are not permitted to go out for production until sample propellers have first been made from them and given severe whirling tests and flight trials at the experimental station, McCook Field, to prove their worth and safety. (See chapter on "Destructive propeller testing.")

Presuming that we are to make some propellers according to some drawing which we have selected, or have been furnished, and granting that we have made a thorough study of it, and of all of the above specifications which will apply to the manufacture of this particular kind of propeller, we will now take up a few points on woods and wood selection.

*Woods.*—Several kinds of wood have been approved for airplane propeller construction. To begin with, it is of great importance that a propeller stand up, keep its shape, keep in perfect balance, and resist wear as much as possible. The propeller is the highest speed piece that enters into the make-up of an airplane, yet it is made of wood, due to the fact that no satisfactory substitute has yet been found. This will give the reader some idea of the importance of selecting and preparing his materials with care, as well as being careful throughout the process of manufacture.

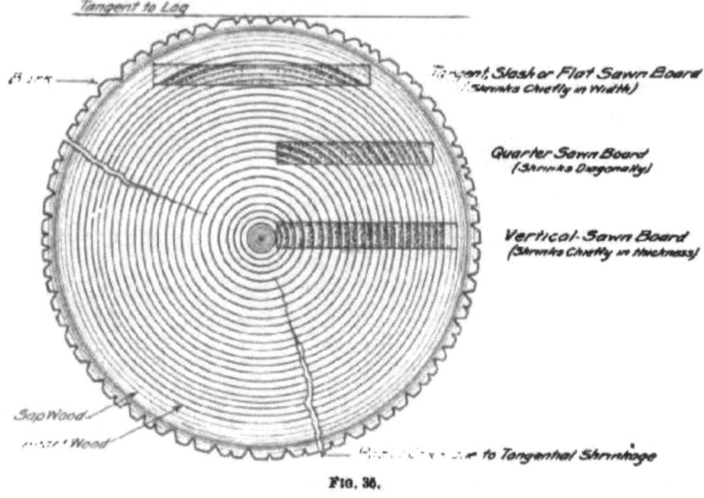


Fig. 36.

In order to understand woods, we must know how they grow. Few people realize that trees grow only on the outside. The small bush puts on a new coat each spring which grows strong during the summer and lies dormant during the winter. Just under the rough bark is a tender white film of bark covering the smooth wood beneath. That smooth wood grew the year before and now serves as a frame on which to build this year's wood, which will thus be found as a layer over the whole tree just beneath the tender bark. Cut the trunk of the tree crosswise and you will see that it is composed of layers of wood almost like the structure of an onion, excepting that the layers of the onion are not attached to each other. These circles on the end of the log are called "annual rings," since they are the cross sections of the layers of wood which have been formed over the tree each year and the age of the tree can be determined by counting these rings. (See fig. 36.)

*Wood inspection.*—For the purpose of classification, all coniferous or cone-bearing trees, such as pine, fir, spruce, cedar, etc., are called soft woods and all others are called hardwoods, while in reality many of the so-called "soft woods" are really quite hard and many of those classed as hardwoods are quite soft. We have really classed them according to the structure and arrangement of their small cells which we study through a microscope rather than according to the way the wood cuts or how hard it is to drive a nail into. This is confusing for a while, but when you once see how the terms are intended to apply you will be able to properly construe our language and not be misled. We can not go too much into details in this brief chapter, but the above explanation will help you to understand various discussions on wood which you may read from time to time, should you care to make a further study of them, and especially of those used in the manufacture of airplanes.

In the soft woods the cells are all alike and appear different in the annual rings merely because of the fact that they begin in the spring by growing quite open and light, and as the season progresses the later cells build on more closely packed and flattened and partially filled with resin, which gives them a yellowish or brown appearance and consequently makes the later summer wood much harder than that which grew first in the spring. This is especially plain in the yellow pines, although less pronounced in white pine and spruce. Of course, the summer wood is the stronger, since it is more dense. (See magnified end wood, fir and spruce, fig. 37.)

The so-called hardwoods differ from the coniferous in that while they, too, have their more open spring growth and denser summer wood in each annual ring, by the aid of the microscope we find that the wood is built up of more than one kind of cells, which makes its structure decidedly different from that of the coniferous. Both classes of woods have countless groups of cells called medullary rays rotating from the heart of the tree to the bark. These are especially large in the oaks, and are known as "flakes" on quarter-sawn lumber. They have a tendency to prevent the wood from shrinking radially or toward the heart, and also make it difficult to split wood in the direction of the annual rings. (See fig. 38.)

The texture of woods varies widely. Some are quite solid; others are so porous that one can blow through a piece of several inches long. Some woods apparently have very large pores, yet neither air nor water can get through them, due to the fact that the pores contain deposits of resin or gum at short intervals. (This is illustrated in the white oak block, fig. 38.) These little obstructions are called tyloses. Red oak does not have these, hence it is too porous for propeller construction. Many of the soft woods have scattered but prominent resin ducts which show as brown or yellow lines running through the spring wood in the direction of the grain. Since it is of the utmost importance that propellers have great strength and evenness of texture throughout, it is best to select wood with quite an even distribution of its cells throughout the annual ring, provided that the other characteristics of the wood are permissible or desirable.

*Wood shrinkage.*—Going back to the log that we sawed off (fig. 36), let us watch it from day to day as it dries out at the end. We know that green wood contains a high percentage of water, hence it must of necessity shrink when it loses this water. The first marked effect that we notice is the appearance of cracks running from the bark toward the heart of the log. They prove that as the log loses its surplus water it shrinks more in the circular direction than it does toward the center. We have names for these different shrinkages. If we lay a straightedge across the log, we say it is "tangent" to the circle represented by the end of the log. Therefore, all shrinkage around the log in the direction of the run of the annual rings is called "tangential shrinkage." A line from the outside of a circle to the center is called a radius. Therefore, all shrinkage toward the center of the log, or crosswise, of the annual rings is called "radial shrinkage." All woods have a greater tangential than radial shrinkage, but in some this difference is not as great as in others. Study table in Specification No. 15020B for a comparison of some of our principal woods.

If this shrinkage difference is too great, the wood will warp so much that a propeller made from it would never hold its shape. This is due to the fact that wood will always take up or lose moisture as the weather changes, causing the propeller to be constantly changing in pitch and in track.

By experimenting we have learned that we can neutralize some of this shrinkage difference by quarter-sawing our logs so that the annual rings run diagonally across the end of the board. In that way a part of each kind of shrinkage is distributed both ways in the board and thus made to equalize to a certain extent.

If, in the sawing of lumber, the boards are cut tangentially from the log, the rings run practically flat across the end of the board and the lumber is said to be flat-sawn or slash-sawn, and its greater shrinkage to the inch will be in the direction of its width and its lesser in the direction of its thickness. If the board was sawn straight through the log from the heart to the bark, the rings will be seen to run crosswise of the end of the board and the board is said to be

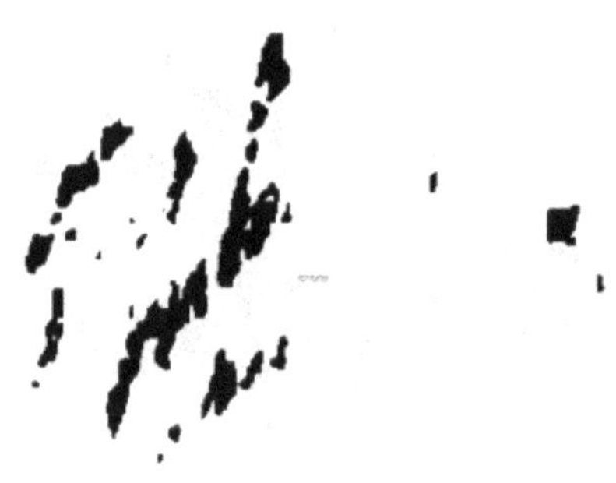

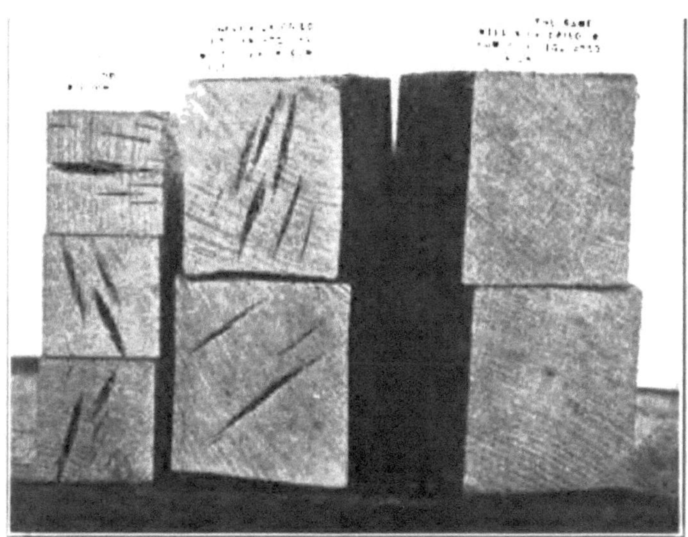

Oak stock honeycombed by air-drying and improper kiln-drying. Also similar stock properly dried.

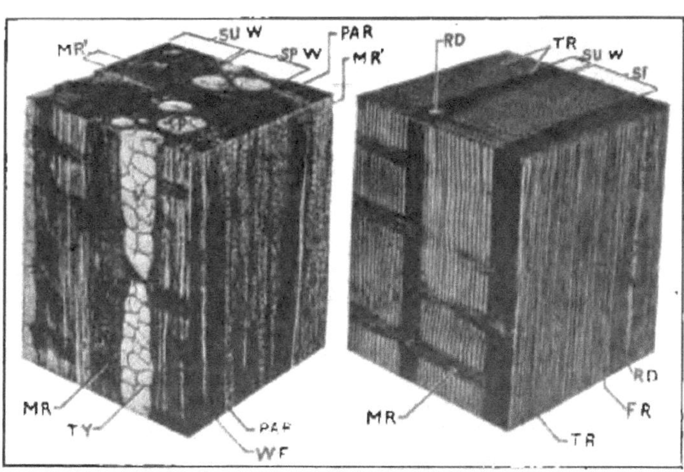


WHITE OAK. SHORTLEAF PINE.

(Photographs of cubes of wood magnified about 25 diameters.)

In each cube the top view represents the transverse or end surface, the left view the radial or "quartered" surface, and the right view the tangential or "bastard" surface. SP W, spring wood; SU W, summer wood.

To the left a hardwood showing: V, vessels or pores; TY, tyloses in a vessel; PAR, parenchyma cells; the dark areas, WF, wood fibers; MR, small medullary rays; MR', large medullary rays.

To the right a coniferous wood showing: TR, tracheids, which comprise the bulk of the wood; RD, resin ducts; MR, ordinary medullary rays; FR, fusiform rav containing a horizontal resin duct.

The medullary rays are continuous from the starting point to the bark, and the vessels are continuous longitudinally, although the illustrations show them interrupted.

FIG. 38.

"vertical sawn." Its greatest shrinkage will be in the direction of its thickness. If you should dry these two boards ever so well, and then glue them together in a propeller, the flat sawn board would try to get wider and the vertical sawn board would try to get thicker as soon as they began to take up a little moisture. This would at least cause warping and the stress thus set up might be great enough to break the glue joint holding them together. Therefore, when it is necessary to use lumber which has been cut in these two ways, you should use only flat sawn lumber in one propeller and only vertical sawn lumber in the other so as to avoid these internal stresses.

*Propeller wood selection.*—As stated above, quarter-sawn lumber is preferred and has been used almost exclusively during the war, excepting in cases of walnut and mahogany, whose shrinkage in the two directions is not really serious. Experience in the flying fields has proven white oak to be about the most durable propeller wood for all-around service, especially where weeds, brush, cotton fields, etc., are to be contended with, but even this wood fails to stand up long under such conditions when not protected by a sheet-metal edging and tipping to take the brunt of the wear. Relative merits of different tippings is discussed in the chapter on "Destructive whirling testing and research." Birch holds second place for general service. It is a tougher wood and is to be preferred for the manufacture of propellers of thin design.

As stated above, walnut and mahogany are rated high from the fact that they hold their shape quite well. However, most of the demand for propellers of these woods is due to the mere preference for the name and appearance of the higher-priced cabinet woods. They will not stand up as well as oak or birch under rough usage.

Cherry and maple have been used with fair satisfaction and yellow poplar has been used a great deal in southern Texas and on the Mexican border, where the air is hot and dry. Poplar is not a strong wood and has not been a success in damp climates, due to its tendency to warp when subjected to change of moisture content.

Philippine lauaan, variously known as "Philippine mahogany," "bataan," and "tanguili," has been used quite extensively as a substitute for mahogany by the Lang, Curtiss, Starr Piano, and Dayton Wright Companies in filling propeller contracts during the war, and it stood up very well when protected by metal tips. It was too soft, however, to give much service when not so protected. Here again is an apparent discrepancy, for while this so-called "Philippine mahogany" was rated and paid for by the Government as a hardwood, the writer knows, from six years' experience in woodwork in the Philippine Islands, that this wood is really the softest, cheapest, and lowest-grade lumber that is milled for the market over there, and that it is never used where a long or durable wood is needed, nor is it used in cabinet work. The use of this wood has been ordered discontinued by the Technical Section.

Although the propeller maker will, as a rule, be told what wood to use, the above discussion will be of some assistance to those who may be obliged to choose for themselves. Let us assume that we are to use oak, since that is one of the standard service propeller woods. If any other wood is to be used, the reader will apply the specifications which cover it and proceed accordingly.

*Purchasing lumber.*—Since it is impossible to get air-dried lumber that can be depended upon for evenness of moisture content, we will get the freshest and greenest oak that we can buy, preferably right from the mill. Weathered lumber is always uneven, usually warped, and often cracked more or less, hence for best results we will take our lumber green. In knowing that short grain, knots, worm holes, and all other defects which may injure the strength or lasting qualities of the wood will eventually be rejected by the inspectors, it is important that we exercise great care in the original selection of our lumber at the mill and thus avoid buying waste. (Study Specifications 15029B, 15041A, and 29500D.)

Lumber which is exceptionally heavy or uneven in weight and that which is unusually light should be rejected at once, since it will interfere with the balancing of the finished propeller and cause its rejection. Sapwood in oak is also worthless and can not be used.

After having studied the contour lines on the drawing you know how large the boards must be in order to cut the desired laminations without undue waste. (See fig. 39.) It is permissible to splice laminations when necessary. (See Specification No. 29500D for kind and direction of joint.) Figure 40 shows a section through a serrated spliced joint, a very good type. All this should be taken into consideration by the man who buys the lumber, otherwise a large stock of high-priced lumber will be accumulated which can not be used for the propellers and thus delay the work and increase the cost by kiln-drying and milling useless material.

Having purchased our lumber, we will store it in a good shed, where it will be protected from sun and rain. It should be open piled on 1 by 1 inch crossers, placed about 2 feet apart, and placed one directly above the other, to allow a free circulation of air and at the same time to hold the lumber straight.

According to Specification No. 29500D, our lumber must be dried down to a moisture content of from 7 to 9 per cent, but according to the writer's field and factory experience and to the experience of the Department of Propeller Experiment and Design at McCook Field, the moisture content should range between 5½ and 7½ per

161988—21——5

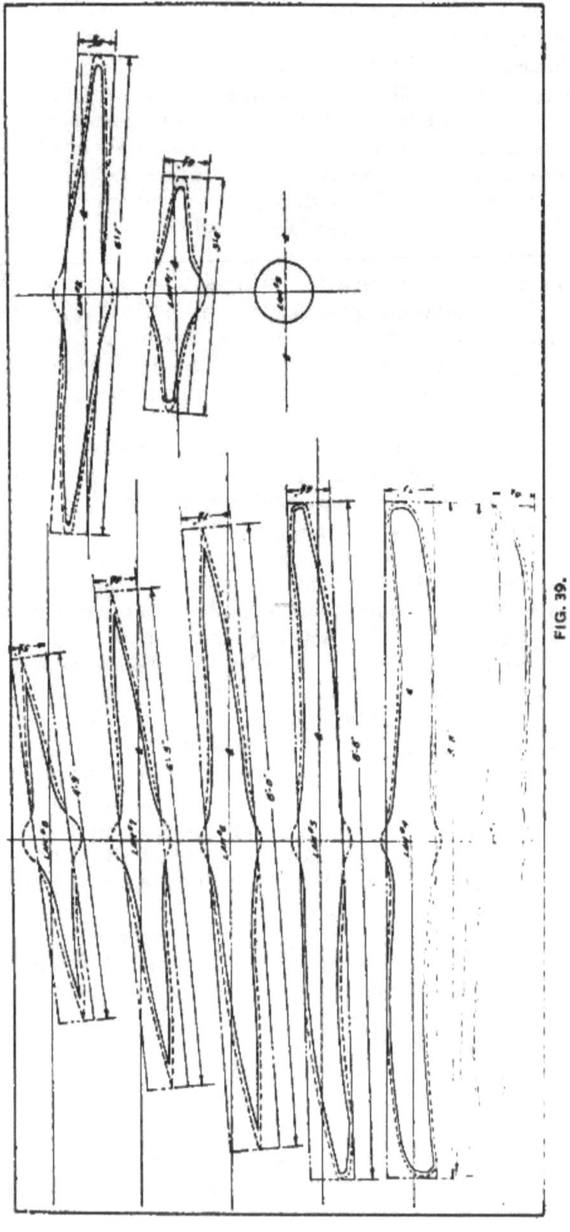

FIG. 39.

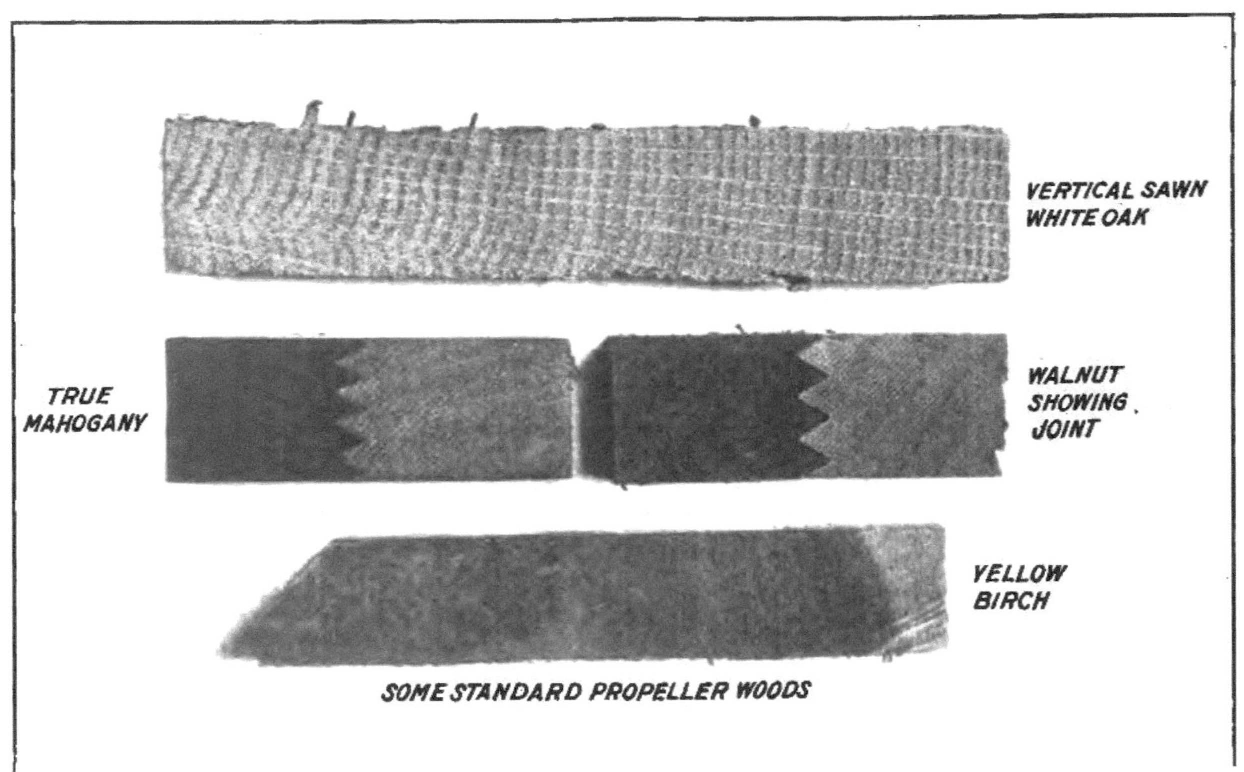


SOME STANDARD PROPELLER WOODS

cent, or possibly as low as 5 per cent, for the best results. Do not risk drying below 5 per cent, for there is danger of baking your lumber, thus making it brash. It is especially important that all the laminations in one propeller have approximately the same moisture content so that all will shrink or swell alike in case of future drying out or taking up of moisture. This protects against warping to a certain extent.

Many people insist that under ordinary circumstances propellers will eventually attain a moisture content of from 9 to 11 per cent, hence they contend that the propeller should be given that much when made. Actual experience does not bear this argument out. In many climates, such as those of southern Texas and Mexico, propellers will lose moisture in service if they contain more than 7 or 8 per cent when they go out.

While kiln-drying the lumber, we must not forget the fact that some moisture will be added to the propeller with the glue.

Hence we must dry the wood lower than the desired moisture content of the finished propeller.

The arguments for drying lumber to a low moisture content are in part as follows: In the first place wood must be dried down fairly low before it will take a permanent set and hold its shape. For that reason it should be dried to around 6 per cent, even if you wanted it to have 9 later. It can then take up an additional 3 per cent and remain more stable than it would have done had it been dried to only 9 per cent in the first place. Furthermore, it is always better for a propeller to take up a little moisture than to lose a little because of the mechanical fact that wood shrinks in losing moisture and, since the moisture near the surface would be the first to go, the surface wood would shrink first and either pull the glue joints open or actually cause cracks in the wood before the center wood began to shrink. This would allow moisture to be taken up readily by these cracks from day to day or be given off as the weather changed, and soon the glue joints would be broken deep into the propeller, thereby causing great weakness. Furthermore, if the wood has been dried lower than it will remain in service, it will take up moisture. Naturally the surface wood will take on moisture, first causing it to swell. Hence, the surface would have a tendency to become more compressed rather than to open as in the other case and no cracks would start. When the moisture finally got out to the center the propeller would still be sound throughout and no harm would have resulted. When propellers lose moisture there is also a tendency for the hub bolts to start cracks, while in case a little moisture should be taken up the same bolts would have a tendency to hold the propeller more firmly.

*Kiln-drying lumber.*—Returning to our lumber, we find it necessary to provide a dry kiln in case we do not have access to a good one. Here let us bear in mind that all green or wet wood is swollen, hence its cells are stretched or distended, which means that they will shrink when the water is taken out if not prevented from doing so. If prevented from shrinking, they must either break or dry in their stretched condition. This explains why boards warp, shrink, and often crack when dried, especially when dried rapidly in the sun or in an overheated kiln where the surface dries first.

We are in a hurry to have the lumber dried so that our manufacturing process may begin, yet there is danger of ruining our lumber if we dry it too rapidly. Therefore we must handle this process with care. The moisture in the wood is largely bottled up in little cells and the only way it can get out is by working its way through the cell walls. This is a slow process if we are to avoid breaking the cells and splitting the wood internally. This excess moisture must be evaporated from the surface of the lumber and be carried off by the air. In order to do this we must surround the lumber with dry air. The mechanical process of doing this is called kiln-drying. If we start this operation too rapidly the moisture will be taken out of the surface wood faster than that at the center of the board can work its way out. In this case the outer shell of the board will either crack from its tendency to shrink while the central part is yet swollen or else it will become stretched and dry hard in that form. Later when the excess moisture has finally gotten out from the center there is a stress in that part of the board due to its tendency to shrink as it dries. This will either cause internal cracks, called honeycombing, as illustrated in figure 38, or the sides of the board will cave in (called collapse) more or less like the sides of a pasteboard box. Then it will set in that form. In either case there will still remain a stress in the central wood. If the board is sawed through the center edgewise to form two thin boards, the two halves will curve inward. This condition is called casehardening and the lumber is unfit for propeller use. (See fig. 41, items 1, 3, 5, 7, and 8.) Oak is especially bad about this, hence it must be dried slowly. It requires from three to four weeks to dry green oak properly to 6 per cent. Specification 20500A should be studied thoroughly before putting any lumber into the kiln. It should then be followed carefully until the charge is taken out. Keep your kiln running steadily.

If the casehardening has not been severe enough to produce honeycomb internal cracks it can be relieved by turning live steam into the kiln for a while before removing the charge of lumber. Steaming softens the outer surface or shell of the boards and allows it to shrink to agree with the central wood. Care must be taken not to steam too long, as oversteaming may set up a reverse condition, which is

FIG. 41.

quite as bad as the first. Test some pieces before removing the charge. (See cuts in Specification No. 20500A; also fig. 41, item 4, for oversteamed samples.)

The safest way in starting a new kiln charge is to turn live steam into the kiln long enough to thoroughly heat your lumber through before any drying of the surface is begun and then begin running your relative humidity down very slowly, as per instructions, and gradually raising your temperature. In the Tiemann kiln you run the relative humidity down by using cooler water in the sprays, or by running cold water through the wall condenser pipes, while in the Cutler kiln it is done by allowing freer circulation of the outside air against the canvas walls. Do not rush this part of the drying while there is so much excess moisture in the lumber, for it can not come out rapidly without causing internal cracks. Keep this in mind, for the writer is speaking from actual experience in drying oak for propellers. We must have control of our heat in the kiln, for it directly affects the rate of drying. A cubic foot of air at a certain temperature will hold just a certain amount of water vapor and no more. If we raise the temperature of that air it will take up more vapor, and so on, always increasing with the amount of heat we add. When the air at a certain temperature has all the moisture it will hold, it is said to be saturated, and we say it has a "relative humidity" of 100 per cent. But if we heat this saturated air without allowing it to get any more moisture, it is no longer saturated and its relative humidity is no longer 100 per cent, but it is something less, because it would take more moisture if it could get it. Relative humidity is the relation between the amount of water vapor present in the air at a given temperature and the amount of water vapor it takes to saturate it at that temperature. Thus the "relative humidity" is constantly changing as the temperature of the air changes. If the air at a specified temperature has all the water vapor it can hold, it has, as above stated, a relative humidity of 100 per cent. If it has only half the amount of water vapor it can hold at that same temperature, its relative humidity is only 50 per cent, and so on. Thus you can always *lower the relative humidity* of the air by *raising the temperature*, or *raise the relative humidity* by *lowering the temperature*. Remember this when you find conditions wrong in your kiln and make your adjustments at once before the lumber becomes seriously affected.

This point has been treated in a very simple way, due to the fact that it is not clear to many kiln operators. This will also help to explain the working of the wet and dry bulb thermometers, which, when used in connection with a graph chart, sometimes called a hydrodyke, can be used as a hygrometer. A hygrometer is an instrument with which to determine the relative humidity of the air. In the thermometer type of hygrometer the dry-bulb thermometer shows the actual temperature of the air, no matter what its moisture content; the wet-bulb thermometer gives a temperature somewhat lower. The difference between the two is called the wet-bulb depression. This difference is a measure of the relative humidity. The low temperature of the wet-bulb thermometer is produced by the evaporation of the water from the wick. The drier the air the more rapid will be the evaporation and the lower the temperature. Hence, the greater the difference between the wet and dry bulb thermometer readings, the lower the relative humidity. When there is no difference between the two readings, there is no evaporation of water around the wet bulb. In this case the air has all the moisture it will hold, and is said to have a humidity of 100 per cent. To obtain the relative humidity from the wet-bulb depression, reference must be made to some standard chart, several of which are in use. (See fig. 59.) These curves are based on computations too technical for the purpose of this discussion. In order to get a reliable reading from the wet-bulb thermometer, it is necessary to fan the air or otherwise keep it moving past the thermometer at a rate of not less than 15 feet per second for about five minutes or longer, and then to make the reading quickly before taking the thermometer outside the kiln. Otherwise it will change and give you the wrong temperature. It is also imperative that the thermometer wick be kept clean and that the bottle be kept full of clean distilled water, so that the wick will not clog up nor have to draw the water more than an inch high to get it to the bulb. The wet-bulb thermometer may remain in the kiln for constant reference, but in this case it is imperative that its water supply and the movement of the air be noted each time before taking a reading. Either *low water* or *slow air* will allow mercury to run up and show relative humidities too high.

Since the relative humidity changes with every change of temperature and with every addition of steam or other moisture, the reading must be taken often in order that steam may be admitted into the kiln in case the relative humidity is found to be getting lower than is permitted by Specification No. 20500A. Of course, it is easy for you to watch the dry-bulb thermometer and keep the temperature right. This *must run steady day and night*, for the sun and the clock have nothing to do with the operation of a dry kiln. They belong to no union and never shut down until the charge of lumber is ready to come out.

There is still another point to be observed. We have discussed the temperature of the air and the relative humidity of the air, but now we must keep this air moving steadily through the spaces between the boards and through every part of the kiln in order to maintain the same conditions everywhere. Otherwise the air will be hot and dry in some places, especially up the passages between the stacks of lumber in the Tiemann kiln and near the ceiling in all types of kilns. At the same time it will be cooler and very humid in the stacks of lumber between the boards where the evaporation from the lumber has filled the air with moisture and at the same time taken some of its heat. This produces unequal drying conditions and some of the lumber will become too dry before that in the middle of the stacks becomes half dry enough.

Thus we have learned that in order to dry lumber evenly and safely we *must have* a dry kiln in which we can absolutely control the *temperature*, the *relative humidity*, and the *circulation of the air*.

Here let us note that wherever drying takes place cooling also takes place. This is just as true for lumber as it is for men. When the air becomes still and ceases to carry the vapor away, evaporation or drying stops and so does the cooling. Therefore, although we may keep the temperature the same, in order to keep all the lumber drying alike we must keep the air moving on all surfaces of every board, so that it will carry the saturated air away and bring drier heated air to furnish more heat to the lumber and to carry away more moisture.

*Types of kilns.*—There are several types of kilns on the market. Only a few of these are worth having, and only one really meets all of the above conditions. This is known as the Tiemann improved water-spray humidity-regulated kiln. It was designed and patented by Mr. H. D. Tiemann, an engineer in the United States Forest Products Laboratory, at Madison, Wis. Though this kiln is patented the public is granted permission to build it free of charge and without any royalty whatever.

Other kilns frequently mentioned are the Grand Rapids, the Morton, the Cutler, etc. The Grand Rapids and the Morton can be changed over to the Tiemann system of control in case they have already been installed, but the Cutler is entirely different. The Cutler kiln is the simplest and by far the most inexpensive to install. It is quite an efficient kiln and will do good work if properly operated. This kiln is patented by the Cutler Co., of Buffalo, N. Y.; nevertheless we will explain its operation.

*The Cutler kiln.*—All that is needed is one wall, a ceiling, a floor space large enough to accommodate the length of your lumber, and 12 or 14 feet wide. The other three walls and even the ceiling when necessary are formed by tent canvas, which may be raised when the lumber is being put in or taken out. This makes it very convenient. Over by the wall are two or more ordinary steam radiators, according to the length of the kiln, and in front of the radiators are common electric fans at about 4-foot intervals. The fans drive the air through the radiators as it comes from the lumber pile. Sheet-iron shields or baffles behind the radiators cause the air to shoot upward and over along the ceiling toward the top of the lumber pile. Plenty of space must be left above the lumber and a clearance of a couple of feet must be left down behind the lumber stack. Very little chance should be allowed for the air to come under the stack of lumber. The lumber is stacked parallel with the wall in front of the radiators and at a distance of about 6 feet from them, or possibly a little more. Crossers, 1 by 1 inch, should be used at intervals of about 2 feet under each layer of lumber in order to allow free passage of air through the stack and at the same time prevent warping. These crosses should be placed over each other in columns. The air passes through the radiators, up the wall, back across the ceiling, and down behind the lumber, from whence it finds its way through the pile of lumber and back to the fans. A diagram of the air circulation is given in figure 42. The air moves in a continuous positive circuit all the time. A thermostat regulates the supply of steam to the radiators, thus keeping the temperature of the air to the desired degree.

The regulation of the humidity is yet to be cared for. Let us trace it. The moisture evaporates from the lumber and is carried away by the air, which in turn comes in contact with the canvas walls which serve as a wick to the outside air. The canvas takes up the moisture from the inside air and allows it to evaporate into the outside air. Then the inside air continues to take more moisture from the lumber, and so on until the lumber is as dry as is desired. Thus it is seen that the same air is used over and over and it is not necessary to allow any to escape or to bring any fresh air in. But suppose that the inside air is still too dry—that is, if our instruments show us the relative humidity of the inside air is too low while the temperature of the air is correct, what shall we do to correct the humidity? There are two ways to do this. Either allow a little steam to escape into the inside of the kiln or shut the outside air away from the canvas walls, so as to prevent too rapid evaporation to the outside air. This can be done by dropping a second set of curtains outside the first or by closing doors if the kiln is inside a building, thereby checking air circulation around the kiln. We can see that the weather outside would have a lot to do with the steady operation of this kiln. For this reason the Tiemann kiln is far better.

One bad feature about this kiln is that the moist air, fresh from the lumber, should go first to the curtain to give up its excess moisture while at its lowest temperature, and then go to the radiators for more heat. In that way it would lose its surplus moisture at a more rapid rate. Some people try to get this effect by driving the air the opposite way, but this again is unnatural, since the cooler vapor-laden air coming out behind the lumber does not like to go up the curtain and across the ceiling to get back to the radiators, since cool air is heavier and tends to fall.

There is a recording thermometer and a hygrometer in addition to the thermostat on a shelf above the radiators. The fans keep up the circulation; the recording thermometer shows how the thermostat has kept the temperature; and by frequent readings of the hygrometer you are able to keep track of how the relative humidity is running. You should check up on the hygrometer at least once per day by means of your wet and dry-bulb thermometers and your chart. Also, see how your recording thermometer readings compare with your dry-bulb thermometer.

The reason for this is that automatic instruments sometimes fail to operate properly. By checking up on them you can locate any trouble before material damage is done to your lumber.

*The Tiemann kiln.*—Let us now take a look at the Tiemann kiln and see how it arranges to take care of the three important points as we have discussed them. This is best done by referring to the description of its construction and operation as written by the designer of the kiln and set forth in the following paragraphs.

*The improved water-spray, humidity-regulated dry kiln.*—It has already been shown that the three fundamental factors necessary in the drying of a pile of lumber are circulation, humidity, and temperature. In an endeavor to produce a commercial kiln in which each of these elements could be regulated independently of the others, the author designed for the United States Forest Service the kiln described in this circular. The principle of the forced circulation and humidity control by means of the sprays of water which was the basis of the first patents taken out in 1912 is still the main feature of the present kiln, which has been greatly improved through five years of constant study and experiment, so that the kiln is now well adapted to commercial work. Moreover, the recently discovered principle of the downward circulation through the lumber pile has been taken advantage of in the newest form by so arranging the piles that the air may descend diagonally through them into the spray chamber.

*General description of the kiln.*—It is not intended to offer here a working plan or specifications, but merely to give a sufficient description to make its construction and operation plain. For the best results, each case should always be worked out to suit the particular requirements and a design made accordingly.

In figure 43 is shown a cross-sectional front elevation of this kiln in one of its simplest forms. Spray chambers *B B* are placed on the sides. These are 6 or 7 feet in height, 12 to 16 inches in width, and extend the entire length of the kiln. They are thoroughly waterproofed on the sides. The top of these flues are open and may be arranged for a footpath or runway. Near the tops are placed the series of sprays *F F*. At the bottom are gutters which drain to the end of the kiln and thence to a well. The bottoms of the flues open into the space beneath the heater coils, but the air is obliged to pass through zigzag baffle plates *D D*, which separate all fine mist from the air but allow the air to pass through freely in a saturated condition. These baffles may be made up of boards in convenient sections. Copper nails or wooden dowels should be used. They should fit tightly, as any leakages will allow the spray to get through to the steam pipes, which would spoil the humidity regulation. *H* represents the heating pipes, which are concentrated toward the center. *K* is a floor or shield of loose planks, which serves as a bottom to walk on and also shields the lower course of lumber from direct radiation of the heating pipes. The lumber is piled as indicated with a flue in the middle about 12 inches wide. While the inclined pile as illustrated will give the best results, flat piling, arranged in the same manner with the flue in the center, will also work well. With inclined piling the boards may be placed solid edge to edge, but with flat piling it is advisable to leave cracks between them so that the air may descend as it cools, passing through in a downward and outward direction. Curtains are hung from the roof to the edges of the piles as shown, to prevent the air from passing over the piles and thus "short circuiting" them. Condensing piles are placed just above the spray chambers at *G G* for use at the end of the drying operation when not so great a circulation is needed. Steam sprays are placed at *M M*, with the holes pointing downward, for use in removing casehardening. The diagram is drawn to scale for a kiln 13 feet wide, 12½ feet high, and of whatever length desired.

The water sprays consist of small brass nozzles fitted by means of a ¼-inch pipe "gooseneck" to the supply pipe, as shown. The adjustable "vermorel" type of nozzle, such as is used in horticultural work, which will deliver 2.5 to 3 pounds of water per minute at about 45 pounds pressure, has been found satisfactory. These should be spaced about 3 feet apart. They should give a spray of water and not a mist.

The temperature of the water is regulated in a very simple manner as follows:

The water flows by gravity from the gutters into a suitable well. From this well it is pumped by a suitable small rotary pump direct to a special temperature-regulating, mixing valve. Cold water from the supply main is also tapped into this valve at approximately the same pressure, which should be adjusted to between 30 and 50 pounds. By merely moving a small lever this valve will deliver water at any temperature between the two extremes and hold it steady within a degree or two. A steam pipe should also be arranged to discharge into the well for heating the water when very high humidity is called for. A strainer is placed on the water line to prevent the sprays from becoming clogged. Should any one become clogged, however, it can be readily cleaned by its small adjusting screw in the top. The exhaust steam from the pump may be advantageously used for heating the kiln by passing it through a separate coil of pipes.

The condensing pipes are also connected to the same circuit through a separate valve from the pipe line leading from the special temperature regulator, so that either sprays or condensers may be operated by the same system by merely opening and closing the respective valves.

*Construction.*—Any kind of building construction may be used, but it should be water and moisture proof. All cement work except the foundations should be waterproofed. This may be accomplished by using a 5 per cent solution of alum and 8 per cent of soap in the water used for mixing concrete, or the walls may be thoroughly coated with high-temperature asphaltum varnish.

In cold climates either standard wooden studding construction or hollow tile is preferable on account of insulation. A certain amount of radiation from the side walls, however, is desirable, since it increases the efficiency of the condensers and the sprays. The heat given to the walls does not need to be removed by the condensers.

Do not use galvanized pipes or galvanized metal in a dry kiln, as the fumes rapidly decompose the coating. Use plain wrought-iron pipes and paint them with a good high-temperature asphaltum varnish or black baking japan. Condensation is liable to occur on the roof when high humidities are used and drip upon the lumber. This may be overcome by placing steam pipes along the ceiling, spaced 3 or 4 feet apart.

*Operation.*—The operation is very simple. The heated air rises in the flue between the two piles of lumber. As it comes in contact with the piles, parts of it are cooled and forced to pass outwardly through the piles to the spray chambers. Here the velocity of the descending column of air is greatly augmented by the sprays. It then passes out through the baffle plates into the space immediately beneath the heaters. Here it is in a saturated condition and its temperature is therefore manifestly the *dew point* of the air after it becomes heated in passing through the steam pipes. This may, therefore, be termed the *dew-point method* of humidity control since this dew-point temperature is easily controlled by the temperature of the spray water.

Only two stationary thermometers are necessary for determining the humidity and temperature of the air entering the lumber, and therefore for operating the kiln, one in the baffles at *D*, which thus records the dew point, and the other in the flue between the piles of lumber. No wet bulb is needed nor any hygrometer. It is very convenient to use the recording type of thermometer, having long, flexible tubular connection with the bulb, and to have both hands recording on the same dial. By means of the humidity diagram the humidity may thus be quickly determined at any time.

The temperature of the entering air may be controlled by any good form of thermostat, of which there are many on the market, or by means of a reducing valve on the steam line.

This type of kiln requires very little attention when properly operating; once a day, or even once every three days, has proved sufficient.

The form of the kiln may be varied from that shown. For instance, a single truck may be used. This form would be represented by dividing the diagram, figure 43, vertically into two parts. In another form the spray chamber may be placed in the center and the lumber flat piled or sloped in the opposite direction. The air would then rise next to the side walls and descend in the center. For a separate kiln, however, the form illustrated is preferable on account of the cooling effect of the outside walls. Ventilation may also be used, which reduces the amount of fresh water required, an outtake flue being placed immediately above the spray chambers and an intake beneath the heating coils.

For further details of construction correspondence should be taken up with the Forests Products Laboratory, Madison, Wis., with a full detailed statement of (1) kinds and conditions and quantity of lumber to be dried, (2) purpose for which dried, (3) present method of handling, (4) kind and amount of present losses, (5) discussion of present difficulties and what it is desired to accomplish, (6) steam available, whether boiler pressure or exhaust from engine, (7) quantity of water available, (8) kind of man available for looking after kilns, (9) any other information which might assist in designing a kiln for the purpose.

While the Forest Service does not act in the capacity of construction engineer and does not furnish working designs and specifications of dry kilns, it is prepared to give such information to parties who desire to build a dry kiln of this type or to remodel their old kilns, so as to enable them to install and operate a kiln of this kind. It does not, however, advise any particular type of kiln without a full knowledge of the requirements and conditions in any specific case.

[U. S. Forest Service, Forest Products Laboratory, Madison, Wis. Revised copy, Apr. 25, 1917.]

All who wish to go to the bottom of the subject of kiln-drying lumber for any purpose should write the Bureau of Agriculture, Washington, D. C., for Bulletin No. 509, entitled "The Theory of Drying and its Application to the New Humidity Regulated and

Recirculating Dry Kiln," and Bulletin No. 104, with humidity diagram. These may be had for 5 cents each.

*Operation of kiln.*—We have discussed the kiln, but it yet remains for us to learn how to keep track of our progress in drying and to know when to take our lumber out. In stacking our lumber in the kiln we must leave about six boards loose enough to be pulled out. There should be one at each side near the bottom of the charge, one at each side near the top, and one at each side about halfway up the stack. These should be representative boards, some of which should be of the very wettest in the stack, so that when they are dry, all are quite certain to be dry. You should be quite careful to put only lumber of about equal moisture content in at one time. If a few fairly dry boards should be put in with a charge of green lumber they would probably be ruined before the rest are ready to take out, and if a few green boards should be put in with a charge of fairly dry lumber they would not be ready to come out when the rest are dry enough. All lumber with moisture content above 25 per cent is considered wet and may go in together, but 12 per cent should not go in with 25 per cent lumber. Differences between the lower percentages are most important.

*Test samples.*—When the charge of lumber is in the kiln no steam nor heat must be turned in until all samples are taken and weighed for record. To do this we pull out the six loose boards and take them to the band saw. Take each board separately and saw off a piece about 3 feet long and mark both pieces plainly with a blue pencil so that you will know them when the lumber is ready to take out. Weigh the long pieces and put them back into their places in the kiln with their weights plainly marked on them. Now take each short piece and after sawing 6 or 7 inches from the weathered end saw off from the other end three or four cross-section pieces about a half inch or less thick and extending the full width of the board. Weigh these immediately on scales that will weigh to one-hundredth of an ounce before they have a chance to gain or lose moisture. Mark each one with the number of its board and with its piece number, as 2A, 2B, 2C, etc., so that it can not become confused with any others. Our object in weighing each little sample separately is that they may serve as checks upon each other. Each of the 2-foot boards should now be ripped down the center, so as to give a sample for each end of the kiln. Paint the ends with asphalt varnish to prevent end evaporation. The pieces will then each be weighed very carefully to the nearest tenth of an ounce, or even closer than that, if possible, and be plainly marked with their respective weights and board numbers.

It is absolutely essential that every piece of your sample boards be so marked and placed that it can not possibly become confused with any of the others. Otherwise the samples will be worthless. Be sure to brush off all sawdust and loose fragments from each sample before weighing it. Otherwise this will fall off later and be counted as lost moisture.

These pieces will now be inserted on crossers in the stack, each *with its respective board*, the one-half being at one end of the kiln and the other at the opposite end, so that they may be subjected to the same conditions as the board from which they were cut. These pieces are your "daily" test samples, and you will weigh them *very quickly* each day and put them back so as not to interfere with their drying. From their weights you will be able to tell how rapidly your lumber is drying out and whether all parts of the charge are drying alike. If you do not have good control of your kiln they will not all dry at the same rate. In this case you will have to get things corrected at once or have the whole charge of lumber condemned by the inspector.

*Moisture test and calculations.*—The small pieces which we weighed to the hundredth part of an ounce are very important and must be handled promptly and very carefully in order to get results. You will dry them, preferably in a steam-heated oven at a temperature of between 212° and 220° F. A temperature of 215° F. is fine. This is for the purpose of driving out all the moisture in order that you may find out what the wood alone weighs. Do not heat above 220° F., for you may evaporate volatile matter other than the water if left too long in the oven. In fact, a temperature of 212° F. is safest, because water boils at 212° F. Weigh your samples at intervals and continue to dry them until they cease to lose weight. When they cease to lose weight you can be certain that they are dry, if you have kept your oven at 212° or 215° F.

It will take 24 hours or more to dry these samples. When you have finished with them they should be labeled and kept for future reference. *Do not discard them* until the lumber is *out of the kiln*, for people sometimes make mistakes. You may wish to weigh your samples again and repeat your calculations. (See Specification No. 20504B for this operation.)

Subtract the original net weight of the sample from the lightest or final dry weight and your remainder is the weight of water that you dried out of the piece. The first weight was *wood and water*, while the last weight is only *dry wood*, therefore the *weight lost* was water. Now divide that water weight, or *weight lost*, by the *dry wood weight*, using ordinary decimal fractions, and carry your division just *two places* beyond the *decimal point* and put the remainder on as a common fraction. *Read* your result as so much per cent moisture content, based on the overdry weight. Suppose that your figures show that your lumber contains 40 per cent water, which is quite ordinary for green oak, and suppose that you wish to dry the lumber down to 7 per cent, then you must take out 33 per cent water in order to get down to 7 per cent.

When we say that wood has 25 per cent moisture content *we do not mean* that one-fourth of the total green weight is water. We mean that if all the water were taken out of the wood it would weigh just one-fourth as much as the dry wood that remained. Therefore we find that the water was just one-fifth of the *total green weight*.

If your small samples show 40 per cent moisture, it is assumed that the whole boards from which they were cut contained the same proportion of water, and so did your short "daily samples." Then by dividing the green weight of the long boards in the kiln by 140 per cent, or the decimal "1.40," you will obtain the bone-dry weight of the boards. Then add 7 per cent of that weight and you will find how much the board should weigh when it is ready to take out of the kiln. Perform this calculation for each of the long boards according to the moisture showing of the small samples taken from them. Of course you will not expect them all to come out at just exactly your figures, but if you have made no mistake and have done all weighing promptly and with care your dried lumber should check up fairly close to your calculations. As a check you could make some new moisture-content tests from samples cut from the boards after they are taken out of the kiln.

After finding the moisture content of the little samples you will calculate the *bone-dry* weight of each of the respective "daily samples" from which they were cut. This will be done by dividing the *green weight* of the "daily sample" by 140 per cent, or whatever you found the little sample to be. If it had 40 per cent water, then the wood is 100 per cent, of course, and the green weight of wood and water must be 100 per cent, plus 40 per cent, or 140 per cent, which equals 1.40 in decimals.

Divide the green weight of the "daily sample" by the 1.40 and you will get the bone-dry weight of the wood. Whatever it weighs over that, each day, is all water. It is most convenient to reduce the total weight of each "daily sample" to ounces and hundredths of ounces so that you can write them down in decimals. You will then not be bothered with reducing them to pounds and ounces each day and your work will be much simpler. This reduction should be made before the above division is performed, so that you can do the whole calculation on the slide rule by setting the 1.40 on the slide immediately over the green weight in ounces on the lower scale of the rule.

The "1" on the slide will then stand above the *bone-dry weight* of the sample on the lower scale. Now, just leave your slide rule in that position and read what the "daily sample" should weigh at 5, 6, 7, 8, 10, 12, 14, 17, 20, 25, 30, and 35 per cent. These several

161988—21——6

weights may be read on the lower scale of the rule under 105, 106, 107, 108, 110, etc., on the slide, all at the same setting. If you can not use a slide rule, figure them out by ordinary percentage. First, make a chart, as shown in figure 44, and fill in the various weights as shown in the sample chart.

Fill out this blank for each of the "daily samples" and then fill out another one of the same kind for your long-board samples in the kiln so that you may take them out and weigh them as a check on the showing of your "daily samples" before shutting off your heat or stopping the kiln. If found to check, take the charge out; otherwise continue for the correction.

In this way we will avoid pulling a charge of lumber before it is ready. Of course, the inspectors will test our lumber later, as they work it up, anyway, but we want them to find it right. By making out the tables as indicated above, we will not have to make any more calculations for moisture during the run, since we have calculated the weights for each sample with each percentage of moisture, and it is only necessary to weigh the samples daily, or every two days if things run well, and check the weights wherever they are found on the chart. Do this for each sample each time and then draw a red line through these checks from top to bottom of chart each day and we have what we call a "moisture graph" of the charge for that day. By this we can judge how evenly our kiln is running and also how our temperature and relative humidity should be run for the next day or so. We will probably find this graph quite crooked at first, owing to the fact that some of our boards started with 35 per cent moisture, others with 40 per cent, and others with more or less, as shown by our samples, but if the kiln is run properly they will gradually even up and the graph will become straighter from day to day, even though it will not be entirely straight at the last, but our specifications will allow us a variation of 2 per cent, which will get us by easily. This chart was devised and used by the writer and it was a great time saver, since any ordinary workman who could do a reliable job of weighing could check up the graph for the day and then it was very easy to tell him how to run his temperature and relative humidity for the next day.

*Closing the run.*—When the chart shows the charge ready to pull, check it up by weighing the long board samples and by running a moisture graph on their chart. If they do not agree with the showing of our daily samples we should finish the run by the long samples, for we know they are the same as the rest of the lumber. The daily samples might be a bit different at the end of the run, due to the fact that they were smaller and were removed from day to day for weighing. When finished, saw some samples from the long boards and try for case hardening, according to Specification 20500A, and steam, as it says, if trouble is found. (See fig. 41, which shows actual tests made by the writer.) Do not expect to get perfect results, but be sure that you get good results. Do not remove the

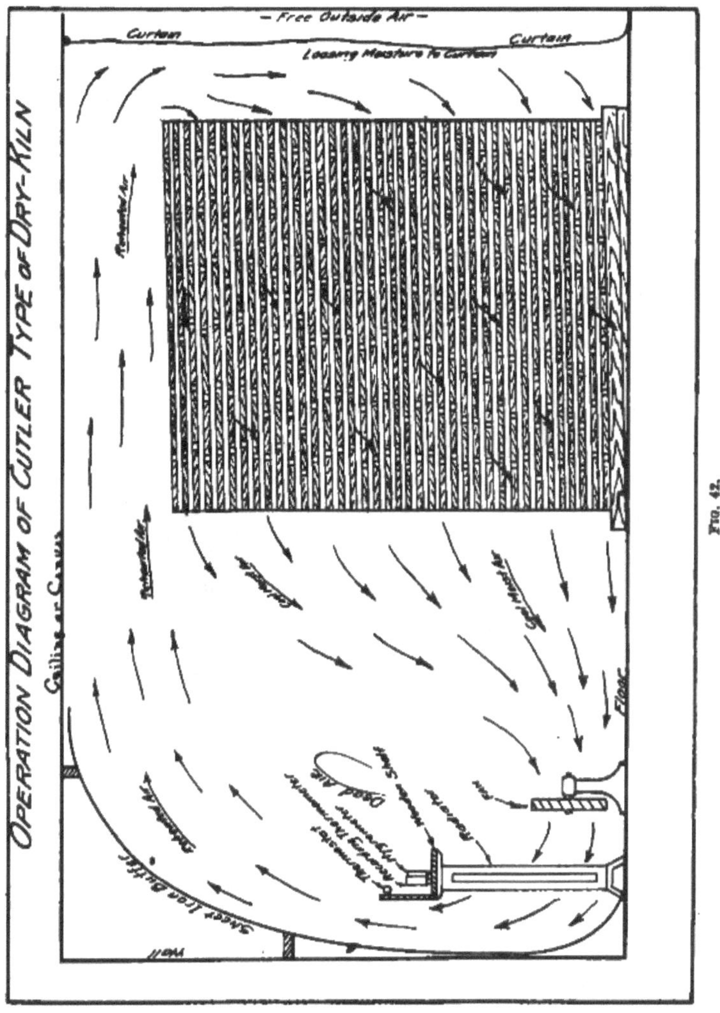


FIG. 42.

charge until the lumber is practically cold in order to be sure that it has taken its set form before being taken out of the pile. Lumber removed too soon often warps in cooling.

*Seasoned lumber storage.*—If our truck tracks run to a closed storage room, it is best to shove the truck out of the kiln and leave the lumber on it for about four days in the conditioning storage before starting to work it up. This will save a lot of extra handling and will give us straighter lumber. All kiln-dried lumber will remain in this conditioning storage room until taken out for use. The temperature and the relative humidity of this room must therefore be controlled, both day and night, in order to keep the lumber the same as it was when removed from the kiln.

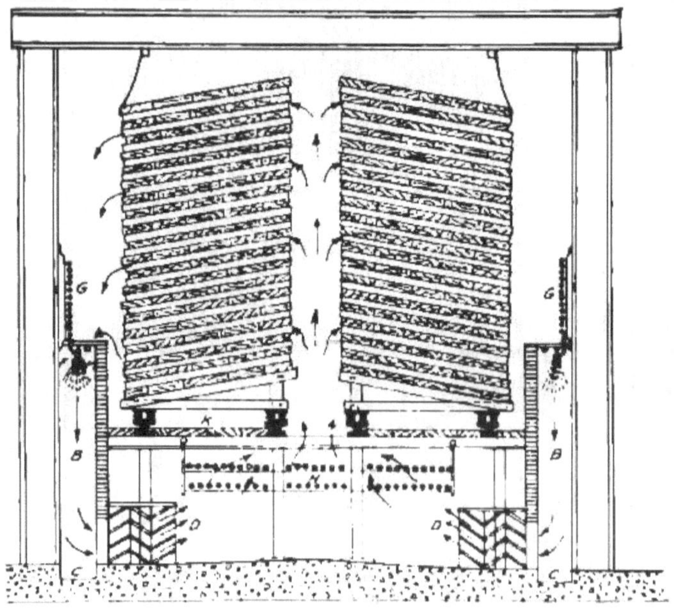


FIG. 43.

*Milling lumber.*—The lumber will next go to the mill room, where it will be surfaced on both sides and inspected. The best way is to run one side over a hand jointer to take out all humps and then run the board through a single surface to plane the other side. This gives us straight lumber.

*Laying out and cutting laminations.*—At this point we need a good mechanic, who is also a good judge of lumber. He must know Specification 29500D and be able to judge his board quickly and safely. If he is slow, it is a loss; if he rejects good lumber, it is a loss; and if he cuts laminations from lumber which will later be rejected, it is a loss. (See drawing of set of laminations, fig. 38.)

Some people still glue straight boards into square solid blocks and then carve out their propellers. This method is wasteful of both labor and material and should not be followed. It is the regular

KILN NO. DATE STARTED

KILN END — MOISTURE GRAPH CHART — DATE STOPPED

LUMBER CHARGE — SHOWING PROGRESS OF DRYING — STEAMED: at start hrs; at finish hrs

| | SMALL SAMPLES | | | DAILY SAMPLES | | | CALCULATED WEIGHTS BASED ON SMALL SAMPLES | | | | | | | | | | | | | ACTUAL |
|---|---|---|---|---|---|---|---|---|---|---|---|---|---|---|---|---|---|---|---|---|
| FROM BOARD NO | ACTUAL GREEN WT OZ TOTAL | ACTUAL OVEN DRY OZ TOTAL | ACTUAL MOISTURE % AVERAGE | ACTUAL GREEN WT OZ | CALCULATED OVEN DRY OZ | MOISTURE ACCORD TO SMALL SAMPLES | OZ AT 5% | OZ AT 6% | OZ AT 7% | OZ AT 8% | OZ AT 10% | OZ AT 12% | OZ AT 14% | OZ AT 17% | OZ AT 20% | OZ AT 25% | OZ AT 30% | OZ AT 35% | OZ AT 40% | WTS. CKD. |
| 1 | 41 75 | 29 75 | 40% | 65 25 | 46 60 | 40% | 483 | 494 | 498 | 505 | 512 | 522 | 531 | 543 | 555 | 582 | 606 | 629 | 682 | |
| 2 | 32 65 | 24 20 | 35% | 71 25 | 52 75 | 35% | 554 | 559 | 565 | 570 | 580 | 591 | 601 | 617 | 635 | 660 | 686 | 712 | | |
| 3 | 43 40 | 33 40 | 30% | 61 75 | 47 50 | 30% | 499 | 504 | 508 | 513 | 522 | 532 | 541 | 561 | 570 | 594 | 617 | | | |
| 4 | 39 80 | 27 85 | 43% | 72 50 | 50 72 | 43% | 532 | 537 | 542 | 547 | 558 | 568 | 578 | 594 | 608 | 634 | 659 | 685 | 710 | 735 |
| 5 | 30 45 | 22 40 | 36% | 67 25 | 49 50 | 36% | 520 | 524 | 529 | 530 | 544 | 554 | 564 | 579 | 594 | 618 | 643 | 667 | | |
| 6 | 34 25 | 26 55 | 28% | 60 50 | 47 25 | 28% | 497 | 501 | 506 | 511 | 520 | 529 | 539 | 555 | 567 | 591 | | | | |

NOTE: Fill out separate charts for samples at other end of kiln and then compare the two charts daily to see that both ends of kiln are running about alike.

REMARKS: It is assumed that the "daily samples" had the same moisture content as the "small samples" since they were cut from same boards at the same time. Therefore, calculate "oven dry" and various percentage weights for daily samples from "green weight", assuming moisture content as found in small samples. Weigh samples daily and check weights where found in calculated table. Now draw red line graph through check marks.

FIG. 44.

practice to have patterns a little larger than each lamination has to be in the finished propeller. These patterns are therefore of various sizes and shapes, so that a man of good mechanical judgment can save lumber by choosing his pattern according to his board and thus cut out defects and use the good part of the board.

In fact, the man who makes or lays out the laminations rarely does the cutting. Usually a couple of other men saw them out on a common band saw. They are then stacked in their respective piles and shoved over to a boring machine, where a center hole of about an inch or a little more in diameter is bored in each one. Next they are shoved through a tooth planer, which shaves them to their exact final thickness by means of a bearded blade, which leaves a combed-like surface for gluing. Care must be exercised in sawing out laminations, due to the fact that all are to be weighed and balanced before being glued up.

*Conditioning and assembling laminations.*—The laminations are next carted off to a conditioning room, where they are slipped into open racks in an edgewise position and left for a period of five days, as per Specification 29500D. While these laminations are being cut, the inspector gets samples for his special tests for moisture, as per Specification 20504B, and for his density tests, as per Specification 20505A or 20506, so that he may throw out such laminations as are beyond the limits before they are glued up. He stamps all the approved laminations on the edge of the hub part by means of a steel stamp so that he may inspect them after they are glued up into propeller blocks and before any carving is done. This is to make sure that no faulty or uninspected laminations have been put in by accident or otherwise. After the work gets well started, there will be fresh laminations coming in all the time, while those which have been conditioning long enough will be going out in propeller units or groups to be glued up. A full set of laminations of about equal density and moisture content as specified will be grouped and hung up by the center holes on a peg or on a shaft across a balancing stand and tried for balance, both horizontal and vertical. When they are properly arranged by turning first one and then another over until the *group as a whole* balances fairly well, both horizontally and vertically, they are marked for position and sent to the gluing room.

*Gluing up propellers.*—The assembled group of laminations is placed temporarily into a hot box at a temperature of from 120° to 140° F., and left there long enough to get warmed through. Specification 29500D says that they shall remain in the hot box 10 minutes, but experience has taught the writer that oak, especially, should remain for a longer time in order to become heated to the center. A half hour or more is not too much, for it must be remembered that, since glue and water are applied to both surfaces of the laminations, the surfaces should be dried to a lower moisture content than the center so that the water from the glue will merely equalize the surfaces with the center, rather than set up a tension by putting more moisture into the surfaces than the center contains. Furthermore, when cold oak is first put into the hot box, sometimes errone-

2279

ously referred to as the caul box, it chills the air that touches it and causes a light film of moisture from the hot air to condense over its surface. This is due to the fact that air at the temperature of the oak will not hold as much moisture as air at 140° F. will hold. Thus while the air at 140° F. might be dry enough to have a relative humidity of only 50 per cent or thereabouts, it would be quite wet or humid if cooled quickly to 80° or 90° F., which is the case with the air which touches the oak laminations when they are first put in. Then it takes some time to warm the surface, due to the fact that the center of the lamination is constantly taking the heat away until it, too, is fairly warm, and it will take the heat away from the glue and cause the glue to set prematurely if an attempt is made to glue the propeller up too soon after putting the laminations into the hot box. Some factories had serious trouble in this way during the war. Some even turned free steam into the hot box to try to keep the humidity high while the laminations were getting their 10 minutes, with the result that all their propellers thus glued up came open in the glue joints after a few days.

It is far better to warm laminations three hours at 120° F. than to warm them only 10 minutes at 140° F.

*Presses and clamp.*—There are several ways of gluing up and clamping propellers to dry. Most manufacturers use heavy steel presses, with a row of large screws with which to apply the pressure. They take a set of scrap cuttings or trimmings from the several laminations and nail them in proper form on a heavy bottom plank and on a heavy top plank, so that they virtually form a caul, or mold, which receives the set of glued laminations. Sometimes instead of using a set of lamination trimmings, stacks of small blocks are built up like stair steps to apply the pressure from the crossties to the different laminations at regular intervals. This is illustrated in figure 45. As the hot laminations come from the hot box they are quickly given a coat of glue and assembled on a center bolt, which passes through their center holes and then through the two planks. We next apply the top half of the caul and screw down the center bolt. The group is then pressed down solidly to a pressure of approximately 150 pounds per square inch, and tie clamps are screwed up at short intervals. There is some difference of opinion as to how much pressure should be applied, but all agree that it must be sufficient to insure a tight joint throughout. The whole bulky unit is then removed from the press and stacked to one side to dry until the next day, when the tie clamps may be removed. Thus, the same press may be used over and over for as many propellers as the factory can make per day, providing there are plenty of cauls and baling clamps on hand. Other manufacturers assemble the laminations on a big flanged hub bolt as the glue is applied and then screw a big nut down securely, clamping the laminations together at regular intervals both ways from the hub by means of crossties and "C" clamps (fig. 45). Small blocks are built up on the various laminations in order to bear on the crossties and hold all joints securely. Be careful to lap the laminations just right before clamping. Otherwise, although they may be sawed correctly, you will not be able to make a propeller from them if they are not adjusted properly when glued up.

Both methods are successful when properly used. The former is the better for quantity production, but the latter is cheaper for the plant which expects to make only a few propellers. The American Propeller & Manufacturing Co., one of the largest propeller factories in the country, uses the latter method. The important point is that the work be glued and clamped quickly and securely in order to insure tight, solid joints and to leave as little surplus glue in the joints as possible. A temperature of approximately 90° F. should be maintained in the glue room in order to be sure that the glue does not chill before the work can be properly clamped. The relative humidity is also very high, due to the evaporation from the glue. For this reason the glued propellers should be removed from the glue room to the conditioning room as soon as possible after gluing.

*Glues.*—There are different kinds of glues, including casein cold glue, which may be used; but we advise the inexperienced propeller man to stick to first run hide glue which has been inspected by the Government. Hence, this article will deal only with this one kind. Government certified glue may be purchased from any of the big packing houses direct. This glue should be clear, like varnish, when ready to use and should spread with a brush about the same as varnish. Beware of opaque glues.

A well-equipped steam glue heater like the "Advance" is to be preferred. It has a thermometer attached, so that the operator may know exactly what temperature he has. This thermometer should stand at about 140° to 150° F., since that is enough to melt the glue and keep it right for use. We all know that meat becomes very tender when boiled, and that it falls apart if cooked enough at a high temperature. Hide glue is an animal product. Hence, it will do the same if cooked. We do not want our glue to become tender and fall apart. We want it to remain tough, so as to hold the wood together. Therefore, we should heat it enough to melt it and no more. This old idea that glue must be boiled or even cooked is all wrong. The dry glue should be weighed and then mixed with about two and one-half times its weight of cold water. Some glue may require a little more or a little less water, but this is usually about right. Allow this cold glue to soak until soft if it takes all

FINISHED
AFTER ROUGH CARVING
AFTER GLUEING
ASSEMBLED LAMINATIONS AFTER GLUEING

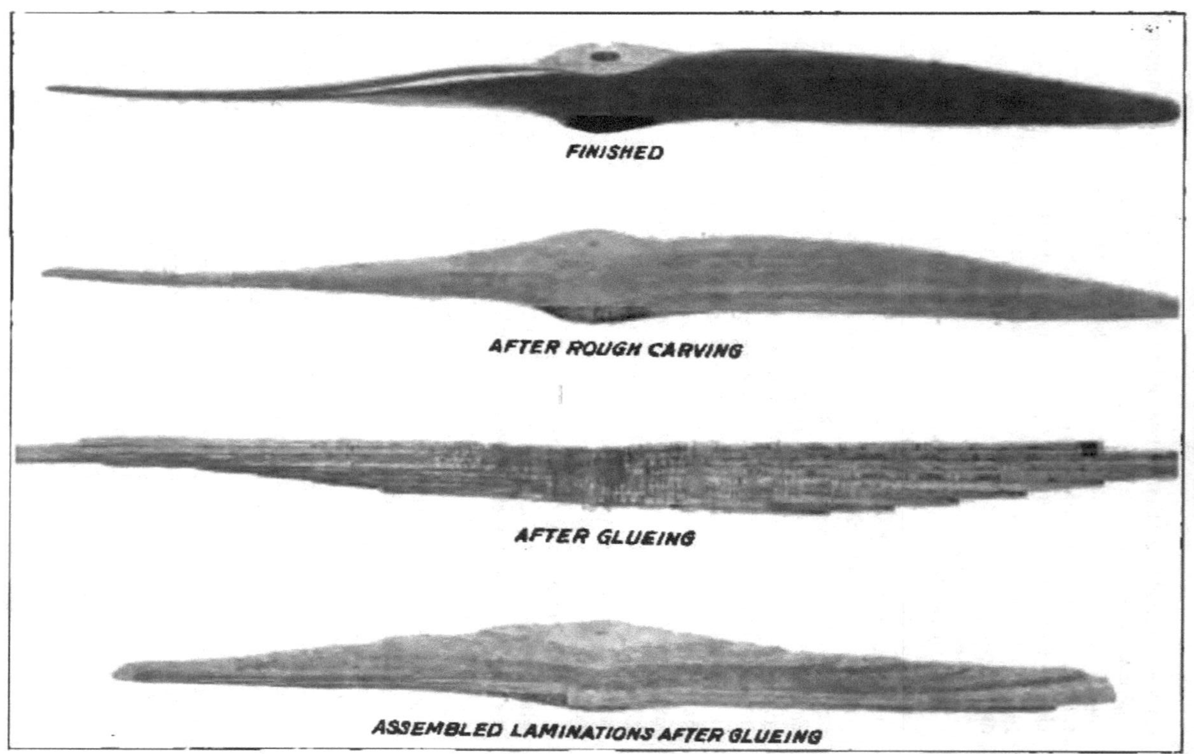

FIG. 46.

night, and then put it into the heater; or, rather, allow it to melt and fall through the screen into the heater tank. It should be ready to use in about three minutes. Do not put in too much water and make soup instead of glue, as many people do. Neither should the glue be so stiff that it will not enter all the rough places on the surface of the board. The glue room may be large enough to answer also as a storage room for the freshly glued propellers, which will cool off during the night and the glue will set sufficiently, so that the clamps may be removed in the morning and be used on the next run of propellers. This will save a lot of heavy handling. In this case cold-water pipes, or other condensing equipment, should be installed for condensing the moisture from the air as the glue room cools, and thus prevent it from being absorbed by the freshly glued propellers. After the clamps have been removed the propeller blocks should be taken to the rough carving room and left to equalize internally for not less than 48 hours before being roughed out. It is always best to give them as much time as possible, so as to render them less liable to warp after being roughed out. We have discussed the solid laminations only. In case it is found advantageous or necessary to use spliced laminations simply follow Specification 29500D. A serrated joint is a knitted or saw-tooth joint made by a special cutter on a shaper. This type of joint is illustrated, both before and after gluing, in figure 40. When splices are made, the surface tooth planing must be done afterwards in order to insure level surfaces and an even thickness. Experience and tests have proven that spliced laminations are as good as any, providing that the splices run parallel with the grain and are well made.

*Machining or rough carving.*—We now come to the rough carving stage, where rough, unsightly glue-smeared blocks are to receive their first shaping, after which they begin to look like propellers. (See fig. 46.) There are many ways of doing this work. The regular factories all use some sort of duplicating machines for this job. In this case the machines take care of the proper shaping. The man who must do it all by hand has a long, tedious job, which must be done carefully. Judging by the personal experiences of the writer, the Defiance propeller lathe is probably the most satisfactory. However, a number of other lathes have been giving good results. This machine is merely a copying lathe of large size. It has a large ring chuck at the one end through which the propeller block is passed and in which it is securely bolted through the hub while the end which is being worked is held by an ordinary tail center. The other end is left free to stick out the other way until the one end is finished. The block will then be taken out and turned so that the other end may be worked down. The wood is worked down by a cutter head on a moving carriage which is pushed to and

FIG. 47.

94

from the revolving stock by a blank wheel of the same size as the cutter head and acting on a revolving propeller model, which revolves at the same rate as the propeller block and serves as a cam as the carriage travels from one end of it to the other. For best results this model should be a half inch or more thicker all around than the propeller is to be. It may be made of wood and sheathed with sheet iron to prevent wear. The machine is very successful when operated by a good man who knows how to line-up his work. It will rough out an ordinary propeller in less than an hour if operated properly. All of the propeller-carving machines follow some sort of a model. Some factories use the ordinary multiple-spindle variety carver. This machine carries four high-speed cutter spindles and one similar blank spindle on a swinging frame. A finished propeller and four glued-up propeller blocks are clamped to a fixed vertical frame. The four cutters work down the propeller blocks as the blank spindle traces the contours of model propeller. This machine must be operated by a careful carver.

*Roughing out propellers by hand.*—The simplest way to cut a propeller out by hand according to a given drawing is first to have your stock glued up full width in a solid rectangular block. Plane this block to the proper thickness to agree with the desired hub thickness. Next draw a center line down the top face and bore the hub hole. Now, take a square and mark lines across the top face of the block at 6-inch intervals out both ways from the center of hub to serve as station lines. Then take a pattern which we have made from the plan of the drawing and lay it on top of the block, with its hub hole over the same bolt which passes up from the bench through the block and with its center line exactly on the center line as drawn on the block. Draw a definite line on the block around this pattern and then turn the pattern around and mark the face of the other end of the block in the same manner. Now take the block to the band saw and cut it straight through all around and exactly to this line. This gives us the *plan form* or characteristic shape of the propeller. For this laying-out job we should have a level surfaced iron topped table or bench with a vertical spindle at one end on which to mount our block while working on it. Such tables are shown in figure 47. We will next mark vertical lines down both sides of the block from the ends of each 6-inch station crossline. It is best to mark the crosslines while the block is square before sawing it to plan form. These are our station lines, and we are now ready to locate the leading and trailing edges, and thereby the flat or *working face* of our propeller. We will now set our surface gauge halfway up the side of the hub or, according to the center line, on the side elevation on our drawing, and then mark our center line all around our block by sliding the surface gauge along the surface table and by turning the propeller block around. Now, at each 6-inch station both the trailing and leading edges are located by measuring from the center line according to the dimensions given on the drawing (fig. 6). The method of locating these points can probably be better visualized by referring to figures 33, 34, and 35.

We will next connect all of these points, ranging from the rear top edge of the block diagonally down around the end and back toward the hub near the lower front edge of the block. The next step is to cut out this top wood and dress down to the face of our propeller. The edges will require more or less rounding, but this can be done later. Do not round face edges until back of propeller is cut to template form. Now we get our station back templates and turn our block over. We have a big job of cutting the backside of this block down until the back templates will fit. Do not cut it all down at first, for we will need some wood for balancing, and furthermore it is always better to leave a propeller a little heavy than too light. Now we round our hub off to its proper shape before bothering with the problem of balancing. Some people will tell us to balance by shaving wood off the hub, but they are mistaken. We will get the best results by shaving and scraping farther out on the blade nearly halfway to the tip. We should be extremely careful to get the outer two-fifths of the blades exactly right in shape, pitch, and template fit, but the inner three-fifths of the blades can vary somewhat without materially interfering with the preformance of the propeller so long as they are not weakened by cutting too much away. From two to four weeks should elapse between the first roughing and the final shaping and balancing. This is to permit the propeller to do all of its warping and adjusting before the last shaping is done in order that it may have the right pitch and track when finished and remain that way when it goes into service.

*Laying out of shaped lamination propeller blocks.*—After seeing how to mark and cut out a propeller from a squared block it will be easier to see how to cut one out of an apparently shapeless block which has has been glued up of a series of various shaped laminations, as shown in figure 46. In this case we should first bore the center hole large enough to fit the center bolt on our workbench or iron-topped table. Some men prefer to bore the hole full final size at this time, but this is not the best way, due to the fact that the propeller is to be allowed to season nearly a month before it is finished and varnished, hence there is a possibility that the center hole might become either flat or oversize later on. For this reason it is best to keep the hole a little undersize until the propeller is finished or nearly so, and then rebore it to its proper size. We do not have a broad flat top surface now as we had in the other case, hence it will be best to get a smooth straight board and make a full plan form or pattern for the propeller according to the propeller drawing. Nail this to the top of the glued-up block with short nails.

Now take our block to the shaper and turn it pattern side down on the shaper stand. The shaper has been rigged up with a high straight cutter head similar to a vertical planer head and with a gauge ring of the same diameter at the bottom just above the surface of the stand. Now run the pattern against this gauge ring and our shaper will cut the block to the proper plan form. If we have no shaper we may be able to do this work on a band saw by being very careful.

Next we will mark our horizontal and our vertical propeller center lines in both ends of the block.

Get our vertical lines by using a square on the laying-out surface table and extending up to the center line on our pattern which is still nailed to the block. Mark this vertical line plainly and then locate the horizontal center line by means of a surface gauge on the table. Mark these points by little holes so that they will not become lost.

*Rough carving.*—Remove the pattern and carefully work the hub faces down absolutely parallel to each other until the hub is of the proper thickness. Then by means of the surface gauge locate the *leading edge* and the *trailing edge* at each 6-inch station.

The first rough carving must not be done sooner than 48 hours after the gluing. After the first roughing to approximate form it must be hung in a conditioning room for a period in order that its moisture content may become thoroughly equalized and distributed before the final shaping is done. (See Specification 29500D.)

The same procedure in this respect must be followed whether the work be done by machine or by hand.

Referring again to the machine method, we find that the center holes shall be bored to near their final size either just before or just after the first roughing out and before the conditioning storage period. This is to allow the hub moisture to become adjusted before the hole is bored to final size. This ruling was made due to the fact that the Army had quite a bit of trouble in 1917–18 with hub holes which changed after manufacture.

By following this rule we hope to avoid most of this trouble in the future. The conditioning period was decided upon due to the fact that hundreds of propellers that were rushed through the factories proceeded to do more or less warping and changing of pitch, track, etc., after leaving the factory. Many such propellers have been condemed at the various flying fields during the past two years from no other cause. This is extremely wasteful and expensive as well as very annoying.

After the conditioning period the propeller will be put back into the machine for a second working to very near its finish size, but before doing this the bench men will see that the hub faces have been properly dressed parallel to each other and that the hub is of the proper thickness and the hub hole will be re-bored to finish size. It is now necessary to re-center the ends of the blades due to the "working" the propeller has done since it was in the machine for its first roughing. This alone proves the wisdom of requiring a conditioning period between the first roughing out and the final shaping. In order to be sure of the time that has elapsed we always write the date of the roughing plainly on the propeller with a blue pencil as soon as it is taken out of the machine. We do the same thing when it comes out of the machine the second time and then let it hang a little while again before doing the final bench work on it.

*Second final shaping and carving.*—Inspect very carefully for cracks and open joints when the propeller comes from the conditioning room, for sometimes a poor job of gluing was done, and if so it will probably show up by this time. In that case it is best to condemn such propellers now before putting any more work and expense into them.

Always bear in mind that no matter how many inspections the material or the propeller has passed, it is still subject to rejection at any time that sufficient cause is discovered, and this should be so, since a faulty propeller might ruin a very expensive airplane, or kill some good man later, and we are sure that no person could wish to be the blame for either.

After the second time in the machine, the propeller will be finished by hand exactly the same as though it had been started by hand excepting that it will only be necessary for the workman to smooth up the face side to the correct pitch angles at the several 6-inch stations outward from the hub and work to the proper depths at the tips according to his end center marks. These he will establish according to the drawing by means of a straightedge lengthwise over the top of the propeller and a square from the table up to it and by using his surface gauge with which to get the height of the center point above the table. While the long straightedge is still clamped in position, mark center lines the full length of both blades by means of a try-square and a pencil. The machine left our edges a little wide for safety, hence we must do all face measurements from our center line in order to play safe. By all means be sure that our center line is really straight and in line from tip to tip. Otherwise our two blades will not be in line and we will never be able to balance the finished propeller in the vertical position, although it may balance perfectly horizontally. One manufacturer had many propellers returned to him from the flying fields in 1918 because of this one trouble alone. They were beautiful in every other respect. In some of them the running unbalance was so bad that they tried to jerk the motors out of the planes. The writer inspected a lot of them in the field and condemned them to be returned to the factory.

Instead of measuring out from the hub center to locate our stations, as we did when we began with the squared block, we will transfer our measurements up from the table which has station marks scratched into it. By means of a square it is very easy to find your stations at any time. On some tables these marks are made on radii from the center of the vertical spindle which holds the propeller. The lines should be square across the table, so that when the center line of our propeller is directly above the center line of the table we can locate all of our stations correctly by marking at both edges against a square. These points should then be connected by straight cross lines. Then we will take the dimensions from the vertical center line on our cross-section drawings to the edges in the direction of the pitch or slope, and step them off from the center line which we marked on the face of our propeller. Now work to these points all around and our propeller will have the correct profile. It is really best to work the leading edge to shape first and not dress the trailing or thin edge until we have fitted all the back templates. In this way we always have a little room in case we make a mistake. It is easy to dress the trailing edge later. Of course, if we are making several propellers from the same drawing we will make a face template from a piece of sheet metal. With this it is only necessary to locate the center line in the regular way and then mark around the metal face template while it is correctly placed relative to the center line. The tips of the blades must be cut to the proper length before the face profile lines are drawn. Otherwise we will have trouble. It is easy to get the tip measurements from the table and when one tip is cut, turn the other around and cut it so that when turned they will both show the same against a square erected on the table.

After shaving the back side of the propeller pretty nearly to a template fit, turn it over and measure your pitch again at each station with your protractor, as described in the chapter on Inspection. Sometimes these angles change a little after we have cut down the back.

In this case it is best to make the discovery while we still have enough wood to make the correction.

*Final shaping and balancing.*—Take the propeller to the balancing stand quite frequently when nearly done in order that you may cut only where we can spare the wood and remember that all balancing goes by the law of levers, which means that an ounce 4 feet from the hub will cause as much trouble as 4 ounces 1 foot from the hub. That is, for comparison we always multiply the weight by its distance from the center of the propeller. (For description of balancing stand, see chapter on Inspection.)

This shows us why it does more good to remove a small amount of material from the outer end of the blade than it does to cut a handful of shavings off the hub. Therefore give the hub the proper size and

161988—21——7

shape before starting to balance and then make your corrections by shaving, scraping, and sanding farther out as we go along.

It is always best to be absolutely sure that our blades lie in the same straight line running through the center of the hub from tip to tip. Then work the outer two-fifths of the blades to their full template fit, the same as we worked the hub. All balancing can now be taken care of on the inner three-fifths of the blades. Do not forget that the outer two-fifths of the blades do the work and also cut the air with by far the highest velocity. That is why they should be made alike in every respect. The two blades must balance in weight, alignment, track, pitch, length, pattern form, and cross-section form.

If we get the outer end of one blade a little heavier than the same part of the other it will require a great deal of surplus wood on the inner half of the other blade to hold a balance. In that case it will be very noticeable and will look bad, and although our propeller may balance while on the stand, it will not balance when revolving at high speed, which is most important. In trying for balance be sure that the balancing spindle fits the hub hole securely, for if there is a little shake balancing will be impossible. We have seen men work for hours because of this trouble before finding out the cause. We have seen propellers continue to roll over and over on the balancing stand, which would seem to indicate that first one end and then the other was the heavier. Men have most trouble when the propeller hangs at an angle, but will hang neither vertically nor horizontally. This trouble is usually due to the fact that the lower blade is the heavier and that the surplus weight is nearer the hub than the tip and that it lies to one edge of the blade rather than near the center line. This is especially true if the blade hangs from vertical. In this case first try our templates to see that we can spare the wood and then shave principally near the lower edge.

Before we do any cutting we should inspect the upper blade because sometimes, although the lower blade is certainly the heavier, the sidewise disturbance which prevents the lower blade from hanging straight down is due to the fact that one side of the upper blade is heavier than the other. The workman who solves the problem of balancing can make two propellers to the other fellow's one, and do it easily.

*Hanging propellers in shop or in storage.*—Another point to be observed is that of how to hang up propellers. It must *not* be left with one end much higher than the other, especially in the conditioning room or overnight while being worked on, due to the fact that the upper end will soon become lighter than the lower if they were balanced before. This is because the upper air in a room is always warmer and drier than that nearer the floor, due to the fact that the warmer air is lighter and its relative humidity is the lower. Therefore, either the upper blade will lose moisture and become lighter, or the lower blade will take on moisture and get heavier. If the whole propeller has too much moisture and is drying out the top will dry the most. If the whole propeller is drier than the air of the room will permit, and is therefore taking up moisture, the lower blade will take on the most. This is why propellers should always be hung horizontally. They should always hang on a peg or shaft through the hub hole in order that no pressure that might cause slight warping may be exerted on either blade. They should also be turned occasionally to prevent one edge from becoming lighter than the other.

In a factory last year an old Frenchman who was an excellent workman had worked for hours without success trying to balance a propeller. When the shop closed he marked the heavy end and leaned the propeller, heavy end up, against the wall. Next morning before shaving any wood off he tried again for balance in order to see where to cut, and to his surprise the other end was now the heavier. This man believed that the change in weight was caused by the moisture in the propeller running down to the lower points during the night. This belief is shared by many who should know better.

*Time and labor estimates.*—It takes a good workman about two days to finish carving and balancing a propeller after the first roughing, or a little over a day to do the same after the second machining. They can often do better, but not much. Since it takes a good man from two to three days to shape and balance a propeller it will take about 25 carvers to average 10 propellers per day, if the first roughing out is done by machine. A good man on the machine for the second roughing can save nearly half of the carver's time by doing a close accurate job.

Five men can take care of the handling and assembling of the laminations and of the gluing up of 10 propellers per day. Three men will copper tip 10 per day and one man will varnish them. Two men will run the dry kiln, and two will do all the planing, marking, sawing, etc., of laminations. Two more men for the crating, handling, and odd jobs will nearly make up the force necessary to turn out 10 propellers per day.

*White or first general inspection.*—Returning to our unfinished propeller we find it finished so far as the woodwork is concerned, and although it has been inspected more or less all along, it must now be thoroughly measured and inspected for pitch, track, length, width, shape, template fit, alignment, hub thickness, hub fit, balance, materials, and workmanship. This general inspection is sometimes known as the "white" inspection, due to the fact that the wood is yet bare. (The details in this connection are given in the chapter on Inspection.)

*Tipping.*—After passing the "white" inspection the propeller goes to the tipping room, where it receives some sort of reinforcement. The type and extent of tips as well as the materials will be specified on the drawing. For a long time propellers were merely varnished and put into service with no further protection, but experience soon showed that the bare wood would not last long if the plane had to take off or land where there was any grass, weeds, brush, or field crops, especially cotton, to be contended with, nor would it last long in a little rain. Several varieties of materials have been tried as protection on the propellers, chief among which have been fabrics and copper. Fabric (cotton or linen) is usually put on as a sheathing extending from the tips to within a few inches of the hub. It adds much strength to the propeller and especially tends to prevent splitting.

*Copper tipping.*—Presuming that we are to apply the copper tipping, the first step after the white inspection is the routing or thinning of the propeller to receive the copper so that the finished job will be the same size and shape as the wood before the tip was applied. This must be done very carefully in order not to weaken the wood too much. The copper is cut so as to boot the last 4 or 5 inches of the tip and sheathe the leading edge up about 40 to 60 per cent of the radius toward the hub. It is cut with a graceful curve so as not to make a weak line in the blade. One piece fits the face of the blade and the other fits the back or rounded side and the two are lapped and soldered around the blade edges. The face side piece should be continuous and form the leading edge strip without a seam. The sheet copper is cut to pattern and then shaped by pounding and rubbing on a hollowed-out wood block. It is then tacked onto the propeller temporarily so that the tip may be marked for the limits of the routing. It is then removed and the surface which it is to cover is cut down to a depth equal to the thickness of the copper. This must be done very carefully in order to prevent an unsightly edge fit. The tipping is again tacked on the holes and drilled through copper and wood by means of a portable electric drill, and copper rivets are put in, as per Specification 29500D and blue print. An attempt is made to keep the rivets out of line with the grain of the wood in order to prevent splitting. It is important that the rivets fit the holes and that a very light hammer (not over 3 ounces) be used against a heavy back stop for riveting them, in order to avoid buckling rivets internally and thus splitting the wood. Next the edges of the face copper are turned over around the back side of the propeller about ¼ inch all around and soldered down. Both ends of all rivets are also soldered. All soldering must be done quickly and carefully in order to avoid scorching the wood and thus ruining it. The excess solder is then scraped off and the tip is polished. The propeller must now be balanced again and, although the copper on both ends was cut by pattern and should weigh the same, the riveting and soldering always make some difference, hence the tip must be scraped until perfect balance is again secured. When scraping and dressing tips, be careful not to thin the edges, for they must be strong to take the wear. A mill file, slightly hollow ground, is an excellent tool for this scraping job. In the tipping room, as elsewhere, the propeller is held on the bench by a big bolt through the hub. For convenience special benches are sometimes made with bolts which may tilt to different angles. They are also sometimes provided for the carving room and are quite a help. We should have stated that the filler should be applied to the wood before going to the tipping room in order to keep the propeller from becoming discolored with lead, soot, etc., and thus injuring its appearance. The part which is to be covered with copper should also receive a good coat of shellac to serve as a waterproofing.

*Pigskin tipping.*—In case pigskin is to be applied, the propeller is routed down the same as for the copper tips and to practically the same shape and extent. The skin, which has been cut to pattern and sized with very thin glue, which is now dry, is now given a good coat of fresh hot glue and smoothed into place with a rounded piece of iron, which crowds out the surplus glue and all air bubbles. Before the glue was applied the surface of the wood was warmed with a hot smoothing iron to prevent it from chilling the glue. When the leather has been thoroughly rubbed down it is ironed smooth with an electric smoothing iron. The edges from the face side are lapped back about ¾ inch on the back side and glued down. When all is dry the leather is smoothed down by sanding a little, and then the propeller goes to the finish room and is finished the same as all others.

*Fabric sheathing.*—The application of fabric sheathing is the easiest of all. When finished in the "white" the propeller goes to the tipping room, where a coat of our standard hot glue is applied to the back of the blade. A sheet of fabric is stretched over this from the leading edge to the trailing edge and ironed down with a small or medium-sized electric smoothing iron. The propeller is turned over and the fabric is laid on around the blade in the same manner. It is then trimmed off, leaving a margin of about an inch. The propeller is again turned over and this margin of fabric is glued up over the leading edge to the back of the blade. The excess fabric is neatly trimmed to a graceful curve and all is ironed down smooth. Some also add a thin glue sizing to the outside over the sheathing by means of a sponge, and iron dry.

The propeller is now allowed to dry thoroughly, and then it goes to the finishing room to be finished as any other would be.

*Moisture proofing.*—All propellers being built by the Air Service designers are being protected with aluminum leaf finish. This is by far the best protection against moisture absorption yet devised.

*Finish room.*—The propeller goes next to the varnish room, where it receives five coats of varnish, as per Specification 29000D, and is carefully balanced for each coat. All excepting the last coat are rubbed down with steel wool or sometimes fine sandpaper, but the last is a flowing coat, which is finally rubbed with pumice and oil to deaden the luster and thus prevent undue reflection when in flight. Great care must be exercised in the varnish room in order to keep the propeller in balance, due to the fact that a little too much varnish some place will make a big disturbance. Do not use alcohol shellac under the varnish, since it will cause the varnish to blister and peel off when exposed to the sun. Keep the finish room neat and clean, and keep the temperature and the relative humidity of the air under control both day and night for satisfactory results.

*Final inspection.*—When the last varnish is dry the finished propeller is given a thorough final inspection in which it is thoroughly examined in every respect by the manufacturer's inspector, and if found to come within the limits of the specifications it is turned over to the Government inspector, who also goes over it thoroughly. If he accepts it he stamps the required data around the outer face of the hub, where it may be seen when the propeller is on the plane.

It is then securely bolted down in a crate containing felt-covered stays a couple of feet out from the hub both ways which give no support other than to prevent the edges from striking the sides of the crate in shipping.

A square piece of board is under the hub and another on top of it in the crate, and a bolt passes through a round centering block in the hub hole and out through both sides of the crate. When this is drawn up securely the propeller is held firmly between the two blocks, and thus receives no support excepting at the hub.

An inspection report is filled out in triplicate—one copy to go in the crate with the propeller, one for Washington, and one for the inspector's file. (For a copy of such a report see chapter on Inspection.)

*Conclusion.*—We have intentionally left out the descriptions on most of the necessary equipment, since there is nothing unusual about it, excepting that used by the inspector. You will find that fully described and illustrated in the chapter on Inspection.

Much unnecessary labor and loss of time can be saved by properly planning the whole plant in the beginning so that the material will take its natural routing from start to finish and thus save much needless handling.

While coming through the factory we have not mentioned the fact that every part of it is supposed to be conditioned constantly as per Specification 29500D. This is especially true of all parts or rooms where the lumber or the unfinished propeller must remain for any time as long as overnight.

The relative humidity can be kept safe by heating the room if too damp and by some system of sprinkling if the air is too dry. We are most particular about the conditioning rooms for the laminations and for the roughing out of propellers.

Some people have been so foolish as to carefully kiln-dry their lumber and then stack it in the open air or in open sheds where it was only protected from the rain. Then they could not see why the inspector should object to the same lumber later on. It is plain to see that lumber once dried right must be kept under proper conditions or it will not stay at the required moisture content.

Aside from ordinary breakage and wear due to grass, weeds, cotton stalks, rocks, and mud, etc., our greatest trouble has been with propeller warping and getting out of pitch, track, and balance, due to improper seasoning of the lumber and improper conditions in the factory, all of which were avoidable to a very large degree.

The arrangements for heating the rooms should be near the floor, while the arrangements for adding moisture, and thus raising the relative humidity of the room, should be near the ceiling.

*Outline of suggested equipment.*—The equipment that we have found necessary for an estimated production of 10 propellers per day is as follows:

*Sample-testing room.*—One pair balances true to one-hundredth of an ounce or to one-tenth gram; one pair scales true to one-tenth ounce and with capacity up to 10 pounds or over; one drying oven, steam or electrical, for drying wood samples; thermometers, charts, etc., for making tests and for checking dry-kiln.

*Mill room.*—Planer, single surfacer; planer, hand jointer; band saw medium; boring machine, ordinary; tooth planer (this work can be done by hand); shaper (making lamination joints or shaping profile of propeller).

*Lamination room.*—Scales, light platform, very accurate to one-quarter ounce; balancing stand, for trying and assembling laminations.

*Glue room.*—Glue heater, automatic temperature regulator, equipped with thermometer; hot box for heating laminations; glue press; cauls or forms (10 sets).

*Gluing stand.*—"C" clamps and supply of blocks (about 300 for 10 propellers per day).

*Roughing room.*—Duplicating machine for rough shaping; heavy boring machine, for hub boring (some are equipped for hub surfacing); arbor press.

*Rough-conditioning room.*—Rack system for hanging 14 times as many propellers as the daily factory output is to be (due to the fact that all propellers must hang 14 days).

*Carving or bench room.*—Work benches, special, with center bolt for holding propellers, preferably adjustable to different angles (two and one-half times as many benches as daily propeller output is to be); two balancing stands and arbors and enough more to accommodate the workmen so as not to keep them waiting (three or four for 10 per day); two surface or laying-out tables for laying out and for proving and inspecting work, as many more as daily output makes necessary (four for 10 per day); bench protractors, for trying pitch angles, one for each surface table; surface gauges, one for each surface table; vertical square or straightedge set in corner of heavy square base, one for each surface table; steel station templates, two set for inspectors; aluminum station templates, set for each carver.

*Tipping room.*—For metal (for capacity of about 10 per day)—two benches for tip routers' use (with big center bolt to hold propeller), two benches for riveters' use (with big center bolt to hold propeller), two benches for solderers' use (with big center bolt to hold propeller), one bench for dressers' use (with big center bolt to hold propeller), one emery grinder, one burnisher, two electric drills (hand portable); for fabric (for 10 per day)—two benches for tippers, one glue heater (advance of other with equal control), two electric irons (medium or light).

*General equipment.*—One balancing stand, one set of racks for hanging propellers.

*Finishing room.*—One stand for sanding, five stands for varnishing, two balancing stands and arbors, rack system for 60 propellers.

*Multibladed propellers.*—Although our whole discussion has been confined to the manufacture of the standard two-bladed propeller built of plain, flat laminations, there are several other types of propellers in use. Of those built up of plain flat laminations many have three and four blades and a few have more. The making of these propellers differs from the making of similar ones with two blades only in the joining of the various laminations in the hub.

It is a very tedious and expensive job to make these many-bladed propellers. They really are not very efficient performers, hence it is not worth while to use them, excepting where space will not permit the use of a two-blader of sufficient diameter to do the work. It is the usual practice now to simply match two two-bladed propellers crosswise on each other and put them on the single shaft where a four-blader is required. This not only saves time and money in the manufacture of the propellers, but it saves time and money in the crating as well. Furthermore, in case of an accident to one of the blades it is only necessary to replace one single two-bladed propeller instead of a whole new four-bladed one.

*De Grandeville process.*—There is still another method of building up wooden propellers known as the "De Grandeville" process. The finished propellers look very much like any others of the two-bladed class. They are built up from many very thin laminations or layers warped and clamped down to a forming block or mold, one upon the other, as the glue is applied. The laminations do not have to be in single pieces or even edge glued. The pieces are simply cut to match each other so as to make complete layers each time and then glued down; thus narrow or scrap lumber may be used so long as the quality is standard. The forming block to which the laminations are warped and clamped as glued has a twisted surface approximating the face side of the finished propeller, so that when the propeller is afterwards carved out the fibers of the wood nearly follow the twist of the blade, thus retaining great strength.

A casein glue, which is used cold, is usually chosen for this process of manufacture, due to the fact that the time required for fixing each lamination as it is applied is so great as to allow hot glue to chill before the job is finished. Sometimes a loose woven fabric is laid between the laminations to prevent shatter from gunshot.

When the glued-up block is dry the propeller is carved out and finished in the same manner as those built by the common method. This is a French idea, and while great merit is claimed for it, very few have been made in this country, largely due to the time and care required in their manufacture.

Many specially-designed propellers have been proposed and built, most of which have proven impracticable. We will not attempt to discuss these here, since it is our purpose to go into the manufacturing details of only the standard type of propeller.

We hope that this brief chapter may prove of interest and assistance to many who have had more or less to do with the manufacture, use, or upkeep of propellers, but have never had the opportunity of visiting a regular propeller factory.

*Propeller data sheet.*

SERVICE PROPELLERS.

| Airplane. | H. P. engine. | Propeller. | Number of blades. | Standing r. p. m. | R. H. or L. H. T. or P. | Diameter. Ft. | Diameter. in. |
|---|---|---|---|---|---|---|---|
| Albatross | Mercedes | X-1014 | 2 | 1,425 | R. H. T | 9 | $1\frac{1}{8}$ |
| Bristol | 190 Rolls-Royce | X-7453 | 4 | 1,050 | do | 9 | $4\frac{1}{2}$ |
| Caproni | 410 U. S. A. 12-A | X-4875 | 2 | 1,380 | 2 R. H. T; 1 L. H. P | 6 | 6 |
| DeHaviland-4 | do | 34555 | 2 | 1,400 | R. H. T | 9 | 2 |
| Do | 375 Rolls-Royce | X-3968 | 2 | 1,020 | L. H. T | 10 | 6 |
| DeHaviland-9 | do | X-3968 | 2 | 1,020 | do | 10 | 6 |
| Handley-Page | 410 U. S. A. 12-A | 14163 | 2 | 1,340 | R. H. T | 10 | 6 |
| Do | 435 U. S. A. 12-B | X-7033 | 2-2 | 1,080 | do | 11 | 0 |
| H. F. John multiplane | 410 U. S. A. 12-A | X-5795 | 2 | 1,390 | do | 10 | 6 |
| LePere | do | 36167 | 2 | 1,470 | do | 9 | 7 |
| L-W-F | 275 U. S. A.-8 | X-487 | 2 | 1,430 | do | 8 | 4 |
| G. L. Martin (3-seater fighter). | 410 U. S. A. 12-A | 37603 | 2 | 1,570 | 2 R. H. T | 10 | 0 |
| Nieuport Scout | 110 LeRhone | 8-8 | 2 | 1,150 | R. H. T | 8 | 1 |
| Do | 150 LeRhone | X-6727 | 2 | 1,150 | do | 8 | $4\frac{1}{2}$ |
| Ord. Eng. Corpn. S-2 | 160 Gnome | X-4341 | 2 | 1,150 | do | 8 | 9 |
| Rumpler | 260 Mercedes | X-5565 | 2 | 1,450 | do | 10 | 6 |
| SE-5 | 180 Hispano-S | 36945 | 2 | 1,675 | do | 7 | 11 |
| Do | 150 Hispano-S | 36945 | 2 | 1,650 | do | 7 | 11 |
| Do | 225 Hispano-S | 8-48 | 2 | 1,350 | L. H. T | 8 | $2\frac{1}{16}$ |
| S P A D | 200 Hispano-S | 8-48 | 2 | 1,350 | do | 8 | $2\frac{1}{16}$ |
| Do | 180 Hispano-S | X-5903 | 2 | 1,675 | R. H. T | 7 | $8\frac{1}{4}$ |
| Thomas-Morse MB-2 | 430 U. S. A. 12-C | X-6856 | 2-2 | (1) | L. H. T | 8 | 8 |
| U. S. A. C-1 | do | X-4908 | 2 | 1,090 | do | 11 | 0 |
| U. S. B-2 | 275 U. S. A.-8 | X-6389 | 2 | 1,700 | R. H. T | 8 | 6 |
| U. S. B-1 | 320 Hispano-S | 36461 | 2 | 1,560 | do | 9 | 6 |
| U. S. D. 9-A | 410 U. S. A. 12-A | 34291 | 2 | 1,475 | do | 10 | 0 |
| Do | do | 37593 | 2 | 1,420 | do | 9 | 10 |

[1] No report.

PROPELLER CLUBS.

| Engine. | Club. | Number of blades. | R. P. M. | R. H. or L. H. T. or P. | Diameter. Ft. | Diameter. in. |
|---|---|---|---|---|---|---|
| U. S. A. 12-A | 36950 | 2 | 1,700 | L. H. P | 8 | 1 |
| Do | X-6599 | 2 | 1,200 | R. H. T | 10 | $2\frac{11}{16}$ |
| | | | 1,400 | do | 9 | $2\frac{11}{16}$ |
| | | | 1,600 | do | 8 | $5\frac{1}{16}$ |
| | | | 1,700 | do | 8 | 1 |
| | | | 1,800 | do | 7 | $9\frac{11}{16}$ |
| 300 Hispano Suiza | 37170 | 2 | | L. H. P | 7 | 8 |

TRAINING PROPELLERS.

| Airplane. | H. P. engine. | Propeller. | Number of blades. | R. P. M. | R. H. or L. H. T. or P. | Diameter. Ft. | Diameter. in. |
|---|---|---|---|---|---|---|---|
| Avro | 100 Gnome | X-4557 | 2 | 1,200 | R. H. T | 9 | $0\frac{3}{8}$ |
| Curtiss R-2 | 160 Curtiss VX | 8-6 | 2 | 1,400 | do | 9 | 3 |
| Curtiss R-4 | 420 U. S. A. 12-A | 34555 | 2 | 1,460 | do | 9 | 2 |
| Do | 220 Curtiss (V-2, type 3). | 11612 | 2 | 1,400 | do | 9 | 6 |
| Curtiss JN-4 | 90 Curtiss OX-5 | 13706 | 2 | 1,400 | do | 8 | 4 |
| Curtiss JN-4H | 150 Hispano-S | 34889 | 2 | 1,350 | do | 8 | 6 |
| Do | 180 Hispano-S | 34889 | 2 | 1,425 | do | 8 | 6 |
| Standard E-1 | 80 LeRhone | 37125 | 2 | (1) | do | 8 | 0 |
| Heinrich (scout) | do | X-6391 | 2 | 1,120 | do | 8 | 0 |
| Standard J-1 | 95 Hall-S, A-7A | 13706 | 2 | 1,425 | do | 8 | 4 |
| Thomas-Morse (scout) | 105 Gnome | 14161 | 2 | 1,200 | do | 8 | 2 |
| Do | 80 LeRhone | 13631 | 2 | 1,200 | do | 8 | 2 |
| VE-7 (Vought) | 150 Hispano-S | 34923 | 2 | 1,560 | do | 8 | 6 |
| VE-7 | 180 Hispano-S | 37643 | 2 | | do | 8 | 8 |

[1] No report.

## CHAPTER III.

### INSPECTION.

*Apparatus.*—To obtain uniform inspection results throughout the various manufacturing and military organizations having to do with propeller construction, maintenance, and repair, it has been necessary to standardize the majority of the propeller-inspecting apparatus. Those desiring specific information concerning the details of this standard equipment, can obtain drawings from the Technical Section, Air Service, United States Army. These drawings are listed by title and number in Table 1. A very good idea of the general construction of this apparatus can be obtained from the photographs. The propeller-inspection table is shown in figure 48. It consists of a heavy casting mounted on a wooden stand. The top surface of the casting is planed perfectly smooth and true. At one end, on the longitudinal center line of the table, is mounted the aligning stand, or aligning spindle, as shown in the photograph, figure 48. This spindle is bolted to the table so that its top surface is parallel to the top of the inspection table. It serves as an arbor upon which to mount and hold the propeller in a fixed position while performing the inspection.

A number of other inspection devices are also shown in figure 48. Item No. 1 is a surface gauge. This is used in determining whether or not a propeller "tracks." This process will be described later. Any good make of gauge can be used as standard. Item No. 2 consists of an assortment of hub bushings. These are used to mount the propellers on the mandrel when testing their balance and also when mounting them on the aligning spindle of the inspection table. The tapered bushings are for use after the propellers have been fitted with hubs. Item No. 3 is a pair of parallel blocks. These are often used to raise the "protractor" when measuring blade angles. Item No. 4—Calipers for measuring blade widths, depth of camber, etc. Item No. 5—Machinist's combination square for determining the face alignment. Item No. 6—This is the standard protractor for measuring the blade angles. Item No. 7—Marking dies for stamping the inspection data on the propeller hub. Item No. 8—Scale graduated in sixty-fourths of an inch.

The standard balancing stand is shown in figure 49. Other types of stands have been used with varying degrees of success. One type which was frequently used at first is the roller disk type. In this type the hardened-steel knife edges as used in the standard type are replaced by four hardened-steel roller disks mounted in pairs. The

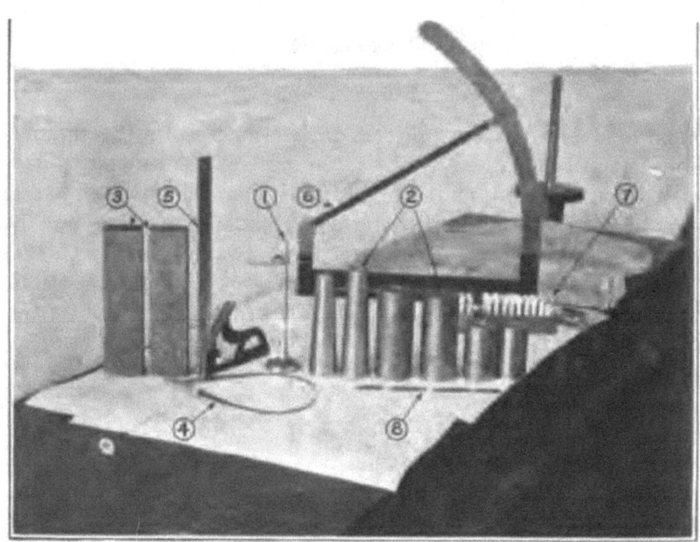


1, Surface gauge; 2, hub bushings, straight and tapered; 3, parallel blocks; 4, calipers; 5, machinists' combination square; 6, protractor; 7, marking dies; 8, 12-inch scale.
Inspection Apparatus.
FIG. 48.

FIG. 49.—STATIC BALANCE OF PROPELLER, BLADE HORIZONTAL.
Note.—The propeller must remain in this position without persistent movement in any direction.

110

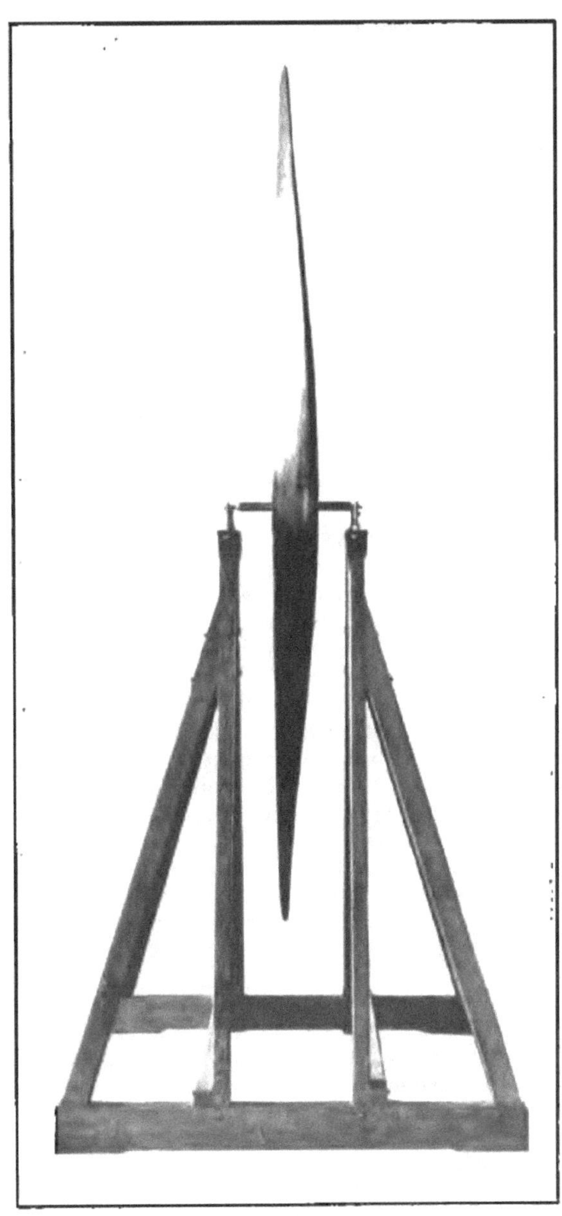

FIG. 50.—STATIC BALANCE OF PROPELLER, BLADES VERTICAL.
Note.—The propeller must remain in this position without persistent movement in any direction.

centers of each pair of disks are mounted in a horizontal line with the edges of each pair overlapping. Thus each pair of disks forms a crotch support to carry the mandrel. In this type of balancing stand leveling is necessary in the direction of the mandrel only. The circumferences of the hardened disks become dented after using a while, due to difficulty in handling heavy propellers or to carelessness. The slightest dent or irregularity will prevent them from working properly. However, this type of balancing stand can be successfully used in making preliminary balance tests. The finished test should be made on the standard type of knife-edge balancing stand.

*Inspection during manufacture.*—All raw materials must pass Government inspection before entering into a propeller. A set of specifications issued by the Bureau of Aircraft Production and covering the use and inspection of propeller materials are appended herewith for ready reference. These specifications are sufficient guidance for both the manufacturer and inspector to turn out a first-class finished product.

Specification No. 29500D is a general treatment of the entire subject of propeller materials, manufacture, and inspection. A thorough study of this specification should be made.

The first duty of the inspector is to see that none but Government-inspected raw material, bearing the inspection stamp, goes into the make-up of any propeller. Excepting lumber, most raw material will have passed Government inspection upon receipt at the plant. Most manufacturers kiln dry their own lumber. It is the duty of the inspector to see that this operation is done properly in accordance with Specification No. 20500A. The subject of kiln-drying is taken up quite thoroughly in the chapter on propeller manufacture.

An inspector's duties begin with the arrival of the first shipment of raw material and end when the finished propellers are crated for shipment. The mere possession of a set of specifications and the ability to perform the duties described therein do not make an efficient or good inspector. There are numerous opportunities occurring constantly for the inspector to exercise judgment on matters not covered in the specifications. Occasions will arise when the work will be held up unless the inspector has the ability to make decisions pertaining to matters not specifically covered in the specifications. On the other hand, occasions will arise where the same sort of decision on the part of the inspector will prevent the manufacturer from turning out work that will result in rejections later. The inspector has a big job, burdened with responsibilities, and has to constantly make decisions which test his knowledge and judgment. He should bear in mind that the quality of the finished propeller can not be consistently good unless he performs his work thoroughly, and that the safety of the men who have to use the finished propeller is jeopardized unless the propellers *are turned out consistently good.*

When the rough lumber is brought into the machine room from the storeroom to be dressed, the inspector should be present to observe the boards as they come from the planer. Doty boards are more easily detected during the planing process than at any other time. The chips from a doty board are much finer than those from a sound board. Boards should be inspected after dressing one side. Many imperfections are exposed after the first planing which result in rejections, making it unnecessary to plane the other side. Defects are often exposed when planing the second side that were not visible before. Frequently a defective board will contain enough clear stock to make one or more laminations. In this case, of course, the clear stock should be salvaged.

At the time of cutting the laminations from the boards, the manufacturers are required to cut small, data samples, one from near each end of each lamination. These specimens are for the use of the inspector in determining the moisture content and density of the laminations. Methods of making the specific gravity or density determination is given in Specifications Nos. 20505A and 20506. The method described in 20505A is preferred. More accurate and reliable data can be obtained by this method.

The method of calculating the moisture content is fully described in Specification No. 20504B. The average values for moisture content and density of each pair of samples are taken as being equal to the actual moisture content and density of the respective laminations from which the samples were taken. These values are recorded for each lamination and are used later in assembling the laminations into propeller groups. This data enables the inspector to prevent the assembling of laminations into propeller groups that vary in moisture content and density within any one group by amounts greater than the limits set by the specifications.

The requirements of the specifications relative to mixing fresh glue daily and to the cleanliness of the glue pot, etc., should be strictly enforced. The gluing up of the laminations is a very important stage of the manufacturing process. The proper heating of laminations before gluing and the maintenance of the proper temperature in the glue room must be thoroughly looked after, otherwise much good material, time, and work will find their way to the rejection heap.

During the process of carving, the inspector should watch carefully for imperfections which are sometimes visible during this stage, but are later, as the job nears completion, cleverly concealed by the workman. Wind shakes are a good example of this. They are readily distinguishable while the wood is being worked, but a good workman can easily finish off the surface perfectly smooth without

FIG. 51.—VIEW SHOWING PROTRACTOR IN POSITION FOR READING THE BLADE ANGLES.

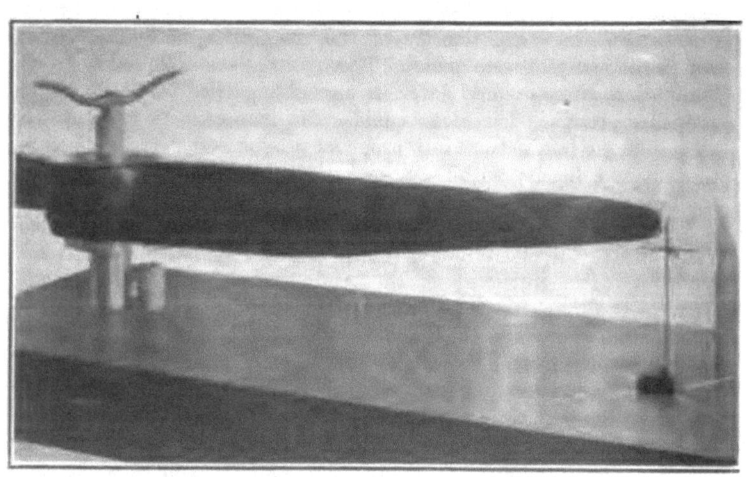

FIG. 53.—CHECKING TRACK OF BLADES.

Note.—Surface gauge is set up for blade No. 1.

the slightest signs of a shake being present. This is but one instance which illustrates the necessity of constant alertness on the part of the inspector.

After the carving is finished, the hub hole bored out, and the bolt holes drilled, the propeller is given a thorough inspection covering workmanship, material, static balance, and conformity to drawing.

First the entire surface of the propeller is looked over carefully for open glue joints and other defects. Open glue joints are hard to detect. The inspector should be provided with a very thin steel-blade probe of only a few thousandths of an inch in thickness. Such an instrument can be readily inserted into open glue joints. Many open joints can be detected in this manner that would otherwise pass unnoticed.

The next thing to be determined is the condition of the static balance. For this test the propeller is mounted on the hardened steel mandrel of the "knife-edge" type of balancing stand. Propellers undergoing this test are shown in figures 49 and 50. The room in which the test is made should be inclosed free from all air currents. To pass the balancing test a propeller must stand at any angle of rotation without showing persistent motion in either direction. Propellers that are slightly out of balance can be corrected by the addition of a very thin coat of varnish. The insertion of weights in the hub is not allowed. (See Specification No. 29500D, pars. 43 to 46, inclusive.)

After passing the balancing test the propeller is mounted on the aligning stand of the inspection table and inspected for hub thickness, hub diameter, diameter of hub hole, diameter and spacing of bolt holes, blade length, track, edge alignment, face alignment, width of blade, thickness of blade, section or air-foil shape, and blade angles. The allowable tolerances covering the above points are given in Specification No. 29500D, paragraphs 29 to 32, inclusive. Tolerances marked on propeller drawings supersede those in the specifications. A propeller mounted for inspection is shown in figure 51. The protractor is in position for measuring the blade angle at a given station. The inspector is shown measuring the blade width at another station. The blade section is checked by the same template from which the woodworker carved the propeller to shape. These templates are made up of sheet metal (preferably duralumin) to conform to the details of the propeller sections at the stations shown on the drawing. Figure 52 shows a type of template suitable for this work. These templates can be used in checking the blade widths and thicknesses.

The top of the inspection table is marked with graduating lines at right angles to the longitudinal center line every 6 inches, beginning at the center of the aligning spindle. These graduations are useful in conjunction with the combination square in locating the positions

161988—21—8

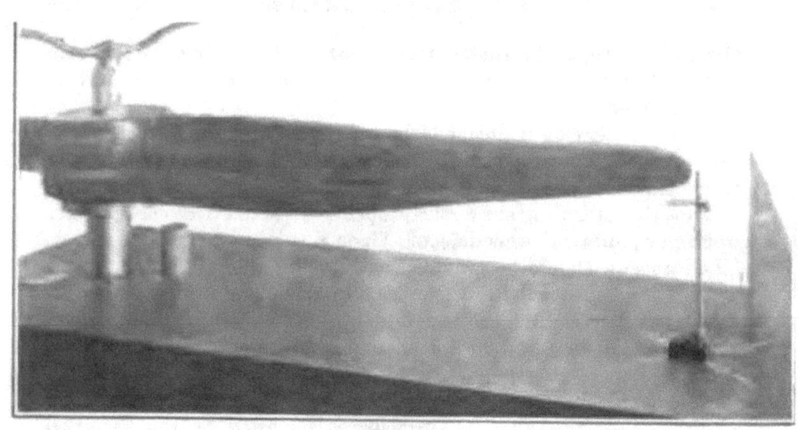

FIG. 54.—CHECKING TRACK OF BLADES.

Note.—Observe that blade No. 2 falls above the point of surface gauge when the propeller is swung around. The propeller is out of track.

FIG. 55.—VIEW SHOWING PROPELLER AFTER HUBBING MOUNTED ON INSPECTION TABLE TO CHECK FOR TRACK.

of the blade sections as shown on the drawing and in measuring the blade length.

Perfect tracking of the blades is very important. The method of determining whether the blades track is illustrated in figures 53 and 54. The propeller in this illustration is badly out of track. To correct this, the hub face is scraped down on the high side until the proper track is secured. If a propeller is so badly out of track that it can not be corrected by this method, it is not fit for service and is rejected.

After passing this first inspection, the propellers are returned to the shops to be sheathed and varnished. A description of the

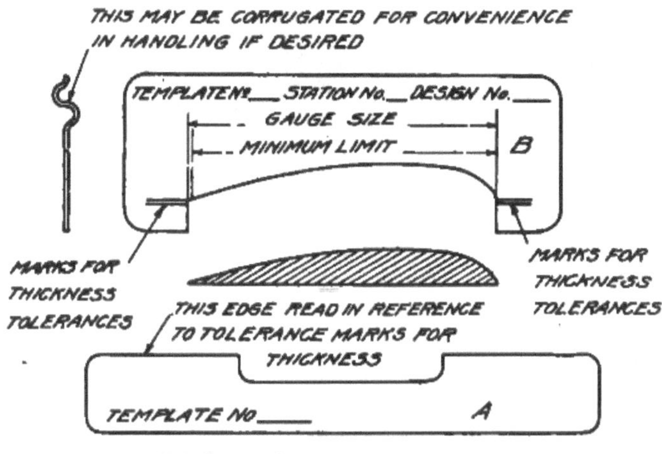


FIG. 52.

sheathing process is found in Specification 29500D, paragraphs 33 to 40, inclusive.

After varnishing, the propellers are given a final factory inspection to check for any changes that might have been produced by the sheathing or finishing processes. This inspection covers balance, track, and angles. Propellers which pass this test are ready to be packed for shipment or to be installed on planes.

A typical propeller inspection report is given on following page.

Packing for shipment, especially for overseas, should be painstakingly done. Specification No. 40538B covers this subject. The inspector should see that these specifications are faithfully followed.

Propellers that are to be kept for any length of time before using should be stored under constant conditions of humidity. This

FIG. 56.—VIEW SHOWING METHOD OF TRACKING PROPELLER ON PLANE.

118

reduces to a minimum the likelihood of their being damaged by warping. (See chapter on propeller storage.)

No further inspection of propellers is necessary until after they are fitted with the hub preparatory to being installed on the plane. The installation of the hub sometimes affects the balance, tracking, and blade angles. Consequently an inspection covering these points must be made after the hub is bolted into place. Lack of balance can be corrected by either scraping off wood from the outside diameter of the hub or by removing light shavings from the blade close in toward the hub. Sometimes it is necessary to remove material from both places. This destroys the finish, which must, of course, be replaced. Tracking is accomplished as before by scraping down the propeller hub face on the high side. Since the metal propeller hub has a tapered hole to fit the engine shaft, a special tapered bushing of the type shown in figure 48, item 2, is provided for this inspection.

Figure 55 is a photograph of a propeller after hubbing, mounted on the inspection table to check for track. It is interesting to note that this propeller is made to drawing No. 34291, which is used to illustrate the method of laying out a propeller in the chapter on design.

*Inspection after installing on plane.*—It is often desired to make an inspection of a propeller on the field without removing it from the plane. It is comparatively simple to check for track. The method is well illustrated in figure 56. The end of a stick is held stationary so that it just touches the trailing edge at a point near the tip of one blade. The propeller is then rotated so that the other blade or blades are made to pass the end of the stick. Thus any variation in track can be observed and measured.

A rough check on the blade angles can be obtained as follows: Choose a section where the blade is fairly wide and at a point about one-third in from the tip. Track both the leading and trailing edges at this chosen radius by the method just described. If both edges track perfectly, or if the leading and trailing edges of one blade are out of track by the same amounts, the angles at this radius are equal. This method can only apply when the blades are of equal widths at the chosen radius. The method only serves to check the size of the angles with relation to each other and does not give the magnitude of the angles in degrees. However, if carefully done the method will be sufficient to determine any serious discrepancies in the angles, and will often be of value on the field where more accurate methods are not available.

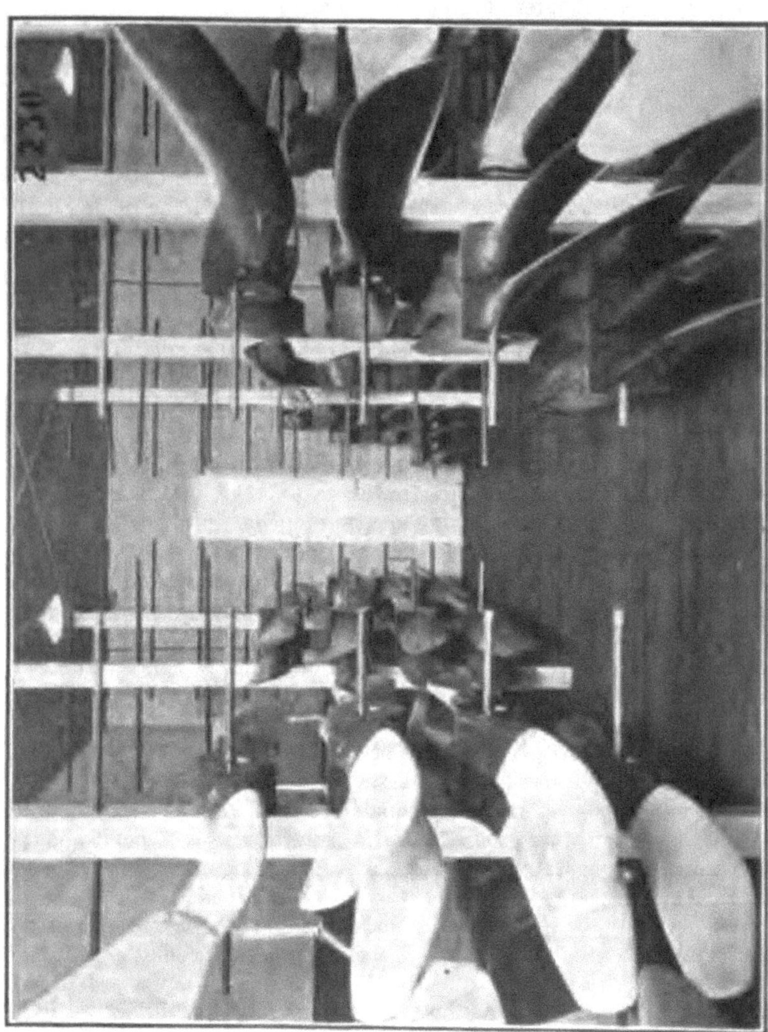

FIG. 58.—STORAGE HOUSE.

TABLE 1.—*Air service drawing numbers of standard propeller inspection apparatus.*

| Item. | Standardized drawing and part numbers. |
|---|---|
| Propeller balancing stand, assembly | 37680 |
| Propeller balancing stand, subassembly | 37681 |
| Propeller balancing stand, subassembly part | 37682 |
| Do | 37683 |
| Do | 37684 |
| Do | 37686 |
| Do | 37689 |
| Do | 37691 |
| Do | 37692 |
| Propeller inspection table |  |
| Propeller inspection table, layout, plate No. 2 | 37667 |
| Propeller inspection table, layout, plate stand | 37668 |
| Propeller inspection table aligning stand, subassembly | 37647 |
| Propeller inspection table aligning stand, subassembly part | 37648B |
| Do | 37649B |
| Do | 37650B |
| Do | 37651A |
| Do | 37652A |
| Do | 37663A |
| Do | 37664A |
| Propeller protractor, assembly | 37626E |
| Propeller protractor part | 37625C |
| Do | 37638A |
| Do | 37627D |
| Do | 37634A |
| Do | 37637A |
| Do | 37630A |
| Do | 37629A |
| Do | 37632A |
| Do | 37631A |
| Do | 37628A |
| Do | 37635A |
| Do | 37636A |
| Do | 37633A |
| Do | 37645A |
| Do | 37644A |
| Do | 37646A |
| Do | N. P. 288 |
| Do | PS1846 |
| Do | RR753 |
| Do | RR569 |
| Do | WE111 |

## Chapter IV.

### PROPELLER STORAGE HOUSE.

*Building.*—The McCook Field storage house occupies three bays of a standard hangar; that is, a space 60 by 66 feet, including the inspection room and apparatus room, as shown on figure 57. The storage house proper is built inside the hangar with 8-inch walls of common brick, leaving an air space of about 2 inches between the brickwork and the corrugated iron siding. It is floored with wood nailed to sleepers embedded in a concrete base, and has a wooden ceiling nailed to joists supported on the roof trusses. The upper side of these joists has a floor of sheathing boards. The purpose of this interior wall and ceiling is to insulate the storage space, so as to require less heat in winter, and also to keep down the temperature in summer.

All doors are 10 feet 6 inches high, so that propellers may be carried through them in a vertical position. It may be noted that the easiest way to carry a propeller is in approximately the position of "right shoulder arms" with the hub of the propeller resting on the shoulder. The doors in the exterior walls are double, to assist in insulation of the storage space.

In the storeroom proper, 6 by 6 inch posts are erected in rows, extending to the ceiling, which in this building is 14 feet high. Five-foot lengths of 1-inch (inside diameter) iron pipe are driven through holes bored in these posts at regular intervals, eight to each post, and projecting half on each side. These form the pins on which propellers are hung by their hub holes. The lower four pins on each post are all that can be used conveniently; the upper four are utilized for dead storage. The general arrangement is shown in figure 58.

Posts and pins may be given letters and numbers, and the location of each propeller kept in a card index.

The space marked "inspection room" on the plan (fig. 57) serves not only that purpose but also as an office, packing room, etc., as may be required. The object of inspection here is to see that propellers are properly hubbed, especially as to balance and track, before being put into service; and also to condemn them when worn or damaged. The equipment used is the same as that described in the chapter on "Inspection".

The effects of moisture and humidity on wood have already been treated in the chapter on "Manufacture". The finished propeller is still sensitive to humidity changes. Although stock of the same initial moisture content is selected for each propeller, it is impossible to take into account the small differences of grain and texture,

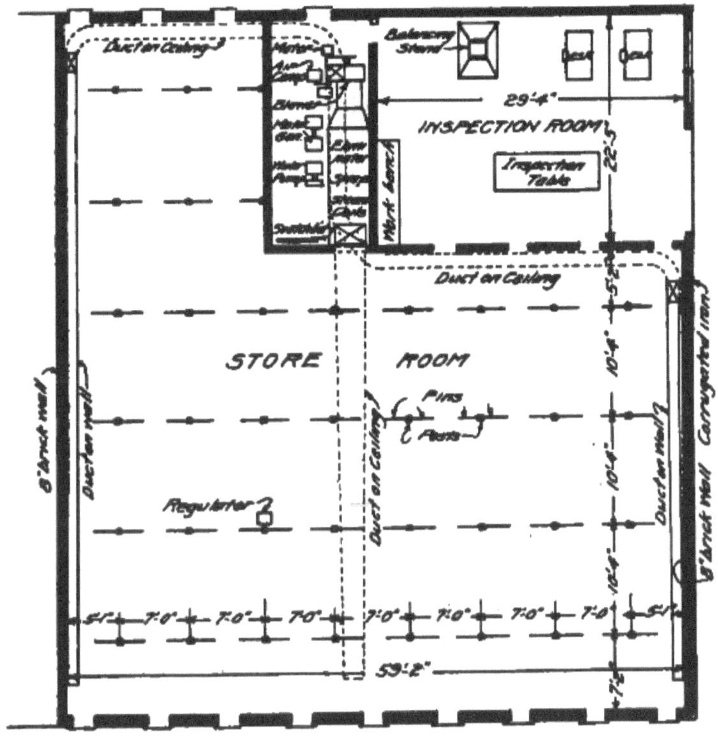


FLOOR PLAN
PROPELLER STORAGE HOUSE
McCook Field, DAYTON, O.
Scale 1/8"=1'-0"

Fig. 57.

which may cause one lamination to absorb moisture more rapidly than the next, with consequent warping or change of pitch. The air in the propeller-storage house should, therefore, be kept in a condition which will not tend to change the moisture content of the propellers, whatever that may have been at the time they first passed inspection.

This condition has been experimentally determined to be 65 per cent relative humidity and 65° F. temperature, though the temperature is comparatively unimportant.

To verify this conclusion, a test was made on a propeller without varnish or covering of any sort, the bare wood being directly exposed to an atmospheric condition of 65 per cent humidity and 65° temperature for 135 days. The effect of this exposure is shown in the following table:

*Propeller inspection data.*

Serial No. S. C. 60511.
Drawing No. 34889.
Inspection Report No. 0-480.

Plane, JN-4-H.
Engine, Hispano-Suiza 150 H. P.
Material, mahogany.

[Manufactured by airplane engineering department, McCook Field.]

(Blade angles measured in degrees.)

| Radius inches. | Specified angles. | Nov. 7, 1918, after carving (blade number). | | Feb. 10, 1919, during test (blade number). | | Apr. 7, 1919, conclusion of test (blade number). | |
|---|---|---|---|---|---|---|---|
| | | 1 | 2 | 1 | 2 | 1 | 2 |
| 30 | 19.70 | 19.70 | 19.70 | 19.75 | 19.60 | 19.60 | 19.60 |
| 36 | 17.00 | 17.00 | 17.00 | 16.90 | 16.90 | 16.95 | 16.90 |
| 42 | 15.00 | 15 00 | 15.00 | 15.00 | 14.70 | 14 95 | 14.60 |
| 48 | 13.70 | 13.75 | 13.70 | 13.55 | 12.40 | 13 70 | 13.30 |

Track........................................................ O.K. $\frac{1}{16}$ inch out. $\frac{1}{16}$ inch out.
Balance...................................................... O.K. O.K. O.K.

The variations in angle and track shown on the last inspection are within the allowable limits and the propeller was in serviceable condition.

In practice, humidity is determined by the "wet-bulb depression," that is, the difference in temperature registered by two thermometers, one normal and the other with its bulb constantly kept moist and in a current of air. The dryer the air, the more rapid will be the evaporation and the lower will be the reading of the wet-bulb thermometer. The relationship between moisture content of air, normal (dry-bulb) and wet-bulb temperatures and humidity may best be obtained from a chart plotted by W. H. Carrier (fig. 59).

Wet-bulb and dry-bulb temperatures may be obtained simultaneously by a device known as a psychrometer. It consists of two thermometers, one with its bulb covered by a wick, which is dipped in water before using. The thermometers are mounted in a frame with a swiveling handle so that they may be spun around in the air. This motion is necessary to get the full effect of the evaporation on the wet-bulb thermometer. An instrument of this kind is useful as a check on the working of the air-conditioning apparatus.

*Air-conditioning apparatus.*—Several concerns are already making air-conditioning apparatus for use in cotton mills, cigar factories, tobacco warehouses, etc., where the requirements are practically the same as outlined above, so that there is no difficulty in obtaining reliable equipment. Among these concerns are the Atmospheric

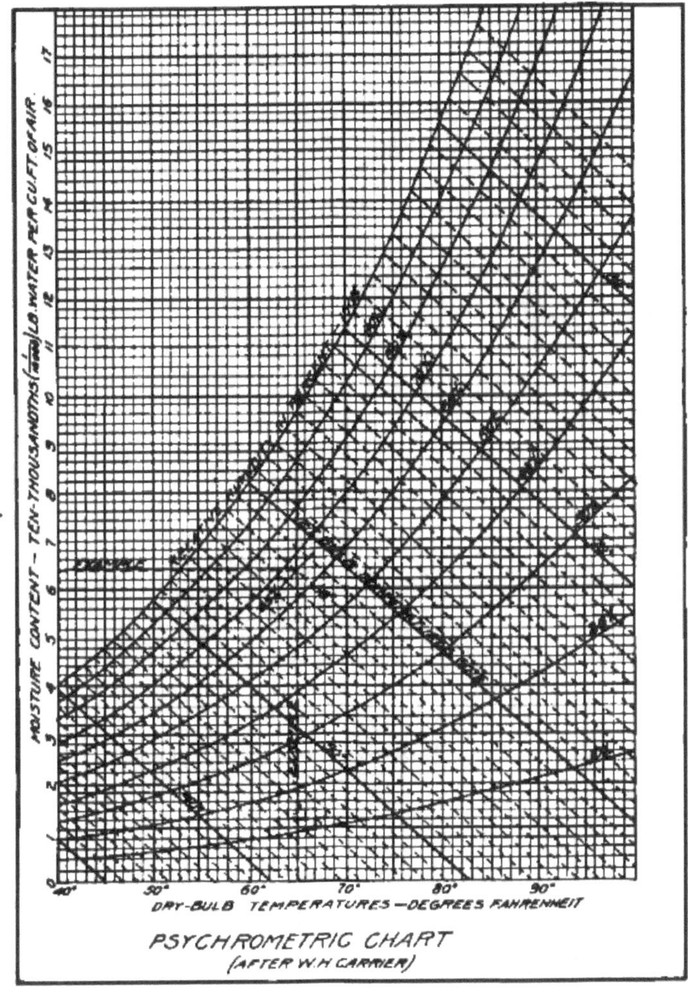


Fig. 59.

Conditioning Co., Philadelphia; Bontz Engineering Co., New York; Carrier Air-Conditioning Co., Buffalo, N. Y.; Stuart Cramer Co., Charlotte, N. C.; and Johnson Service Co., Milwaukee, Wis. All

FIG. 60.—APPARATUS ROOM STORAGE HOUSE.

conditioning outfits operate on the same principle—the air is circulated by a blower and passes through steam coils and water sprays, the differences are in the method of control and in the design and arrangement of the spray nozzles.

The apparatus at McCook Field was furnished and installed by the Stuart Cramer Co. It is in effect a hot-air heating plant, similar to those used in schools and theaters, with the addition of the water spray and controlling devices. As there are never enough people in the storage house to consume any appreciable amount of oxygen the air is usually all recirculated, though a damper is provided through which fresh air may be drawn in. Recirculation saves the heat which otherwise would be required to bring the incoming outside air up to the room temperature.

Air enters the storeroom through ducts along the side walls and leaves through the large duct in the middle on the ceiling, as shown in figure 57. All the heat necessary to make up for the losses through walls, ceiling, etc., must be carried by the incoming air, which may reach a temperature of 85° F., with a resulting humidity of only 35 per cent. Care must be taken that this hot air does not strike directly upon any propellers; on the contrary, it must be deflected so as to flow along the walls and ceiling until its excess heat has been absorbed. The inlet ducts have openings on top and bottom only, and these are fitted with slides which may be adjusted so as to distribute the hot air uniformly.

The general arrangement of the apparatus may be seen in figures 60 and 61. Heating coils, water sprays, etc., are inclosed in a galvanized-iron casing about 4 feet wide and 6 feet high, with a glazed inspection door and a waterproof electric light inside. Air circulation is forced by a 21-inch fan or blower turning 580 r. p. m., and delivering 5,700 cubic feet of air per minute. The fan is driven by a two-horsepower electric motor through a silent chain. Air passes first through three stacks of steam coils with a total of 384 square feet of heating surface. Beyond the heating stacks a vertical partition is introduced in the casing. On one side of the partition are three water pipes, each with six spray nozzles, and the eliminator, a bank of small angle-shaped galvanized-iron strips which prevent any drops of water from passing through with the air. As air which has passed through the sprays and eliminator is saturated or has 100 per cent humidity it is diluted by air from the other side of the partition, which has been by-passed around the spray chamber. The dilution is controlled by a damper. The bottom of the spray chamber forms a well, from which the water is recirculated by a centrifugal pump. The water level is maintained by a float valve.

The automatic regulation is the distinctive feature of the apparatus. (See fig. 62.) The regulator box is mounted on a post in

FIG 61.—APPARATUS ROOM STORAGE HOUSE.

FIG. 62.—CONTROL APPARATUS.

FIG. 63.—SWITCHBOARD.

the storeroom about 10 feet from the floor. It is divided into three compartments—two contain each a vapor thermostat with a mercury thermometer as telltale. One compartment has openings allowing the air to circulate freely through it, and the thermometer and thermostat in it form the dry-bulb side of the device. The other compartment contains also a small water spray, formed by a jet of compressed air impinging on a jet of water and forcing it against the other thermometer and thermostat, which, of course, form the wet-bub side. The two thermostats are connected by linkages to sliding electric contacts, which are mounted in the upper or third compartment. The contacts actuated by the dry-bulb thermostat control the steam coils, and the wet-bulb ones the water sprays. There are two contacts for each thermostat, one slightly in advance of the other, so as to shut off part of the steam or water a little before the desired condition is reached, and so prevent overrunning. Controlling levers are introduced into the connections in such a way that the device may be set for any reasonable degree of temperature or humidity.

The four electric circuits controlled by the sliding contacts each contain a solenoid actuating a valve in a compressed-air line. The compressed air provides the actual force for opening and closing the steam and water valves, the air being admitted into a "Sylphon" diaphragm connected to the valve stem. The air lines to the water valves are also connected to additional diaphragms which operate the by-pass damper mentioned above. The four solenoid valves, mounted on a switchboard, and two of the "Sylphon" diaphragm valves, are shown in figure 63.

Auxiliaries to the apparatus are a Brunner 2-cylinder air compressor driven by a 2-horsepower electric motor, with a tank of 3 cubic feet capacity and an automatic regulator which starts and stops the motor to hold the pressure between 25 and 45 pounds; an Emerson 1-horsepower motor direct-connected to a 15-volt 30-ampere generator, providing direct current for the solenoid valves, and a "Twinvolute" centrifugal water pump driven by a 1-horsepower electric motor, maintaining 25 pounds pressure in the spray pipes. A switchboard carries air and water gauges, voltmeter and rheostat, the four solenoid valves, switches, etc. (fig. 63).

Steam for the heating coils is provided by a central heating plant, connected through a reducing valve which brings the pressure down to 5 pounds. If no such plant was available, a small boiler of the size and kind used for a medium-sized residence would be capable of heating the building.

At the date of writing (May 29, 1919), this apparatus has been in operation about eight months, and during the winter months ran 24 hours a day, with only a few short stops. It has required practically

161988—21——9

no attention beyond cleaning and adjustment, such as would be expected on any similar mechanical outfit. The regulation is very sensitive, holding the temperature, both wet and dry bulb, within 1° variation during the winter. These temperatures were checked by two recording thermometers, sample records from which are shown in figure 64.

It must be remembered that the apparatus can not reduce the humidity of air except by warming it, nor can it reduce the temperature except by increasing the humidity; and although the air in the storage house is recirculated, without taking in any outside air, still the outside conditions will influence the inside conditions by leakage and by transmission of heat through the walls. If the outside temperature is too high, but with a low humidity, say, 75° F. and 35 per cent humidity, the air can be sufficiently cooled by evaporation of spray water to make the required condition; or if the outside temperature is low, with a high humidity, say, 55° F. and 90 per cent humidity, the humidity will be sufficiently reduced by heating. But the combination of high temperature and high humidity can not be corrected by this apparatus. Consequently, in summer the inside temperature will follow to some extent the outside temperature, although not reaching the same extremes. The inside humidity will be kept up to 65 per cent by the apparatus, but may exceed this figure during a spell of damp weather.

It would be perfectly possible to design an apparatus which would maintain a storage house at the desired temperature and humidity, regardless of outside conditions, but it would be much more costly and complicated than that described, which answers sufficiently well. Consideration of such an apparatus would not come within the scope of this manual.

The routine care of the air-conditioning apparatus is simple and can be managed by any engineer with some experience in power-house work. The solenoid valves are rather delicate, and if not kept well cleaned and oiled are apt to become clogged with rust, due to condensation water in the compressed air. If the water used is hard, the spray nozzles in the spray chamber and in the regulator box must be kept open. All bearings and moving parts in the motors, pumps, fan, etc., must be kept clean and well lubricated.

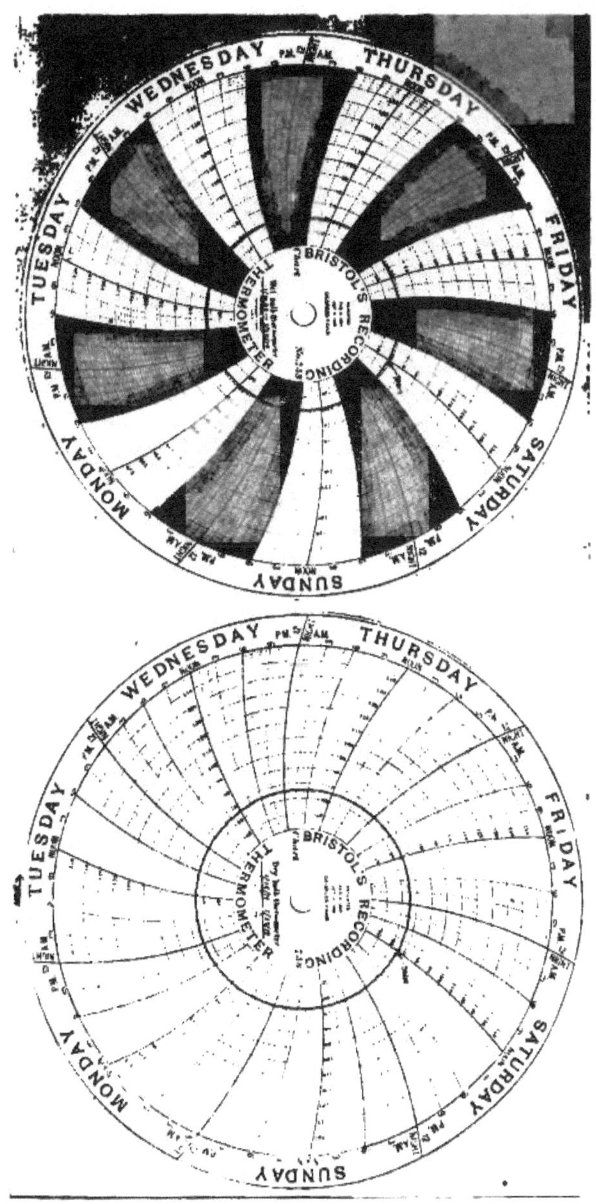


FIG. 64.—TEMPERATURE RECORDS.
(Wet and dry bulb.)

FIG. 65.—POWER HOUSE SHELTERING THE 1,000-KILOWATT MOTOR-GENERATOR SET AND AUXILIARY EQUIPMENT. ALSO SHOWING "TEST RIG" AND "BOMBPROOF" IN COURSE OF CONSTRUCTION.

The bombproof stops flying splinters when a propeller breaks.

FIG. 66.—1,000-KILOWATT MOTOR-GENERATOR SET FOR SUPPLYING POWER AT 550 VOLTS TO THE DYNAMOMETERS.

FIG. 67.—MOTOR-GENERATOR PANELS AND CONTACTOR PANEL FOR STARTING THE DYNAMOMETERS ON REDUCED VOLTAGE. SERVICE PANEL SHOWN ON THE LEFT.

## CHAPTER V.

# INSTRUCTIVE PROPELLER TESTING.

### EQUIPMENT.

The present equipment at McCook Field consists of a power house and a rig on which the tests are made.

*Power house.*—Power is taken from a special 3-phase, 60-cycle, 6,600-volt power line of the Dayton Power & Light Co. Inside the power house this power is supplied through necessary switching equipment to a Westinghouse Electric & Manufacturing Co. 1,420-horsepower, 80 per cent power factor, 6,600-volt, 3-phase, 60-cycle synchronous motor. This motor is direct-connected to a 1,000-kilowatt, 600-volt, 1,663-ampere, compound-wound, compensated, direct-current generator. The direct-current generator supplies power for the motors driving the propeller. (See figs. 65 and 66.)

The exciters for the 1,420-horsepower synchronous motor and the fields of the dynamometers driving the propeller are also located in this building. All resistance grids for armature voltage control in the dynamometers in starting, together with the contactor panels for cutting this resistance into or out of the armature circuits, are located in the power house. (See fig. 67.)

The switchboard for the 1,000-kilowatt motor generator set consists of four panels—one special metering panel belonging to the Dayton Power & Light Co. carrying a Fort Wayne watt-hour meter and General Electric oil circuit breaker; one Westinghouse Electric & Manufacturing Co. synchronous motor-starting panel carrying an indicating kilowatt meter, an ammeter for alternating line current, a power-factor meter, a direct-current ammeter for the synchronous motor field current, a totalizing alternating-current watt-hour meter, a handle for the rheostat controlling the synchronous motor field current, and a double-throw oil circuit breaker for starting the 1,420-horsepower synchronous motor on reduced voltage; one Westinghouse Electric & Manufacturing Co. direct-current generator panel carrying a voltmeter, an ammeter, an air-break circuit breaker, and a double-pole single-throw knife switch; and one exciter panel for the small motor generator set supplying the synchronous motor-field current.

*Test rig.*—The test rig proper consists of a 6-inch nickel-steel shaft mounted in oversize bearing pedestals attached to a heavy steel bedplate. A Kingsbury marine type double-thrust bearing in a special floating housing is mounted on the jackshaft just back of the rear main bearing. This special housing is carried in a forced arm of a 90° bell crank, which is carried by the back bearing pedestal. (See fig. 68.)

The main jackshaft is driven by four Sprague General Electric Co. 200/300 horsepower, 230-volt, 550-ampere, 1,000 to 3,000 r. p. m. dynamometers through a special flexible floating coupling. This special coupling allows the main shaft to move freely in an axial direction under the influence of the thrust of a propeller. It is held very nearly in the same position at all times regardless of the thrust by the mechanisn of the scales used to weigh the thrust. The scale mechanism is connected through rods and bell cranks to

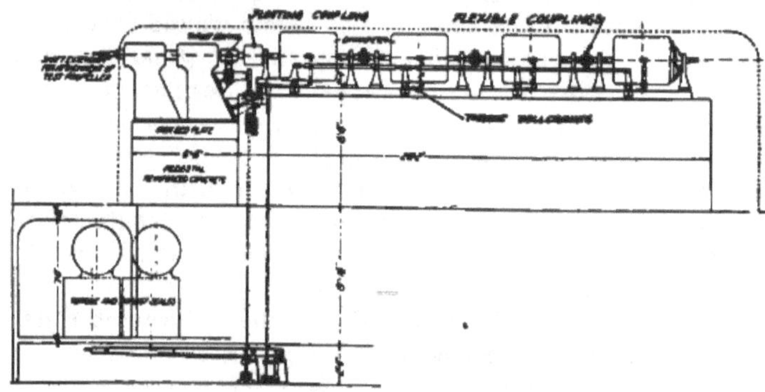

FIG. 68.

one arm of the forked bell crank carrying the special housing for the Kingsbury bearing previously mentioned.

Three Francke flexible couplings rated at 1,000 horsepower at 1,000 r. p. m. are used to connect the shaft of the four dynamometers. All three couplings were made the same in order that, if necessary, large gasoline engines could be tested at the end of the dynamometers opposite the main jackshaft.

The frames of all four dynamometers are connected by a system of bell cranks and rods to a scale which indicates the total torque force in pounds at the 21-inch radius, developed by the motors at any time. (See fig. 68; also figs. 69, 70, and 71.)

A stream-lined sheet-metal housing covers all the dynamometers and the jackshaft and bearings. It extends down to the ground level of the concrete foundation for the motors and mail bearings.

There is a bombproof mounted on wheels which can be moved into such a position as to surround a propeller being tested. This is absolutely essential since pieces of breaking propeller may leave traveling in the plane of rotation at a velocity greater than that of an ordinary small caliber rifle bullet. This bombproof is also arranged to be used as a crane for handling sections of the stream-line hood, the jackshaft, or the dynamometers. A three-quarters view of the above ground portion of the rig is given in figure 78 and also 79.

*Instrument room.*—Underneath the ground, around the reinforced-concrete pier supporting the jackshaft and main bearings, there is an instrument room. In this room are located a control switchboard for the dynamometers, the scales for measuring the torque of the dynamometer fields and the static thrust of a propeller, the optical device for measuring the distortion of a propeller while

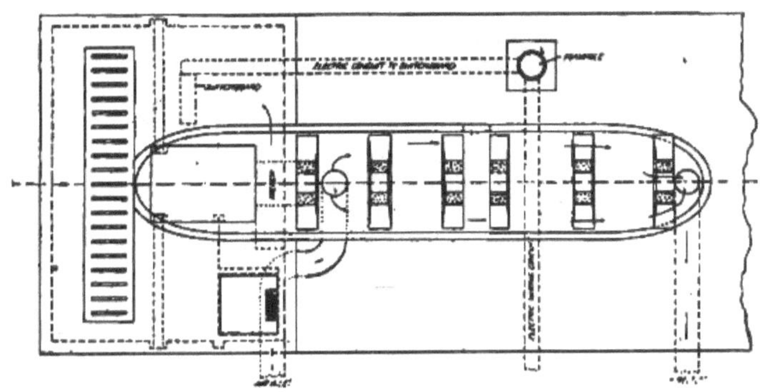

FIG. 80.

running, and the oiling system for the main bearings and the Kingsbury thrust bearing.

The dynamometer control panel was furnished by the General Electric Co. and is made of slate with a black and marine finish. Upon it are mounted two double-pole carbon circuit breakers, one duplex direct-current ammeter for indicating the armature current in each of the two sets of dynamometers, a special dial switch for the remote control of the contractor panel in the power house, the rheostat handles which control the speed of the dynamometers above full field speed, and the rheostat handle, which enables the voltage of the main 1,000-kilowatt, direct-current generator in the power house to be set at any desired value. On the framework back of this panel there is mounted a secondary relay, which is operated by a primary pressure relay on the oil system. This relay is so arranged that the dynamometers can not be started if the oil pressure is below a safe value, and in case the dynamometers are already running it will trip the main direct-current circuit breaker in the power house and open the contractor control circuit should the oil pressure for any reason fall below a safe value. (See fig. 72.)

*Scales for measuring torque and static thrust.*—Two Toledo springless scales of the combination dial and beam type are connected through suitable levers, bell cranks, links, and rods to the dynamometer frames and the thrust bearing previously mentioned. The scale connected to the dynamometer frames indicates pounds torque force at 21-inch radius and has a capacity of 6,000 pounds on the dial or beam alone and 8,000 pounds when both dial and beam are used. As will be seen the horsepower as calculated from the readings of this scale equals, approximately, torque force multiplied by r. p. m. and divided by 3,000. The scale-registering thrust indicates pounds

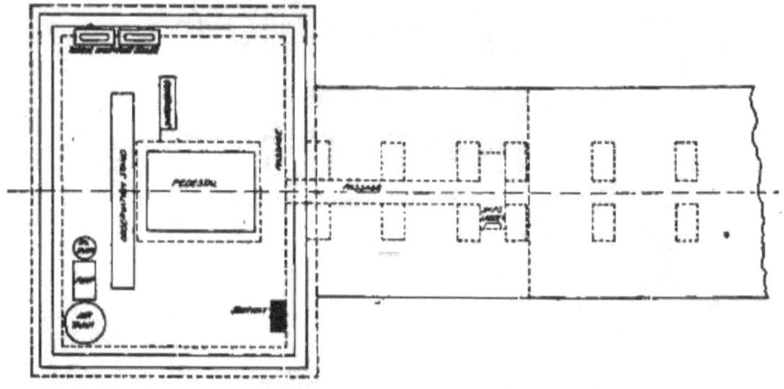

—General dimensions and position of sub-structural raft, pedestal, supporting columns for dynamometer, and approximate location of passageway. The drawing includes arrangement of control switch, torque and thrust dials, forced lubrication system and observation stand in the observation chamber

FIG. 70.

static thrust directly. Its capacity is the same as that of the torque scale. (See figs. 68 and 72.)

*Optical deflection and distortion device.*—This device consists essentially of a transit telescope with totally reflecting prism or mirror so arranged that its line of sight always lies in a plane perpendicular to the propeller axis. The carriage for this telescope is so arranged that it can be set to observe a propeller at any radius, and the telescope itself can be moved so that its line of sight lies in any desired plane perpendicular to the propeller axis. This enables readings of the position of the leading and trailing edges of the propeller at any radius from zero feet to 9 feet to be taken in 1-foot steps. (See fig. 73.)

*Oiling system.*—The oiling system for the main bearings and the Kingsbury thrust bearing consists of a Deming No. 3 rotary pump driven by a General Electric Co. 1½-horsepower, 1,200 r. p. m.,

3-phase, 60-cycle, 230-volt induction motor, and two tanks. One tank serves as a reservoir for the oil draining back from the bearings, and the other, which is under air pressure, insures a flow of oil for a short period in case of a failure of the driving motor.

### OPERATION.

*Power house.*—The 1,420-horsepower synchronous motor is started on reduced voltage and brought up to synchronism as an induction motor. As soon as this synchronism is reached the motor is thrown on line voltage by the starting switch and its field current adjusted for a slightly leading power factor at no load.

The oil pump in the instrument room is then started. It will be remembered from previous description that the oil pressure must be at a safe operating value or neither the main circuit breaker nor

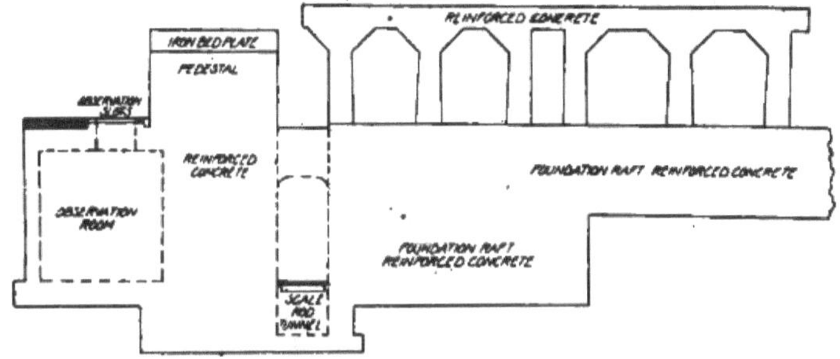


—Vertical section through laboratory structure, showing reinforced concrete raft and observation room, with pedestal monolith

FIG. 71.

the contractor control circuit can be closed. As soon as the oil pressure is at a safe value the main circuit breaker and knife switch on the main generator panel are closed and the field control of the main generator transferred to the control panel in the instrument room.

*Test rig proper.*—The propeller to be tested is placed on an extension of the proper taper, which is bolted to the projecting end of the 6-inch nickel-steel jackshaft with four, six, or eight bolts, the number of bolts used depending upon the size of the propeller to be tested. These bolts are ⅝-inch United States standard thread, and have a ¼-inch hole drilled the entire length of the bolt and a ½-inch hole drilled 1 inch into the head of the bolt. This is done in order to insure the extension jerking off the jackshaft in case a propeller breaks while being tested. (See fig. 74.)

After adjusting the field current in the dynamometers to full field value, they are started by closing the contractor circuit. At starting

FIG. 72.—TORQUE AND THRUST DIALS IN UNDERGROUND OBSERVATION ROOM, WITH DYNAMOMETER CONTROL SWITCHBOARD.

FIG. 73.—TRANSIT TELESCOPE AND MIRROR MOUNTED ON SLIDING BED, FOR OBSERVING DEFLECTIONS OF PROPELLER BLADE WHILE RUNNING.

FIG. 74.—PROPELLER TEST RIG SHOWING A PROPELLER IN POSITION FOR TEST.
Note the observation slots in the floor below the propeller.

the dynamometers the voltage of the main generator is usually about 150 volts. This reduces starting current and enables a very low speed to be obtained for an idle reading of the position of leading and trailing edges of the propeller.

As soon as an idle reading has been obtained the contractor control switch is moved slowly to its running position, thereby cutting out resistance from the armature circuits of the dynamometers. The speed desired, usually 600 r. p. m. for the first set of readings, is adjusted by adjusting the voltage of the main generator. As soon as the desired speed is obtained a complete set of readings is taken consisting of current, voltage, r. p. m., torque, thrust, position of leading and trailing edges of propeller blades, and wind. This procedure is repeated for any desired speed or horsepower.

*Standard destructive whirling tests.*—Propellers for the 80-horsepower Le Rhone engine are usually given a 10-hour test at 250 horsepower and approximately 1,650 r. p. m. Preceding the 10-hour run an r. p. m. power curve is usually taken from 600 r. p. m. to the normal test speed in steps of 200 r. p. m. each. A complete set of readings is taken at each speed.

Propellers for the 90-horsepower OX-5 engine usually are given a 10-hour test at 1,800 r. p. m. At this speed these propellers absorb about 265 horsepower. A power curve is usually taken on these propellers before the 10-hour run.

Propellers for the 150-horsepower Hispano-Suiza engine are given a 10-hour test at 500 horsepower. At this horsepower, these propellers usually turn from 1,950 r. p. m. to 2,000 r. p. m., depending upon the design. An r. p. m. power curve is usually taken on these propellers before the 10-hour run.

Propellers for the 300-horsepower Hispano-Suiza engine are usually given a 10-hour run at 450 horsepower. At this input, these propellers turn about 1,950 r. p. m. An r. p. m. power curve is usually taken on these propellers before the 10-hour run.

Propellers for the 400-horsepower Liberty engine and engines of higher horsepower are usually given a 10-hour test at 50 per cent power overload. An r. p. m. power curve is always taken on these propellers before the 10-hour run.

Propellers built of a material which is not standard are usually given tests of much longer duration than those built of approved woods. These tests are usually of 30 hours' duration in order to determine whether the material will show fatigue.

Propellers of unusual design, such as adjustable pitch propellers, etc., are given special tests in order to obtain design data. In any case in which the propeller is to be used on an airplane it is given an endurance run in line with those specified above.

Whenever it seems advisable any propeller is run to a speed and horsepower input which destroys it. The maximum horsepower available is about 1,350. Propellers for the Liberty 12 engine have been run to this horsepower without any visible damage.

*Typical test data.*—The following is typical of the data usually taken on a propeller of conventional construction.

The primary data is given as observed from the test, and the correction factors are applied to reduce this data to standard conditions. The following description of this test is an exact duplicate of destructive whirling test No. 213 prepared by the propeller section, McCook Field.

## DESTRUCTIVE WHIRLING TEST No. 213.

### A. E. D. PROPELLER FOR THOMAS-MORSE AIRPLANE WITH 300 H. P. HISPANO-SUIZA ENGINE, APRIL 28, 1919.

Drawing No. X-13855. S. C. No. 105228. Insp. No. O-843. Diameter, 8 feet 2 inches. Pitch, 7.75 feet.

*Object.*—To determine the ground characteristics of the above propeller and to assure the safety of a flight test by a 4-hour run at 450 horsepower.

*Preliminary.*—This test was made by and for the engineering division, propeller section, Air Service, McCook Field, Dayton, Ohio.

*Date and place.*—It was run on April 28,1919, at McCook Field, Dayton, Ohio.

*General description—Inspection report.*—This propeller was of the conventional construction and was cotton tipped.

Inspection No. O-843.
S. C. No. 105228.
Drawing No. X-13855.
Radius of blades (1), 49 1/16.
Radius of blades (2), 49 1/16.
Tips, cotton.

Error in track, before covering, 1/16.
Error in track after covering 1/16.
Number of laminations, 9.
Thickness of each, 3/4.
Hub thickness, 6 1/4.

TABLE 1.—*Blade angle.*

| Radius. | Before covering. | | | After covering. | | Angles after test. | |
|---|---|---|---|---|---|---|---|
| | Specified. | 1 | 2 | 1 | 2 | 1 | 2 |
| 24 | 31.2 | ........ | ........ | ........ | ........ | ........ | ........ |
| 30 | 25.9 | 26 | 26.9 | 26 | 25.9 | 26.05 | 25.9 |
| 36 | 22.1 | 22.2 | 22.2 | 22.2 | 22.2 | 22.2 | 22.2 |
| 42 | 19.3 | 19.35 | 19.3 | 19.3 | 19.3 | 19.3 | 19.3 |
| 46½ | 17.5 | 17.6 | 17.7 | 17.7 | 17.5 | 17.5 | 17.6 |

TABLE 2.

| Radius. | Blade width. | | | Blade thickness. | | |
|---|---|---|---|---|---|---|
| | Specified. | 1 | 2 | Specified. | 1 | 2 |
| 12 | 7 3/16 | 7 [illegible] | 7 [illegible] | 3 [illegible] | 3 1/4 | 3 1/4 |
| 18 | 8 | 8 | 8 | 2 [illegible] | 2 [illegible] | 2 1/16 |
| 24 | 8 1/16 | 8 1/16 | 8 1/16 | 1 [illegible] | 1 [illegible] | 1 [illegible] |
| 30 | 8 1/16 | 8 1/16 | 8 1/16 | 1 [illegible] | 1 [illegible] | 1 [illegible] |
| 36 | 7 1/2 | 7 1/2 | 7 1/2 | 1 1/16 | 1 [illegible] | 1 1/16 |
| 42 | 5 [illegible] | 5 [illegible] | 5 [illegible] | [illegible] | [illegible] | [illegible] |
| 46½ | 4 [illegible] | 4 [illegible] | 4 [illegible] | [illegible]/16 | [illegible]/16 | [illegible]/16 |

Manufactured by A. E. D.
Material, walnut.
Remarks, 1/16 after test.
Error in fit of templates, 1/16 inch unless noted otherwise.

Inspected at McCook Field.
Rated diameter, 98.
Weight, 33¾ pounds not bored.

*Procedure.*—An r. p. m. power curve was taken from 600 r. p. m. to 1,868 r. p. m., in steps of 200 r. p. m. each. The propeller was then run for 4½ hours at approximately 450 horsepower.

TABLE 3.

| Time. | | Motor. | Input. | Speed. | Thrust. | Torque force at 21-inch radius. | Horsepower. | Air temperature. | Bar. pressure. | Wind. |
|---|---|---|---|---|---|---|---|---|---|---|
| Start. | Stop. | | | | | | | | | |
| Apr. 28, 1919 (p. m.): | | *Volts.* | *Amperes.* | *r. p. m.* | *Pounds.* | *Pounds.* | | ° C. | *Mm.hg.* | |
| 1.48 | ........ | 225 | 40-34 | 593 | 155 | 91 | 18 | 15 | 749.6 | Low |
| 1.52 | ........ | 298 | 60-70 | 791 | 260 | 132 | 34.8 | 14.8 | 749.6 | Do. |
| 1.56 | ........ | 385 | 86-80 | 1,005 | 370 | 195 | 65.3 | 14.5 | 749.6 | Do. |
| 2.00 | ........ | 462 | 125-110 | 1,197 | 540 | 276 | 110.1 | 14.5 | 749.6 | Do. |
| 2.03 | ........ | 502 | 172-164 | 1,402 | 718 | 375 | 175.3 | 14.5 | 749.6 | Do. |
| 2.07 | ........ | 523 | 250-250 | 1,592 | 970 | 530 | 281 | 14.5 | 749.6 | NE. by N. |
| 2.12 | ........ | 565 | 400-380 | 1,868 | 1,265 | 780 | 486 | 14.8 | 749.6 | Do. |
| 2.19 | 4.19 | 565 | 375-375 | 1,852 | 1,240 | 760 | 469 | 14.8 | 749.6 | Do. |
| Apr. 29, 1919 (a. m.): | | | | | | | | | | |
| 8.00 | 10.30 | 570 | 380-380 | 1,806 | 1,215 | 780 | 470 | 9.4 | 748 | NE. by E. |

*Primary data as observed.*—When shut down at 10.30 a. m., April 29, 1919, the propeller was in good shape in every respect.

FRICTION DATA.

TABLE 4.

| Speed. | Torque force at 21-inch radius. | Horsepower. | Thrust. |
|---|---|---|---|
| | *Pounds.* | | *Pounds.* |
| 625 | 26 | 5.4 | 6 |
| 810 | 29 | 7.8 | 6 |
| 1,015 | 32 | 10.8 | 6 |
| 1,180 | 36 | 14.2 | 10 |
| 1,400 | 40 | 18.7 | 9 |
| 1,610 | 41 | 22.0 | 8 |
| 1,805 | 45 | 27.1 | 10 |

161988—21——10

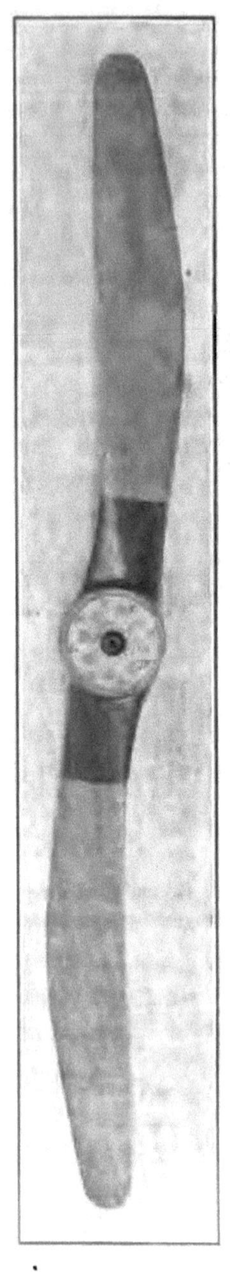

FIG. 77.—VIEW OF THRUST FACE OF A. E. D. PROPELLER AFTER TEST.

Drawing No. X-13855, S. C. No. 105228.

## COMPUTED RESULTS OF TEST.

TABLE 5.—*R. p. m. power curve.*

| Speed (r. p. m). | Horsepower. | | | Cortd. | Torque in foot pounds, "Q." | Thrust, "T." | T/Q. |
|---|---|---|---|---|---|---|---|
| | Total. | Friction. | Absorbed by propeller. | | | | |
| 593.... | 18 | 5 | 13 | 13.1 | 116 | 156.5 | 1.35 |
| 791.... | 34.8 | 7.5 | 27.3 | 27.6 | 183 | 263 | 1.44 |
| 1,005.... | 65.3 | 11.2 | 54.1 | 54.6 | 285 | 373 | 1.31 |
| 1,197.... | 110.1 | 14.2 | 95.9 | 96.8 | 425 | 545 | 1.28 |
| 1,402.... | 175.3 | 18.2 | 157.1 | 158.6 | 594 | 724 | 1.22 |
| 1,592.... | 281 | 22 | 259 | 261 | 861 | 988 | 1.13 |
| 1,868.... | 486 | 28 | 458 | 463 | 1,302 | 1,277 | .98 |
| 1,852.... | 469 | 28 | 441 | 446 | 1,265 | 1,253 | .99 |
| 1,806.... | 470 | 27 | 443 | 440 | 1,289 | 1,206 | .94 |

(See figs. 75 and 76.)

TABLE 6.—*Deflection at 3-foot radius.*

| Speed (r. p. m.). | Shaft position. | Blade position at 3-foot radius. | | Shaft movement. | Deflection at 3-foot radius. | | Blade pitch. | Remarks. |
|---|---|---|---|---|---|---|---|---|
| | | L. E. | T. E. | | L. E. | T. E. | | |
| Idle.. | .770 | 24.32 | 19.94 | 0 | 0 | 0 | 7.75 | Track O. K. |
| 593... | .773 | 24.41 | 19.96 | .003 | +.09 | +.02 | 7.87 | |
| 791... | .778 | 24.49 | 20.03 | .008 | +.16 | +.08 | 7.89 | Slight quivering of edges. |
| 1,005. | .785 | 24.53 | 20.07 | .015 | +.20 | +.12 | 7.89 | T. E. fluttering slightly. |
| 1,197. | .795 | 24.61 | 20.13 | .025 | +.27 | +.17 | 7.92 | Do. |
| 1,402. | .805 | 24.66 | 20.20 | .035 | +.31 | +.23 | 7.89 | Slight flutter both edges. |
| 1,592. | .815 | 24.70 | 20.26 | .045 | +.34 | +.27 | 7.87 | L. E. steady, T. E. fluttering. |
| 1,868. | .835 | 24.75 | 20.30 | .065 | +.37 | +.30 | 7.78 | Intermittent flutter of L. E.; continual flutter T. E. |

## CONCLUSION.

In view of the small deflection and the small amount of flutter present in the propeller, it is believed to be safe for use with the 300-horsepower Hispano-Suiza engine on the plane for which it is designed.

*Sample calculations.*—Torque force at 21-inch radius at 1,005 r. p. m. = 195 pounds.

$$\text{Total H. P.} = \frac{1005 \times 195}{3000} = 65.3 \text{ (approximately).}$$

$$\begin{aligned} \text{H. P. absorbed by propeller} &= \text{Total H. P.} - \text{friction H. P.} \\ &= 65.3 - 11.2 \\ &= 54.1. \end{aligned}$$

$$\text{H. P. corrected to 29.92 inches mercury, and } 16^\circ \text{ C.} = \frac{54.1}{.992} = 54.6.$$

$$\text{Torque force, } F_q \text{ at 1 foot radius} = \frac{\text{H. P.} \times 5250}{\text{R. P. M.}} = \frac{54.6 \times 5250}{1005} = 285.$$

$$\text{Thrust "T"} = \frac{\text{Observed thrust} - \text{No load thrust of bearing.}}{\text{Correction factor for air density.}}$$

$$= \frac{370 - 7.5}{.992} = 373 \text{ pounds.}$$

Deflection at 1,005 r. p. m. on L. E. = (Position of L. E. at 1,005 r. p. m.) − (position of L. E. at idling r. p. m., shaft movement.)
= (24.53 − 24.32) − .015 = .20 inch.

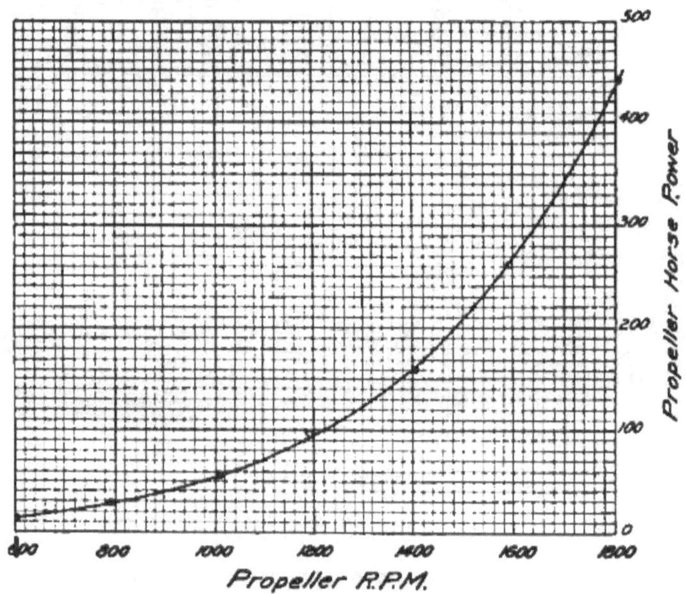


CHARACTERISTICS OF
A.E.D. PROPELLER DRAWING NO. X-13855
INSPECTION NO. 0-843. S.C. - 10522B
FROM GROUND TESTS, McCOOK FIELD, 4-28-19.
DESTRUCTIVE WHIRLING TEST, NO. 213

FIG. 75.

Blade pitch at 1,005 r. p. m. = Pitch idling ×

$$\frac{(\text{Blade position L. E. at 1,005 r. p. m.}) - (\text{Blade position T. E. at 1,005 r. p. m.})}{(\text{Blade position L. E. idling}) - (\text{Blade position T. E. idling})}$$

$$= 7.75 \times \frac{(24.53 - 20.07)}{(24.32 - 19.94)}$$

$$= 7.89 \text{ feet.}$$

*Necessity for testing propellers.*—The one fundamental reason for testing propellers on a destructive whirling test rig, such as that described above, is to be as nearly as possible absolutely sure that no

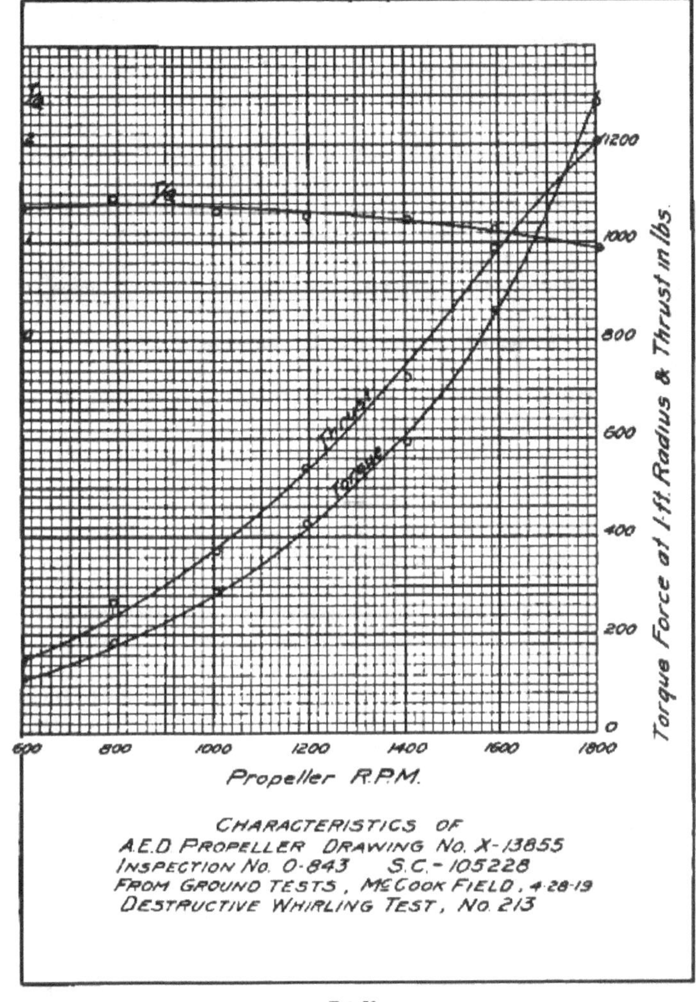


FIG. 76.

propeller of any particular design or construction will break while in service on an airplane. A few comparatively simple calculations will show that the unbalanced forces set up when a propeller breaks badly will be ample to wreck the ship in extreme cases, and in the cases of comparatively small breaks will be sufficient to jerk the engine from its supports before the engine can be stopped. That these forces are tremendous will be at once apparent from the following case: A propeller for the OX-5 engine, weighing 34¾ pounds, broke near the hub while running at 2,000 r. p. m. The unbalanced force set up when one blade was thrown off bent a 6-inch steel shaft 1½ inches out of line and broke a heavy cast-iron cap on one bearing pedestal.

In view of the foregoing it is considered necessary to test the following:

First. A certain number of propellers (usually the first three) of any standard design made by any new manufacturer, to be sure his product is safe for use on airplanes.

Second. All propellers of standard design but of unusual construction.

Third. All propellers of conventional construction but of radical design.

Fourth. All propellers of standard design constructed of any material other than the approved woods.

Fifth. All propellers of any design or construction presented by inventors to the Air Service for approval.

Tests on propellers coming under the first class are necessary in order to discover any defects in manufacture, such as poor gluing or poor manufacturing processes, which can not be discovered by visual inspection.

Tests of propellers coming under the second class are necessary in order to determine the relative merits of different methods of construction. Examples of propellers of this class tested up to the present time are: Propellers having various kinds of splices and in various locations in the laminations, propellers built with twisted laminations, and propellers built with the laminations laid vertical as compared to the conventional method.

Propellers coming under the third class include all propellers built for research tests.

Tests of propellers of the fourth class are necessary in order to prevent any weak material being used in the construction of propellers of approved design. Propellers of this class include those constructed of any wood not previously approved for use; those constructed of rubber or wood covered with rubber; those constructed of composition such as micarta, etc.; those constructed of aluminum, and those constructed of steel or any unusual combination of materials.

Test of propellers of the fifth class are necessary in order that they may not be used on airplanes without previous evidence that they may be used with safety. Under this class come adjustable pitch propellers, automatic pitch propellers, compensating propellers, and any other propellers of unusual types.

COMPARISON OF STRENGTHS OF VARIOUS PROPELLER WOODS AS INDICATED BY DESTRUCTIVE WHIRLING TESTS.

These comparisons of propeller woods are made from the point of view of the strength of the wood alone. In considering the suitability of these woods no account is taken of the weight or tendency to warp. In using propeller wood, however, it is necessary to consider both the weight of the wood and its tendency to warp so that the strongest wood may not be the most desirable when everything is considered. The results of a number of tests are given in Table II. From this data the following conclusions are made:

*Baywood.*—Three propellers out of four constructed of this wood failed when tested. All these propellers fluttered badly during the tests. It has been found that fluttering will cause a propeller to fail sooner or later. In view of the foregoing, this wood is considered unfit for use in the construction of propellers.

*Birch.*—From the results of nine tests on propellers built of this wood, it is believed to be the strongest of all woods tested up to this time. Only three propellers failed, and these failed on account of the radical design and not due to any weakness in the wood itself.

A very valuable quality of the wood is its ability to withstand a continuous flutter without failure. Propellers of radical design which can not be made to stand up when constructed of other woods appear to be satisfactory when constructed of birch.

*Cherry.*—The results of 17 tests indicate that this wood will be satisfactory for use in the construction of propellers of conventional design. Only 1 propeller out of 17 failed in any way, and this due to excess speed. One of these propellers showed a bad flutter. This was to have been expected, since the propeller showed a much greater deflection at normal speed than has usually been found to be the case with propellers of the same design.

*African mahogany.*—One out of four propellers tested did not fail. Defects characteristic of the wood are natural compression failures across the grain, shakes, and cross grain. Due to the foregoing characteristics, it is believed that this wood should not be used in the manufacture of propellers except in combination with Honduras mahogany or in special designs.

*Philippine mahogany.*—Four propellers tested show the same tendency to split at the ends of the blades. This tendency to split combined with a bad flutter, which has been present in all propellers constructed of this wood, is believed to indicate a weakness in the wood. Two of the propellers tested were built on a design which had been found satisfactory when constructed of other woods. This strengthens the belief that this wood is weak and is to be placed in a doubtful class of woods for use in the construction of propellers.

*Honduras mahogany.*—Four propellers constructed of this wood were tested. All of these were in good condition at the ends of the tests. In each case there was very little flutter, and only a small deflection at the tip of the propeller. This variety is believed to be the safest of all varieties of mahogany for use in propeller construction.

*Prima vera (white mahogany).*—As will be seen by examination of the tables, four propellers constructed of this wood (sometimes called "white mahogany") have been tested. In every case the propellers cracked at the tips, and in one test the hub cracked. All the propellers were built from approved and tested designs. Therefore the failure of these propellers is believed to have been due to the character of the wood.

It was found that in three of these tests there was a tendency for the glue joints between laminations to open. This tendency for the glue joints to open is not likely due to some characteristic of the wood. It was probably caused by poor workmanship or conditions in the glue rooms of the companies which manufactured the propellers. The characteristics of this wood with respect to its ability to glue up properly can only be determined by special tests.

In view of the above, this wood is believed to be unsafe for use in the manufacture of combat type propellers.

*Unknown grades.*—Twenty-seven propellers constructed of unknown grades of mahogany were tested. Only six of these propellers stood the test applied without failing. These tests seem to indicate that the average run of commercial mahogany is not a strong wood for use in the construction of propellers.

Examination of the wood in all these broken propellers shows that in practically every case the wood at the point of the breaks has a brash, dead appearance.

*Oak.*—Ten training type propellers constructed of oak have been tested. Of these six were Paragon designs and the others were Signal Corps designs.

The Paragon design is for the Curtis OX-5 engine, and is considered to be normally a weaker design than the Signal Corps design "8-25." All these propellers stood the required test without breaking. As will be noted upon examination of Table II, two of these propellers failed—one at 2,200 r.p.m. and the other at 2,250 r.p.m.

Of the three propellers to Signal Corps design "8-25," all stood the required test without distress.

Six propellers of the advanced training type to Signal Corps design 13279 were tested. As will be seen by referring to the table, all these propellers were tested at speeds above 1,950 r.p.m. None of these propellers failed, although at the end of the 10-hour runs the speed was increased until the input to the propellers was above 750 horsepower. This is 500 per cent normal horsepower. The average weight of this design, as constructed of oak, was 38 pounds.

Seven propellers of the combat type Signal Corps drawing 8–45 have been tested. All these propellers were run at an input of approximately 600 horsepower. Only one of these propellers failed. The failure, as nearly as could be determined, was due to defective wood. All the remaining propellers of this design ran without any signs of failure. The average weight of these propellers was 64.5 pounds.

It may also be well to note that oak propellers seem to stand the effects of a water spray much better than other wood. The damage in a water spray takes place almost entirely on the leading edge. When the propeller tested is constructed of oak the leading edge wears away gradually, whereas other woods show a tendency to split.

From the foregoing it is believed that oak will be a most satisfactory wood for propellers. This applies particularly to propellers for use on airplanes on which weight is not so important a consideration as reliability.

The average weight of an oak propeller to Signal Corps drawing 8–45 is about 64.5 pounds, and the weight of a propeller to the same design constructed of Honduras mahogany is about 50.5 pounds.

*Poplar.*—As will be noted, four propellers made of this wood have been tested. One propeller failed, due to poor glue joints. The other propeller failed on an unusually hard test—750 horsepower continuously. The normal rating of this propeller was 400 horsepower.

The tests which have been run on propellers constructed of this material lead to the belief that it is safe for use in the construction of propellers of any design which is not predisposed to flutter.

*Walnut.*—As will be noted, 30 propellers of this wood have been tested. Of this number eight cases of wood failure are noted. In one case the failure was due to spiral grain and light wood; in all other cases, except one, the failure seems to have been caused primarily by excessive flutter due to poor design. In the one case noted the failure seems to have been due to the fact that the hub thickness had been cut down in order to accommodate a standard hub.

From the results of these tests, it is believed that walnut will prove satisfactory for use in the construction of propellers for any type airplane.

## FLUTTER.

Fluttering is the rapid changing of the pitch of a propeller while running. This change in the pitch of a propeller seems to be caused by a very rapid vibration of the thrust produced by the blade elements and the centrifugal force acting on these elements. As will be evident from the foregoing, fluttering is a result of incorrect design.

Observation of a large number of propellers tested shows that propellers having straight leading edges flutter worse than those having curved leading edges. These observations also indicate that propellers having very narrow blades and thin sections tend to flutter worse than those having wider blades and thicker sections.

The results of these tests indicate that a propeller which flutters is unsafe for use on an airplane for it will break sooner or later. In all but a comparatively few cases of propellers tested which fluttered, failure has occurred. These failures are usually a piece broken off the trailing edge of one blade of the propeller. The loss of a piece off one blade of a propeller is in most cases sufficient to cause enough vibration to pull the engine out of the fuselage.

TABLE II.

BAYWOOD.

| Test No. | Design. | Weight. | Deflections in inches at normal r. p. m. | | | | O. K. | Failed. | Nature of failure. | Flutter. | Remarks. |
|---|---|---|---|---|---|---|---|---|---|---|---|
| | | | At 3 feet radius. | | At 4 feet radius. | | | | | | |
| | | | L. E. | T. E. | L. E. | T. E. | | | | | |
| | | *Pounds.* | | | | | | | | | |
| 10 | S-47 | 38 | +2.70 | +2.40 | | | | 1 hour at 1,600 r. p. m. | Pieces broken off L. E.'s tips. | Slight | Too thin at hub and tips. |
| 11 | X-322 | 62.5 | | | | | | 17 minutes at 935 r. p. m. | All tips splintered | | |
| 32 | X-2180 | 17 | + .40 | + .38 | | | 5 hours at 1,800 r. p. m. | | | Bad | |
| 33 | X-2180 | 17 | + .43 | + .42 | | | | 5 hours at 1,800 r. p. m. | Split blades | do | |
| PRIMA VERA (WHITE MAHOGANY). | | | | | | | | | | | |
| 109 | S-45 | 45 | +0.36 | +0.28 | +0.54 | +0.11 | | 2 hours at 1,700 r. p. m. | Cracked tip and open joints. | Bad | Weak wood and poor gluing. |
| 112 | S-45 | 50 | + .43 | + .34 | + .66 | + .44 | | do | do | do | Do. |
| 124 | S-45 | 48½ | + .11 | + .02 | + .42 | + .09 | | ½ hour at 1,700 r. p. m. | do | Slight | Do. |
| 133 | S-25 | 19.8 | | | | | | 10 hours at 1,800 r. p. m. | Cracked tips and hub. | do | Do. |

TABLE II—Continued.

BIRCH.

| Test No. | Design. | Weight. | Deflections in inches at normal r. p. m. — At 3 feet radius. L. E. | At 3 feet radius. T. E. | At 4 feet radius. L. E. | At 4 feet radius. T. E. | O. K. | Failed. | Nature of failure. | Flutter. | Remarks. |
|---|---|---|---|---|---|---|---|---|---|---|---|
| | | Pounds. | | | | | | | | | |
| 154 | 8-25 | 15.4 | | | | | 10 hours at 1,800 r. p. m. | | | Slight | Copper rivets in tip moved slightly. |
| 167 | 8-25 | 26.0 | | | | | do | | | do | |
| 171 | X-12061 | 34.5 | | | | | 3 hours at 1,770 r. p. m. | | | do | |
| 177 | X-7039 | 44 | | | | | 10 hours at 1,740 r. p. m. | | | Vicious | |
| 206 | X-13833 | 32.5 | +0.45 | +0.03 | | | | 1,848 at 4 minutes | Both blades cracked | Terrific | Radical design. |
| 212 | X-13835 | 33.4 | + .22 | + .32 | | | | 3 hours at 1,727 r. p. m. | | | Do. |
| 116 | X-6233 | 27.6 | + .51 | + .13 | +0.74 | +0.54 | 10 hours at 2,000 r. p. m. | | | Slight | O. K. at 2,260 r. p. m. |
| 132 | X-7035 | 36 | | | | | | 2 minutes at 1,730 r. p. m. | Both blades cracked | Very bad | Radical design. |
| 135 | X-8171 | | | | | | 10 hours at 2,300 r. p. m. | | | | |

CHERRY.

| Test No. | Design. | Weight. | At 3 feet radius. L. E. | At 3 feet radius. T. E. | At 4 feet radius. L. E. | At 4 feet radius. T. E. | O. K. | Failed. | Nature of failure. | Flutter. | Remarks. |
|---|---|---|---|---|---|---|---|---|---|---|---|
| 12 | 8-25 | 21 | +1.16 | +0.59 | | | 10 hours at 1,800 r. p. m. | | | Bad | |
| 15 | 8-25 | 20 | + .56 | + .09 | | | do | | | Slight | |
| 23 | Paragon | 20 | | | | | ½ hour at 1,800 r. p. m. | 10 minutes at 2,400 r. p. m. | Crack at hub | | Excess speed. |
| 25 | do | 20 | + .68 | + .28 | | | do | | | | |
| 27 | do | 22 | + .50 | + .36 | | | do | | | Considerable. | |
| 60 | 13279 | 29.5 | + .25 | + .26 | +0.53 | +0.23 | 3½ hours at 1,600 r. p. m. | | | Slight | |
| 62 | 13279 | 29.5 | + .30 | + .40 | | | 3 hours at 1,600 r. p. m. | | | do | |
| 65 | 13279 | 32 | + .25 | + .20 | + .34 | + .18 | do | | | do | |
| 73 | Curtiss | 30 | + .22 | + .01 | + .45 | + .25 | do | | | do | |

| | | | | | | | | | | | |
|---|---|---|---|---|---|---|---|---|---|---|---|
| 149 | 13631 | 20.4 | | | | | 10 hours at 1,500 r. p. m. | | | Bad | Tested for effect of gun streaks. |
| 150 | 13631 | | | | | | do | | | do | Do. |
| 186 | Pomilio N-103 | | | | | | 5 minutes at 1,496 r. p. m. | | | do | Not considered safe. |
| 74 | Curtiss | 30.5 | + .29 | + .01 | + .47 | — .02 | 3 hours at 1,600 r. p. m. | | | | |
| 85 | Navy | 27.5 | + .32 | | | | ½ hour at 1,800 r. p. m. | | | Slight | |
| 91 | do | 42 | + .89 | + .62 | +1.30 | +1.20 | 4 hours at 1,800 r. p. m. | | | do | |
| 101 | do | 32.3 | + .37 | + .31 | | | 3 hours at 1,800 r. p. m. | | | | |
| 148 | 13631 | 21.4 | | | | | 10 hours at 1,500 r. p. m. | | | Bad | Tested for effect of gun streaks. |

AFRICAN MAHOGANY.

| | | | | | | | | | | | |
|---|---|---|---|---|---|---|---|---|---|---|---|
| 24 | Paragon, 8 feet 3 inches by 5 feet. | 18 | +0.55 | +0.56 | | | 5 minutes at 1,800 r. p. m. | 2,200 r. p. m. | Blades totally wrecked. | Some | Good wood, but cross-grained. |
| 25 | do | 17 | + .73 | + .65 | | | 1,820 r. p. m. | | | Yes | Did not fail at 2,380 r. p. m. |
| 28 | do | 16.5 | + .58 | + .39 | | | 5 hours at 1,800 r. p. m. | 1,800 r. p. m. | Piece broken out along T. E. | do | |
| 34 | do | 18.5 | + .52 | + .54 | | | 5 minutes at 1,605 r. p. m. | 1,700 r. p. m. | do | do | Shot up with machine gun. |

PHILIPPINE MAHOGANY.

| | | | | | | | | | | | |
|---|---|---|---|---|---|---|---|---|---|---|---|
| 111 | L. P. 401-E | 48 | +0.38 | +1.08 | +0.63 | +0.64 | | 10 hours at 1,800 r. p. m. | Brass tips failed | Bad | Wood O. K.; brass tip weak. |
| 83 | L. P. 413-C | 36½ | + .40 | + .52 | +1.55 | + .64 | | 10 hours at 1,845 r. p. m. | Brass tips failed; crack in wood at tip. | Very bad | |
| 67 | 13279 | 34 | + .07 | + .01 | + .18 | — .50 | | 10 hours at 1,960 r. p. m. | Copper tip failed; wood at tip cracked. | Bad | Wood apparently good; Bataan mahogany. |
| 68 | 13279 | 32.3 | + .40 | + .05 | + .50 | — .28 | | 1,942 r. p. m. | Chip out at tip; wood at tip cracked. | do | Cross grain, otherwise good wood; Bataan mahogany. |

TABLE II—Continued.

MAHOGANY.

| Test No. | Design. | Weight. | Deflections in inches at normal r. p. m. | | | | O. K. | Failed. | Nature of failure. | Flutter. | Remarks. |
|---|---|---|---|---|---|---|---|---|---|---|---|
| | | | At 3 feet radius. | | At 4 feet radius. | | | | | | |
| | | | L. E. | T. E. | L. E. | T. E. | | | | | |
| | | *Pounds.* | | | | | | | | | |
| 92 | 8-45 | 49.5 | +0.45 | +0.36 | +1.13 | +0.58 | | 53 minutes at 1,680 r. p. m. | Blade split | Bad | Very poor wood; had been rejected by inspector. |
| 93 | 8-45 | 50.8 | + .43 | + .51 | + .74 | + .56 | | 3 hours at 1,700 r. p. m. | Blade No. 1 cracked in 4 places. | do | Do. |
| 94 | 8-45 | 49 | + .48 | + .42 | + .73 | + .48 | | 9 hours at 1,700 r. p. m. | Tips broke off | do | Cross grain and poor wood. |
| 96 | 8-45 | 46.5 | + .74 | + .51 | + .76 | + .50 | | 50 minutes at 1,600 r. p. m. | Piece out of T. E. at tip of one blade. | Very bad | Very poor wood; full of wormholes and pitch pockets; small knots. |
| 97 | 8-45 | 45.8 | + .47 | + .44 | + .61 | + .35 | 10 hours at 1,670 r. p. m. | | | Bad | Poor wood; wormholes and pitch pockets; small knots. |
| 63 | 13279 | 28 | + .27 | + .13 | + .51 | + .10 | ¾ hours at 1,550 r. p. m. | | | | Wood O. K. |
| 113 | 13279 | | | | | | | | | | Too much vibration to test. |
| 138 | S. C. 5090 | 52 | + .74 | | + .61 | | 10 hours at 1,800 r. p. m. | | | Bad | Double 2-blader. |
| 139 | Charavay | 34 | + .43 | + .19 | +1.72 | +1.03 | | 2 hours at 1,733 r. p. m. | Both blades split | do | Poor design, poor gluing. |
| 112 | Wolseley Viper | 21 | + .85 | + .70 | | | 10 hours at 2,300 r. p. m. | | | Some | Linen covering torn on both blades at end of 10-hour run. |
| 53 | L. P. 413 | 57 | + .70 | + .30 | + .98 | + .53 | | 1¾ hours at 1,600 r. p. m. | Brass tips flew off | Bad | Wood O. K.; riveted tips poorly put on; bad design. |
| 80 | L. P. 413-C | 43½ | + .87 | + .77 | +1.58 | +1.02 | | 3 hours at 1,720 r. p. m. | Cracked hub; brass tips cracked. | do | Wood O. K.; bad design. |
| 81 | do | 45¾ | + .79 | + .90 | +1.50 | +1.15 | | 3 hours at 1,800 r. p. m. | Brass tips cracked | do | Do. |
| 82 | do | 44 | + .64 | + .65 | +1.61 | + .97 | | 3 hours at 1,750 r. p. m. | Hub cracked in splices; blade cracked at tip. | Very bad | Splices too short; wood O. K.; bad design. |
| 121 | do | 47 | + .84 | + .15 | + .67 | + .32 | | 4 hours at 1,800 r. p. m. | Both blades pulled out at hub. | do | Splice failed; splice too short. |
| 86 | 9 feet 6 inches by 5 feet 7 inches. | 37 | + .30 | + .30 | + .89 | + .65 | ½ hour $H_2O$ at 1,800 r. p. m. | | | Bad | Wood O. K. |

| | | | | | | | | | | | |
|---|---|---|---|---|---|---|---|---|---|---|---|
| 107 | L. P. 401-E. | 57 | + .25 | + .97 | + .32 | + .68 | do. | | | do. | All blades worn badly just inside brass tips. |
| 115 | do. | 60 | | | + .67 | | | 4 hours and 5 minutes at 1,780 r. p. m. | Blades 2 and 4 wrecked. | do. | Combination of poor work and short splices. |
| 126 | do. | 57.5 | + .73 | + .65 | +1.06 | + .88 | 10 hours at 1,800 r. p. m. | | | do. | Honduras mahogany; wood O. K. |
| 40 | 8-45. | 53 | + .06 | + .28 | + .20 | + .08 | 4 hours at 1,560 r. p. m. | | | Slight. | Do. |
| 65 | 8-45. | 43 | + .21 | + .12 | + .51 | + .17 | | 40 minutes at 1,690 r. p. m. | Large piece of tip of 1 blade cracked on T. E. | Some. | Wood apparently good. |
| 58 | 8-45. | 50.5 | + .14 | + .29 | + .52 | + .07 | 3 hours at 1,700 r. p. m. | | | | O. K. at 1,7[illegible] r. p. m.; Honduras mahogany. |
| 59 | 8-45. | 48 | + .20 | + .06 | + .41 | + .06 | 3 hours at 1,690 r. p. m. | | | | O. K. at 1,7[illegible] r. p. m.; Honduras mahogany. |
| 75 | 8-45. | 49 | + .18 | + .25 | + .55 | + .29 | | ½ hour at 1,700 r. p. m. | Blade No. 2 split in 3 places, split 6 inches long. | Very bad. | Very poor wood; punky. |
| 145 | Charavay. | 34.3 | + .75 | + .25 | +1.65 | +1.46 | | 10 minutes at 1,712 r. p. m. | Blade split from tip to hub. | Extreme flutter. | Poor design. |
| 152 | X-7039. | 37 | | | | | | 2 minutes at 1,600 r. p. m. | Blade split. | Extreme. | Wood too weak for design. |
| 153 | X-4987. | 51.8 | | | | | 10 hours at 1,600 r. p. m. | | | Very slight. | |
| 175 | SE-5090. | 53.3 | | | | | 10 hours at 1,815 r. p. m. | | | Some. | Brass tips bulge during run. |
| 195 | 8-45. | 42.7 | | | + .48 | + .23 | | 5$\frac{1}{12}$ hours at 1,630 r. p. m. | Blades cracked. | Bad. | Low density wood. |
| 201 | SE-5111. | 53.5 | | | + .21 | + .16 | 10 hours at 1,750 r. p. m. | | | Slight. | Splices O. K. |
| 89 | Chauviere. | 21.8 | | | | | | | | | Burst as speed was being raised; failure due to poor design. |
| 125 | X-5135. | 45.5 | + .35 | + .47 | +1.29 | +1.32 | | 7¾ hours at 1,749 r. p. m. | 1 blade broke. | Extreme. | |
| 136 | X-6233. | 20 | + .76 | + .69 | + .82 | + .98 | | 2 hours 13 minutes at 2,000 r. p. m. | 1 blade split. | Very bad. | Break due to flutter. |

MAHOGANY AND WALNUT.

| | | | | | | | | | | | |
|---|---|---|---|---|---|---|---|---|---|---|---|
| 55 | 8-47. | 51 | +0.07 | +0.15 | +0.62 | +0.19 | 5 hours at 1,600 r. p. m. | | | | |
| 127 | 8-47. | 58 | + .20 | − .11 | + .04 | − .23 | 5 hours at 1,700 r. p. m. | | | Bad. | |
| 122 | X-5135. | 45.3 | + .22 | + .09 | +1.12 | +1.08 | | 10 hours at 1,800 r. p. m. | Filler pieces cracked | do. | |
| 123 | X-5135. | 46 | + .73 | + .67 | +1.40 | +1.24 | | do. | do. | Very bad. | |

TABLE II—Continued.

MAHOGANY AND WALNUT—Continued.

| Test No. | Design. | Weight. | Deflections in inches at normal r. p. m. | | | | O. K. | Failed. | Nature of failure. | Flutter. | Remarks. |
|---|---|---|---|---|---|---|---|---|---|---|---|
| | | | At 3 feet radius. | | At 4 feet radius. | | | | | | |
| | | | L. E. | T. E. | L. E. | T. E. | | | | | |
| | | Pounds. | | | | | | | | | |
| 141 | X-5136 | 48.3 | + .63 | + .68 | +1.47 | +1.12 | | 10 hours at 1,720 r. p. m. | Filler pieces cracked | Bad | Propeller not destroyed at 1,836 r. p. m. |
| 170 | 8-47 | 54.2 | | | | | | 2 hours at 1,800 r. p. m. | Brass tips cracked | | Filler pieces near hub cracked. |
| 172 | X-5125 | 45.2 | | | | | 10 hours at 1,770 r. p. m. | | | Bad | |
| 174 | X-5135 | 45.7 | | | | | | 9½ hours at 1,745 r. p. m. | 1 blade broken off | do | |

OAK.

| Test No. | Design. | Weight. | At 3 feet radius L. E. | At 3 feet radius T. E. | At 4 feet radius L. E. | At 4 feet radius T. E. | O. K. | Failed. | Nature of failure. | Flutter. | Remarks. |
|---|---|---|---|---|---|---|---|---|---|---|---|
| 1 | 8-19 | 49.5 | +0.5 | +0.1 | | | | 8¾ hours at 1,800 r. p. m. | Cracked tip to hub. | | Soft wood. |
| 3 | Paragon, 5 feet 10 inches by 7 feet. | 34 | | | +0.3 | +0.1 | | | | | Shaft end broke. |
| 19 | Paragon | 27 | | + .3 | | | 5 hours at 1,600 r. p. m. | At 2,200 r. p. m. | Copper tips failed | | Wood O. K.; tips failed. |
| 36 | do | 26 | | | + .19 | + .28 | At 1,800 r. p. m. | | | Some | L. E. slightly worn. |
| 37 | do | 23.5 | | | + .25 | + .50 | At 2,378 r. p. m. | | | Yes | Do. |
| 38 | do | 22 | | | | | 10 hours at 1,800 r. p. m. | At 2,250 r. p. m. | Broke both blades | | |
| 77 | 8-25 | 28 | + .35 | + .12 | | | 3 hours at 1,800 r. p. m. | | | Slight | Splice O. K. |
| 78 | 8-25 | 28 | + .28 | + .09 | | | do | | | do | Do. |
| 79 | 8-25 | 28.5 | + .23 | − .01 | | | do | | | do | Do. |
| 72 | 12279 | 38.5 | + .10 | + .17 | + .10 | | 10 hours at 2,000 r. p. m. | | | do | Blades O. K. at 2,180 r. p. m. |
| 87 | 13279 | 36.3 | + .63 | + .39 | + .54 | + .37 | 10 hours at 1,960 r. p. m. | | | Yes | Blades O. K. |
| 88 | 13279 | 36.3 | | | | | 22 minutes at 1,980 r. p. m. | | Pigskin tips worn off by $H_2O$. | | Wood O. K. |
| 129 | 13279 | 38 | + .30 | + .16 | + .37 | − .05 | 10 hours at 1,960 r. p. m. | | | Yes | Blades and splices O. K. |

| | | | | | | | | | | | |
|---|---|---|---|---|---|---|---|---|---|---|---|
| 130 | 13279 | 38 | + .22 | + .06 | + .49 | − .13 | do | | | do | Do. |
| 131 | 13279 | 40 | + .27 | + .21 | + .44 | + .11 | 10 hours at 1,992 r. p. m. | | | | Do. |
| 95 | 8-45 | 67.3 | + .27 | + .41 | + .50 | + .31 | 10 hours at 1,700 r. p. m. | | | Slight | Do. |
| 98 | 8-45 | 67.3 | + .37 | + .31 | + .61 | + .48 | | 41 minutes at 1,700 r. p. m. | Pieces out of tip cracked all way into hub. | Bad | Poor wood; splices O. K. |
| 99 | 8-45 | 67.5 | + .39 | + .20 | + .84 | + .39 | 10 hours at 1,700 r. p. m. | | | Slight | Blades and splices O. K. |
| 103 | 8-45 | 63 | + .19 | + .13 | + .31 | + .06 | 10 hours at 1,720 r. p. m. | | | do | Blades, splices, and tips O. K. |
| 105 | 8-45 | 65 | + .33 | + .61 | + .10 | + .20 | do | | | do | Do. |
| 106 | 8-45 | 66.5 | + .29 | + .21 | + .37 | + .12 | do | | | do | Do. |
| 108 | 8-45 | 58.5 | + .23 | + .15 | + .32 | − .01 | 10 hours at 1,725 r. p. m. | | | do | Do. |
| 84 | Curtiss, 8 feet by 4 feet 7 inches. | 38 | + .33 | + .28 | | | 3 hours at 1,800 r. p. m. | | | do | Tips and blades O. K. |
| 90 | Paragon, 8 feet by 5 feet 3 inches. | 27 | + .12 | − .05 | | | do | | | do | Do. |
| | | | | | | POPLAR. | | | | | |
| 54 | X-3012 | 38 | +0.19 | +0.16 | +0.55 | +0.32 | 10 hours at 1,600 r. p. m. | | | | |
| 69 | 13279 | 29 | + .61 | + .27 | +1.18 | + .65 | 10 hours at 1,950 r. p. m. | | | Bad | |
| 70 | 13279 | 28.3 | + .28 | + .16 | + .41 | + .22 | | 20 minutes at 1,600 r. p. m. | Cracked between laminations. | do | Glue holding tips not dry. |
| 137 | 34555 | 43.5 | + .18 | + .22 | + .36 | + .27 | 10 hours at 1,730 r. p. m. | 5 hours at 1,830 r. p. m. | Piece off T. E. tips of both blades. | Slight | Failure due to overload. |
| | | | | | | WALNUT. | | | | | |
| 2 | 6-12 | 29.5 | +0.1 | +0.1 | | | 10 hours at 2,500 r. p. m. | | | | Tips frayed on L. E. |
| 14 | 8-25 | 21.5 | +1.05 | + .49 | | | 10 hours at 1,800 r. p. m. | | | | Tips frayed considerably. |
| 21 | Hartzel | 22.3 | + .48 | + .41 | | | 7 hours at 1,600 r. p. m. | | | Slight | |
| 50 | X-361 | 19.5 | + .40 | + .35 | | | | 4½ hours at 1,800 r. p. m. | Raised splice | do | 1 place raised at splice 0.01 inch. |
| 51 | X-365 | 20 | + .41 | + .35 | | | | 3 hours at 1,800 r. p. m. | do | do | Do. |
| 52 | X-362 | 20 | | | | | | | Loose hub | | Hub not properly placed at start. |

161988—21——11

TABLE II—Continued.

WALNUT—Continued

| Test No. | Design. | Weight. | Deflections in inches at normal r. p. m. — At 3 feet radius. L. E. | At 3 feet radius. T. E. | At 4 feet radius. L. E. | At 4 feet radius. T. E. | O. K. | Failed. | Nature of failure. | Flutter. | Remarks. |
|---|---|---|---|---|---|---|---|---|---|---|---|
| | | *Pounds.* | | | | | | | | | |
| 56 | S-45 | 52.5 | + .27 | + .17 | +0.48 | +0.38 | 4 hours at 1,700 r. p. m. | | | | |
| 57 | S-45 | 50.5 | + .15 | + .06 | + .36 | + .13 | do | | | | |
| 64 | S-45 | 51 | + .22 | + .07 | + .47 | + .16 | | 1½ hours at 1,700 r. p. m. | Split blade | Bad | Spiral grain. |
| 102 | SE-5000 | 50.5 | + .40 | + .26 | + .93 | + .54 | | 1 hour at 1,700 r. p. m. | Split tip | do | Large vibration in pitch. |
| 104 | Navy, H-16 | 43.5 | + .69 | + .44 | +1.20 | +1.01 | 10 hours at 1,800 r. p. m. | | | do | Slight abrasion due to water spray. |
| 117 | X-6317 | 38 | + .46 | + .41 | + .98 | + .66 | | 2 hours at 1,800 r. p. m. | Split blades | Very bad | Thin sections and large variations in pitch. |
| 118 | Charavay | 45 | + .38 | + .06 | +1.32 | + .51 | | 3 hours at 1,800 r. p. m. | do | do | Excessive variations in deflection and pitch. |
| 119 | do | 45 | + .22 | + .18 | + .96 | + .38 | 10 hours at 1,800 r. p. m. | | | do | Not recommended account excessive weave and variation in pitch. |
| 120 | do | 46 | + .26 | + .06 | + .91 | + .95 | do | | | do | Do. |
| 166 | Olmsted Geared, U. S. A. 12. | 119.3 | | | | | | 5 minutes at 1,227 r. p. m. | | Very slight | Hub thickness cut down to fit standard Liberty hub. |
| 173 | Olmsted, for Navy | 94.8 | | | | | 10 hours at 1,780 r. p. m. | | | Slight | |
| 179 | Pomilio, N-106 | 39.2 | | | | | 3 minutes at 1,500 r. p. m. | | | None | No endurance run made. |
| 180 | do | 43.3 | | | | | 5 minutes at 1,517 r. p. m. | | | Very bad | No endurance run made; propeller unsafe. |
| 183 | Olmsted, for DH-4 | 91.8 | | | + .32 | + .18 | 10 hours at 1,660 r. p. m. | | | Slight | |
| 187 | Pomilio, N-156 | 48.4 | | | + .46 | + .30 | 8 minutes at 1,573 r. p. m. | | | | Weaving badly at 1,500 r. p. m. |
| 188 | Pomilio, N-105 | 40 | | | + .64 | + .16 | 4 minutes at 1,642 r. p. m. | | | Moderate | |
| 189 | Pomilio, N-102 | 42 | | | + .45 | + .48 | 5 minutes at 1,602 r. p. m. | | | Slight | No endurance run made. |
| 190 | Pomilio, N-104 | 38 | | | + .36 | + .28 | 5 minutes at 1,498 r. p. m. | | | do | Do. |

| | | | | | | | | | | | |
|---|---|---|---|---|---|---|---|---|---|---|---|
| 191 | Pomilio, N-101 | 42.7 | | | + .82 | + .46 | | 10 hours at 1,742 r. p. m. | Tip of 1 blade cracked. | do | L. E. badly chewed by being run in rain. |
| 193 | Pomilio, N-153 | 53.7 | | | + .69 | + .63 | 5 minutes at 1,705 r. p. m. | | | do | No endurance run made. |
| 194 | Pomilio | 45.4 | | | + .63 | + .25 | 10 hours at 1,746 r. p. m. | | | do | L. E. slightly worn by rain. |
| 213 | X-13855 | 33.8 | + .37 | + .27 | | | 4½ hours at 1,850 r. p. m. | | | Intermittent | |
| 128 | Charavay | 58 | + .38 | | + .81 | | | 10 hours at 1,825 r. p. m. | Blades cracked | Bad | 4 blades straight leading edges. |
| | | | | | MICARTA. | | | | | | |
| 7 | 8-21 | 48 | +0.3 | +0.6 | | | 10½ hours at 1,800 r. p. m. | 19 minutes at 2,015 r. p. m. | Broke off 1 tip | | |
| 8 | 8-21 | 47.5 | + .51 | + .03 | | | 10 hours at 1,450 r. p. m. | | | | |
| 13 | 8-21 | 48 | + .4 | + .33 | | | | 2,300 r. p. m. | Broke both blades | | Failure due to excessive speed. |
| 9 | 8-21 | 46 | +1.3 | + .51 | | | 9 hours at 1,800 r. p. m. | 23 minutes at 2,200 r. p. m. | Broke off 1 blade, other blade cracked. | | Failure due to excessive speed and power. |
| 16 | 8-21 | 39.5 | + .37 | + .24 | | | do | | | Slight | O. K. at 2,346 r. p. m.; no metal hub; direct keyed. |
| 18 | 8-21 | 34.8 | + .61 | + .24 | | | 9½ hours at 1,800 r. p. m. | 2,000 r. p. m. | Broke off 1 blade near hub. | Badly | Reinforced, twisted laminations. |
| 29 | 8-21 | 67 | + .33 | + .02 | | | | $1\frac{1}{12}$ hours at 1,800 r. p. m. | Downstream; propeller failed at hub. | Slight | Double 2-blader mortised and hubbed; duck reinforced. |
| 30 | 8-21 | | | | | | | | | | Destroyed by an accident before test was started. |
| 31 | 8-21 | 40.5 | + .25 | + .1 | | | | 2,200 r. p. m. | Pulled apart at hub. | | Duck; straight laminations; eight 1-inch holes bored through hub. |
| 35 | 8-21 | 37.5 | + .54 | + .22 | | | | 2,000 r. p. m. | Burst in pieces | Slight | Propeller shot up by machine gun before test. |
| 39 | 8-21 | 36.3 | | | | | O. K.; $H_2O$ and sand. | | None | | Duck; twisted laminations. |
| 41 | 8-21 | 37.3 | + .61 | + .35 | | | 10 hours at 1,800 r. p. m. | | | Slight | Duck; twisted laminations; O. K. at 2,250 r. p. m. |
| 42 | 8-21 | 37.3 | + .78 | + .61 | | | do | | | | Paper centers; duck surface laminations; O. K. at 2,400 r. p. m. |
| 48 | 8-21 | 58 | | | | | | 1,400 r. p. m. | | | Double reinforced; double 2-blader; mortised; test discontinued. |
| 202 | X-5425 | 82 | | | +0.22 | +0.08 | 13 hours, 600 H. P., at 1,720 r. p. m. | 26½ hours, 800 H. P., at 1,806 r. p. m. | Blades pulled apart at hub. | | Run to 2,100 r. p. m., 1,325 H. P.; following 13 hours, run 600 H. P. |
| 17 | 8-21 | 37 | + .32 | + .18 | | | | 2,100 r. p. m. | do | Slight | Paper; twisted laminations; reinforced L. E. |

FIG. 78.—VIEW OF PROPELLER UNDER WATER SPRAY TEST, LOOKING DOWNSTREAM.

FIG. 79.—VIEW OF PROPELLER UNDERGOING WATER-SPRAY TEST, LOOKING UPSTREAM.

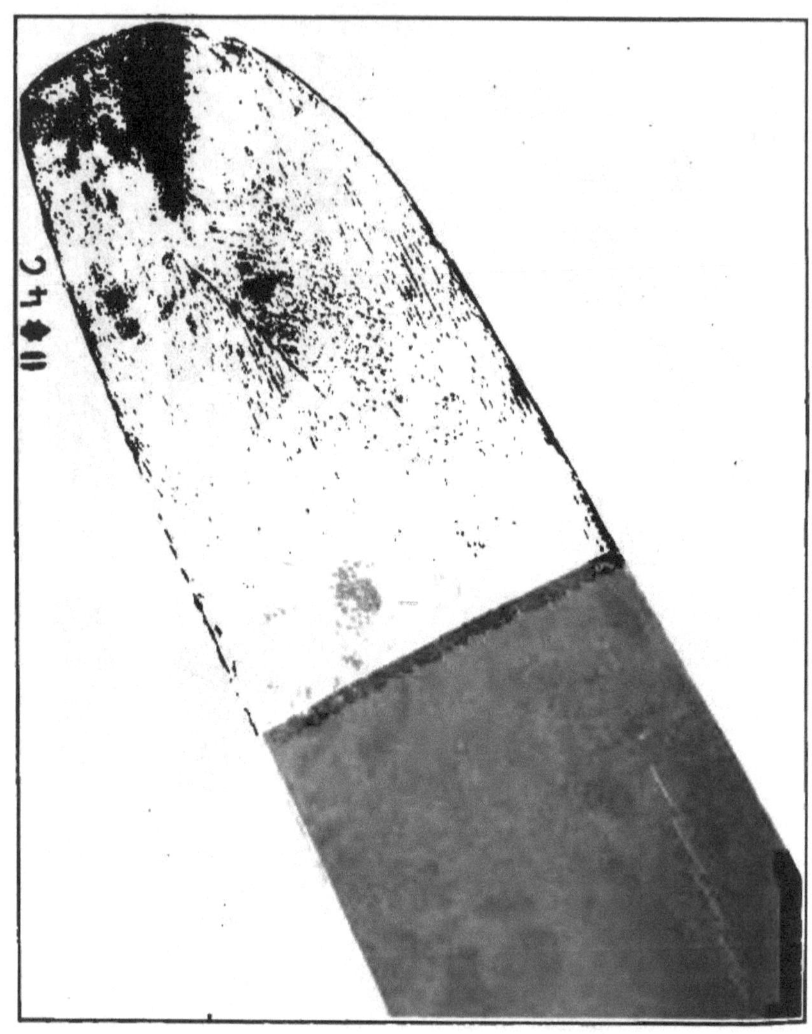

FIG. 80.—THRUST FACE OF LINEN-TIPPED OAK PROPELLER AFTER A 30-MINUTE WATER-SPRAY TEST.

Cambered face of this blade is shown in Figure 81.

FIG. 81.—CAMBERED FACE OF LINEN-TIPPED OAK PROPELLER AFTER A 30-MINUTE WATER-SPRAY TEST.

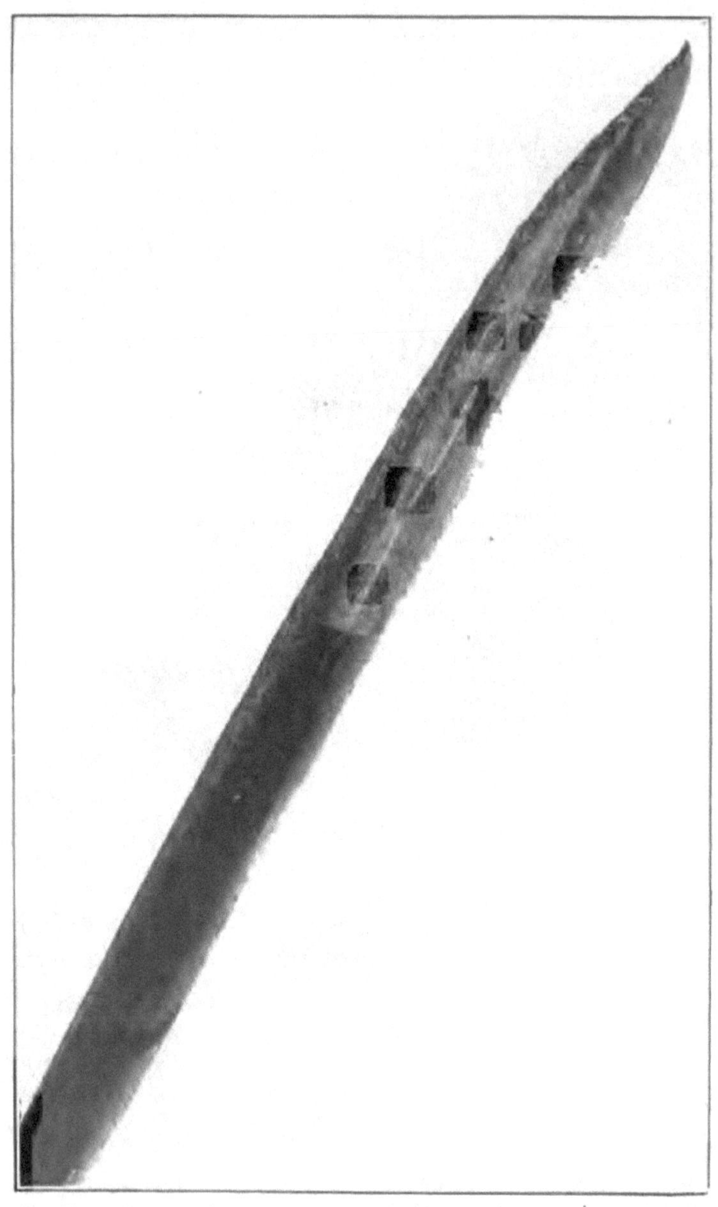

FIG. 82.—UNSHEATHED WALNUT BLADE AFTER A 2½-HOUR RUN IN THE RAIN.
Note the damage to leading edge.

## WATER SPRAY ABRASION TESTS.

A large number of tests have been made in an effort to determine the effect of water spray on propellers. These tests have usually been run at the conclusion of a standard destructive whirling test. The propellers tested have in some cases been tipped and in some cases have been run without tips in order to determine the effect on the wood alone.

In order to obtain comparable results, the same method of introducing the water spray into the propeller draft was used in each test. A 1-inch pipe perforated with a row of $\frac{1}{8}$-inch holes spaced 1 inch apart was suspended above and on the upstream side of the propeller. This pipe was connected to the city water main. The conditions produced with this apparatus are similar to a very heavy rainstorm. The apparatus in operation is shown in figures 78 and 79.

These tests have shown that a propeller without some form of tip will be practically destroyed by a run of one-half hour in a standard water spray. Of all the woods run without tips, oak stands the abrasion of spray better than any other. The leading edge of an oak propeller seems to wear away very gradually without splitting, whereas all other woods tested have shown a decided tendency to split as well as to wear. A one-half hour run on an oak propeller in a standard water spray has been found to wear approximately $\frac{1}{2}$ inch of material off the leading edge. Figures 80 and 81 show the damage done to the thrust and cambered faces of an oak propeller after a one-half hour test.

The effect of rain on a walnut propeller is shown in figure 82. This propeller was run in a light rain for about two and one-half hours. As will be seen from the photograph, approximately 1 inch of material was worn off the leading edge during that time. This propeller was also split at the tip when the test was completed.

All of these tests show that the maximum damage occurs at the tip of a propeller and that this wear tapers to nothing about 15 inches from the hub. It is also found that practically all the wear takes place on the leading edge, and that if the leading edges of the blades are protected no appreciable wear will take place on the faces of the propeller.

*Materials used for tipping.*—Propellers tipped with the following materials have been tested: Copper tips riveted in place, copper tips screwed in place, brass tips riveted in place, brass tips screwed in place, linen tips glued in place, pigskin tips glued in place, rawhide tips glued in place, rubberized fabric tips glued in place, hard rubber covered propellers, hard rubber covered propellers with soft rubber leading edges, and propellers having steel tips.

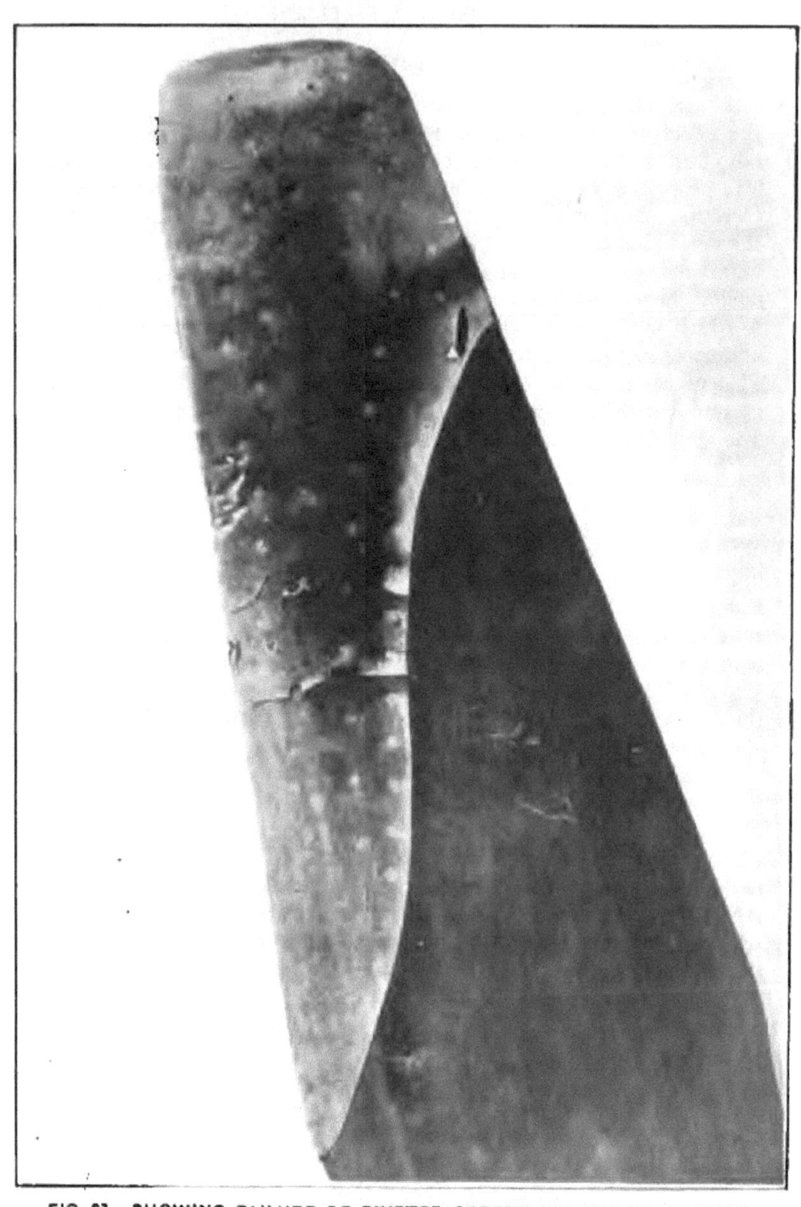

FIG. 83.—SHOWING FAILURE OF RIVETED COPPER TIP DUE TO FLUTTER.

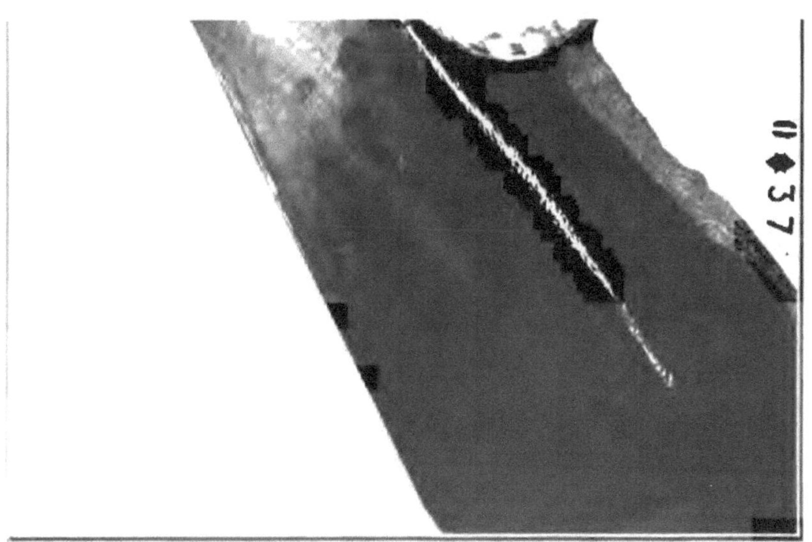

FIG. 84.—A BRASS TIP ATTACHED WITH SCREWS. RUN 30 MINUTES IN WATER SPRAY.

Note damage to unprotected wood and good condition of tips.

FIG. 85.—SHOWING FAILURE ON THRUST FACE OF A LANG PROPELLER WITH RIVETED BRASS TIPS.

Failure partly due to flutter.

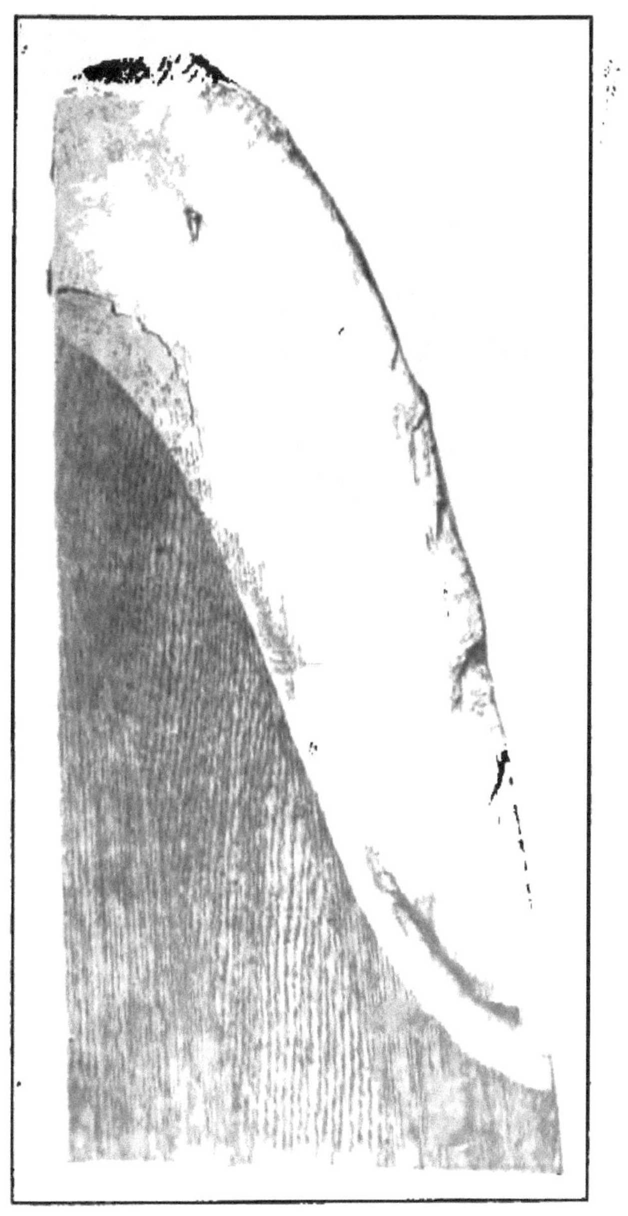

FIG. 86.—THRUST FACE OF OAK PROPELLER TIPPED WITH PIGSKIN. RUN 15 MINUTES IN STANDARD WATER SPRAY.

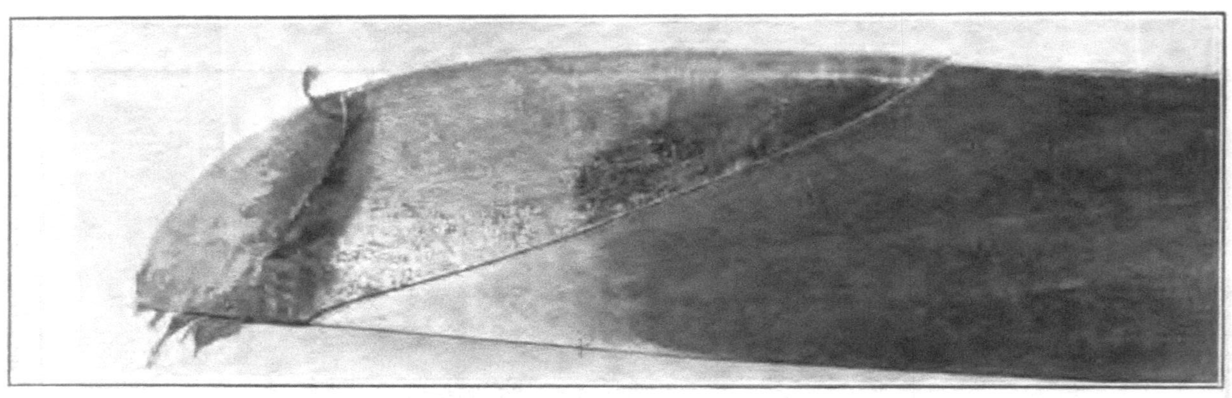

FIG. 87.—RUBBERIZED FABRIC TIP AFTER 2 MINUTES IN WATER SPRAY.

Note the damage to the wood on the unprotected leading edge.

FIG. 88.—LEADING EDGE OF RAWHIDE TIP AFTER AN 8-MINUTE WATER-SPRAY TEST.

*Copper tips.*—These tips seem to afford ample protection to the leading edges of a propeller, but the copper itself seems to be too soft to remain solidly in place when attached either by rivets or screws. (See fig. 83.)

*Brass tips.*—Brass tips seem to be the most satisfactory tips where water spray is likely to be encountered continuously. These tips seem to be more satisfactory when attached with screws than when attached with rivets. (See figs. 84 and 85.)

It has been found, however, that neither the copper tips nor the brass tips will remain on a propeller which flutters to any appreciable extent. When used on such a propeller these tips invariably crack at a point about 18 inches from the ends of the blades.

*Linen tips.*—Linen tips will be cut through by water spray in about two minutes. Figures 80 and 81 show the appearance of a linen tip after a one-half hour standard test.

*Pigskin tips.*—Pigskin tips afford an appreciable amount of protection against the abrasive action of water. The leading edges will be cut through after about four minutes of a standard water spray test. Figure 86 shows the effects of 15 minutes in water spray on a pigskin-tipped propeller. An examination of the photographs shows that the damage started on the leading edge, as has been found to be the case on all other propellers tested. This figure shows very well the character of the damage done to the wood of the blades by the spray after the tip has been cut through.

As has been stated above, these tips afford an appreciable protection against water spray. In view of the considerable amount of skill required in applying them, there is a reasonable doubt as to whether they will be as satisfactory as a linen tip since the latter is comparatively easy to apply.

*Vulcanized rubber fabric tips.*—Only one propeller tipped with this materials has been tested. It was damaged so badly by two minutes in the standard water spray test that it was not considered worth while to continue experiments on tips made of this material. Figure 87 shows the condition of the more damaged of these tips at the conclusion of the test.

*Rawhide tips.*—Rawhide tips appear to afford more protection than linen, rubberized fabric, or pigskin. After an eight-minute test on a propeller fitted with these tips one tip was in fair condition and the other was badly damaged due to a defect in applying the material. The better of the two is shown in figure 88, as at the conclusion of the test. Note the abrasion of the unprotected portion of the leading edge.

These tips are open to the same objection as has been noted above under pigskin tips, viz, they are difficult to apply properly, and to re-

FIG. 89.—VIEW OF THRUST FACE OF ONE BLADE OF A HARD-RUBBER-COVERED PROPELLER SHOWING THE DAMAGE DONE THE LEADING EDGE BY A 30-MINUTE RUN IN WATER SPRAY. MATERIAL WORN AWAY ALONG LEADING EDGE TO A DEPTH OF THREE-EIGHTHS INCH.

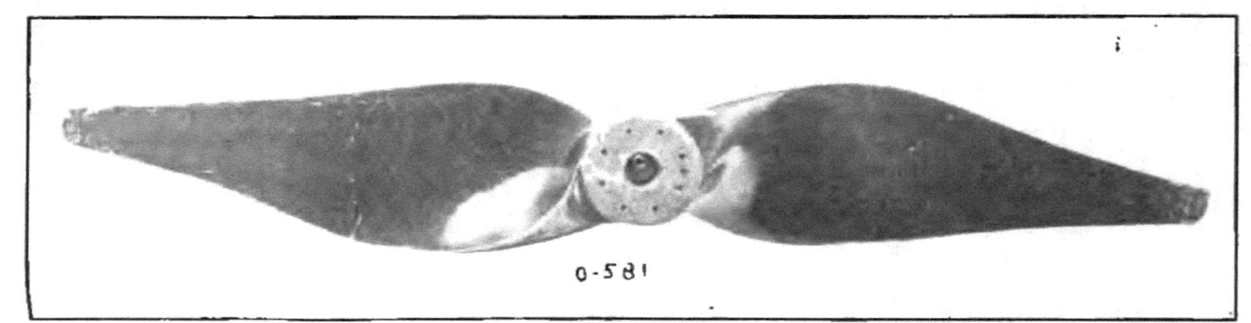


FIG. 90.—THRUST FACE OF A PROPELLER HAVING CHROME-VANADIUM STEEL TIPS. SHOWN AFTER A DESTRUCTIVE WHIRLING TEST RUN IN RAIN.

pair in case they suohld become damaged in any way. Also it may be difficult to get raw material of uniform quality.

*Hard rubber covered propellers.*—Four propellers of this type have been tested. All of these propellers seem to stand the abrasive action of the spray much better than any material yet tested outside of brass. After being run for one-half hour in spray, these propellers have about $\frac{3}{8}$ inch worn off the leading edge. This material seems to have considerable promise as a material for covering propellers, especially those which will not have to be run in a water spray for very long periods. (See fig. 89, which shows the damage to one of these tips resulting from a 30-minute test.)

These propellers are open to the objection that they will be very hard to repair in case they should become damaged in any way.

*Hard rubber covered propellers with soft rubber leading edges.*—Two propellers of this type have been run. Both these propellers stood the abrasive action of the water much better than did the propellers with hard rubber leading edges. At the end of one-half hour in the standard spray only about 1/16 inch had been worn off the leading edges. The wear was characteristic of that shown in figure 89, the only difference being that it was not as severe.

These propellers are open to the same objection as the propellers made entirely of hard rubber, namely, that they will be very hard to repair in case of damage.

*Steel tips.*—Only one propeller having steel tips has been tested. These tips were unaffected by the spray. This type of tip will undoubtedly prove satisfactory where spray is likely to be encountered continuously. The greatest difficulty in connection with these tips, as with all metal tips, is in attaching them satisfactorily to the wood. Figure 90 shows this propeller as at the conclusion of the water spray test.

161988—21——12

## Chapter VI.

### PROPELLER RESEARCH.

#### STRESS ANALYSIS OF A PROPELLER.

The propeller is divided from the 18-inch station out into elements 6 inches long, and the loads of torque, thrust, and centrifugal force are found for each one. These are applied to a horizontal bar at the same point that they act on the propeller. The centrifugal force is resolved into two components parallel to the axes of the bar.

The moments and direct loads are then found about the three axes of this bar for each station, and from these are obtained moments and direct loads about the principal axes through the center of gravity of the propeller cross section.

With these moments and from the section properties the neutral axis is found, after which the extreme fiber distances can be measured.

When these are known, the stress is then calculated by the ordinary methods of $f=\frac{M}{I}d$. This gives normal fiber stress. The steps in detail are as follows: Propeller 34291 is taken as an example. It has standard air-foil cross sections. (See fig. 6.) All dimensions are given in inches, loads are given in pounds, moments in inch-pounds, areas in square inches, and moments of inertia in inches to the fourth power. The exception to this is in table entitled "Data for loading."

1. *Loads.*—In order to find the centrifugal force for each element, the location of the center of gravity and the area of the cross section must be obtained. These are given in Table III, entitled "Section properties." The symbols for items are explained in nomenclature (p. 155). The upper line for each item gives the properties for a similar air foil having a 10-inch blade width and is obtained from curves of "Section properties for standard airfoils for propellers." The reduction table on the instruction sheet (see data sheets, figs. 91, 92, 93, 94, and 95) shows how the properties for the required section are found. The centers of gravity are plotted on a blue print of drawing 34291, where the propeller is shown to full scale, and the principal axes are then drawn on the cross section. The operation is indicated on the sheet entitled "Coordinates of extreme fiber," figure 96.

TABLE III.—*Propeller 34291.*

SECTION PROPERTIES.

| Item. | Section considered. | | | | | | | | |
|---|---|---|---|---|---|---|---|---|---|
| | 12-inch. | 18-inch. | 24-inch. | 30-inch. | 36-inch. | 42-inch. | 48-inch. | 54-inch. | 58½-inch. |
| $b$ | 10 | 10 | 10 | 10 | 10 | 10 | 10 | 10 | 10 |
| | 8.16 | 8.72 | 9.53 | 9.84 | 9.75 | 9.03 | 7.69 | 5.84 | 4.31 |
| $h_U$ | 3.22 | 2.65 | 1.97 | 1.59 | 1.35 | 1.11 | .96 | .856 | .768 |
| | 2.63 | 2.31 | 1.88 | 1.56 | 1.31 | 1.0 | .75 | .50 | .34 |
| $h_L$ | 1.69 | .64 | .33 | .09 | ........ | ........ | ........ | ........ | ........ |
| | 1.38 | .56 | .31 | .09 | ........ | ........ | ........ | ........ | ........ |
| $b_c$ | 4.46 | 4.46 | 4.46 | 4.46 | 4.46 | 4.46 | 4.46 | 4.46 | 4.46 |
| | 3.64 | 3.89 | 4.25 | 4.38 | 4.35 | 4.03 | 3.42 | 2.61 | 1.98 |
| $h_c$ | .63 | .82 | .67 | .625 | .56 | .46 | .41 | .355 | .230 |
| | .51 | .715 | .64 | .615 | .56 | .415 | .315 | .207 | .142 |
| $A$ | ........ | 24.25 | 17.05 | 12.4 | 10.0 | 8.2 | 7.23 | 6.27 | 5.82 |
| | 23.88 | 18.40 | 15.45 | 11.96 | 9.50 | 6.70 | 4.28 | 2.14 | 1.08 |
| $\alpha_p$ | ........ | 2° 33′ | 1° 58′ | 1° 46′ | 1° 33′ | 1° 18′ | 1° 8′ | 1° 0′ | ........ |
| Tan $\alpha_p$ | ........ | .0445 | .0343 | .0306 | .0271 | .0227 | .0198 | .0175 | ........ |
| $I_m$ | ........ | 16.2 | 5.55 | 2.18 | 1.14 | .645 | .445 | .288 | ........ |
| | ........ | 9.36 | 4.57 | 2.04 | 1.03 | .43 | .156 | .084 | ........ |
| $I_M$ | ........ | 146 | 102.2 | 74.3 | 59.6 | 49.0 | 43.3 | 37.6 | ........ |
| | ........ | 84.5 | 84.3 | 69.5 | 53.9 | 32.6 | 15.1 | 4.38 | ........ |

NOTE.—Lower line in each case is properties for X-6306; upper line is for a similar section with 10-inch blade width. Angle $\alpha_p$ is same for both sections.

Required the properties of a standard air-foil section whose—

Blade width equals 7 inches.

Upper camber ratio equals 0.2.

Lower camber ratio equals 0.04.

First. Find properties of similar airfoil with 10-inch blade width.

Second. Apply reduction factor and obtain properties for 7-inch blade width.

REDUCTION TABLE.

| Symbol. | Item. | See propeller data sheet. | For 10-inch blade width. | Reduction factor. | For 7-inch blade width. |
|---|---|---|---|---|---|
| $b$ | Blade width | 94 | 10″ | $\frac{7}{10}$ | 7″. |
| $h_U$ | Maximum upper camber | 94 | 2.0″ | $\frac{7}{10}$ | 1.4″. |
| $h_L$ | Maximum lower camber | 94 | 0.4″ | $\frac{7}{10}$ | 0.28″. |
| $b_c$ | Center of gravity from left end measured along chord. | 95 | 4.46″ | $\frac{7}{10}$ | 3.12″. |
| $h_c$ | Center of gravity above chord | 95 | 0.66″ | $\frac{7}{10}$ | 0.46″. |
| $A$ | Area | 95 | 17.75 sq. in. | $(\frac{7}{10})^2$ | 8.70 sq. in. |
| $\alpha_p$ | Angle between minor principal axis and chord. | 96 | 1° 57′ | 1 | 1° 57′. |
| $I_m$ | Moment of inertia about minor principal axis. | 97 | 6.17 in.⁴ | $(\frac{7}{10})^4$ | 1.48 in.⁴. |
| $I_M$ | Moment of inertia about major principal axis. | 98 | 106.3 in.⁴ | $(\frac{7}{10})^4$ | 25.6 in.⁴. |

FIG. 91.

Section properties for standard airfoils for propellers.

INSTRUCTION SHEET—PROPELLER DATA SHEET NO. 94.

The thrust and torque for each element are obtained as explained in Chapter I. The centrifugal force is calculated as in Table IV, entitled "Data for loading," and is indicated as $\Delta F$. The thrust and torque are also shown in this table and are given by $\Delta T$ and $\Delta Q$.

Draw a curve of center of gravity in both plan view and elevation and do the same for the point of application of the propulsion force, whose components are known as thrust and torque. This force is assumed to act through the one-third point of the chord measured

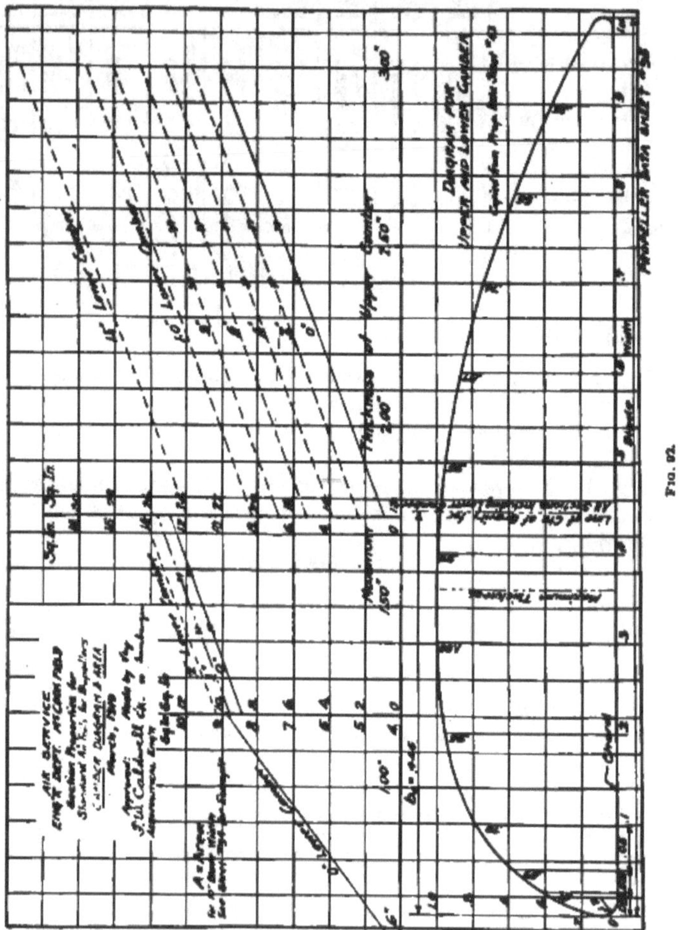

Fig. 92.

from the leading edge for each propeller cross section. Axes of reference are shown as follows: An $r$ axis coming out from the center of the hub in a radial direction; an $s$ axis passing through the $r$ axis at any station and being parallel to the shaft; $a$ at axis perpendicular to these two, making it parallel to a tangent to the rotating disk at the end of the $r$ axis. The centrifugal force which acts in a radial direction in the plan view is laid off from the center of shaft on a line connecting this point with the center of gravity of the element.

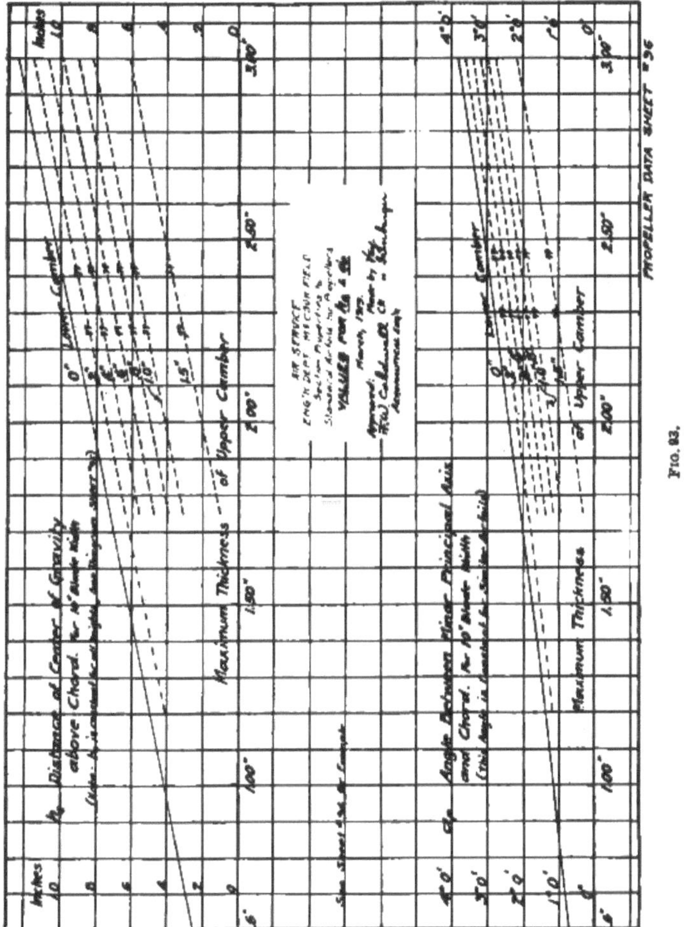

FIG. 93.

The components of this force along the $r$ axis and $t$ axis are then measured and entered in Table IV as $\Delta F_r$ and $\Delta F_t$.

The final elements used are the 6-inch distances between the stations shown on the drawing, but in order to utilize the air-foil cross section on this drawing when obtaining the loads of thrust, torque, and centrifugal force, the element is temporarily considered as the

6-inch length whose mid section is at the given stations. In order to find the loads on the final elements, curves are plotted, as shown in figure 97, entitled "Curves giving increments of load." From these the loads on the final elements may be obtained and are shown by the scratch marks on these curves. These values are given in Table V,

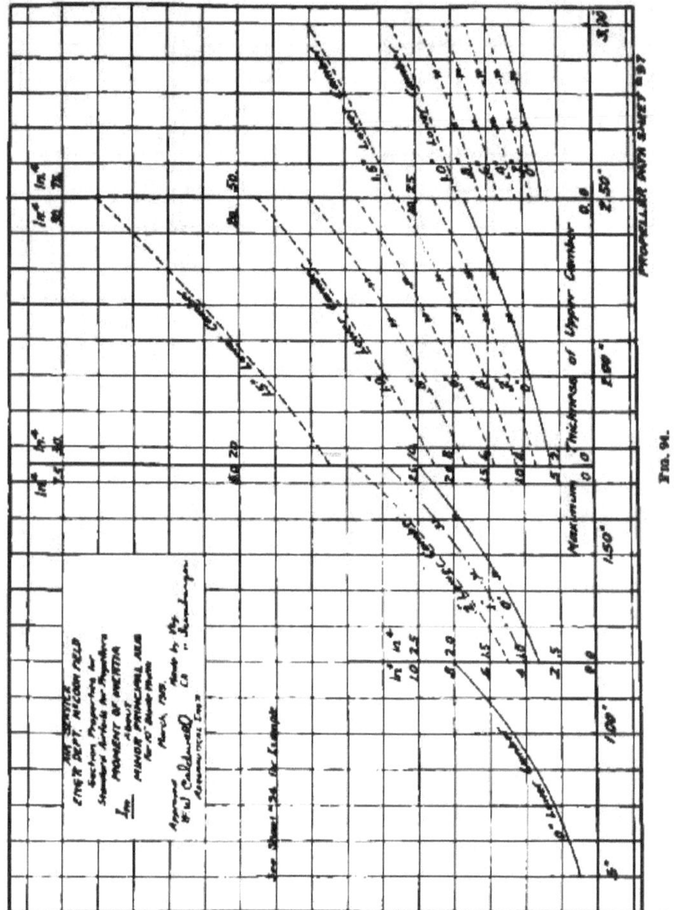

FIG. 94.

entitled "Final loads and coordinates." The coordinates of the points of application of centrifugal force and points of application of the propulsion force are measured from the $r$ axis along the $s$ and $t$ axes. The points are shown by scratch marks between the stations in figure 96, and the values are given in Table V. The signs are geometric and are not to be used as moment arm signs. This table also contains the coordinates of the center of gravity of the cross section at the given stations. The values of these coordinates are shown either on drawing, figure 96, or drawing, figure 98.

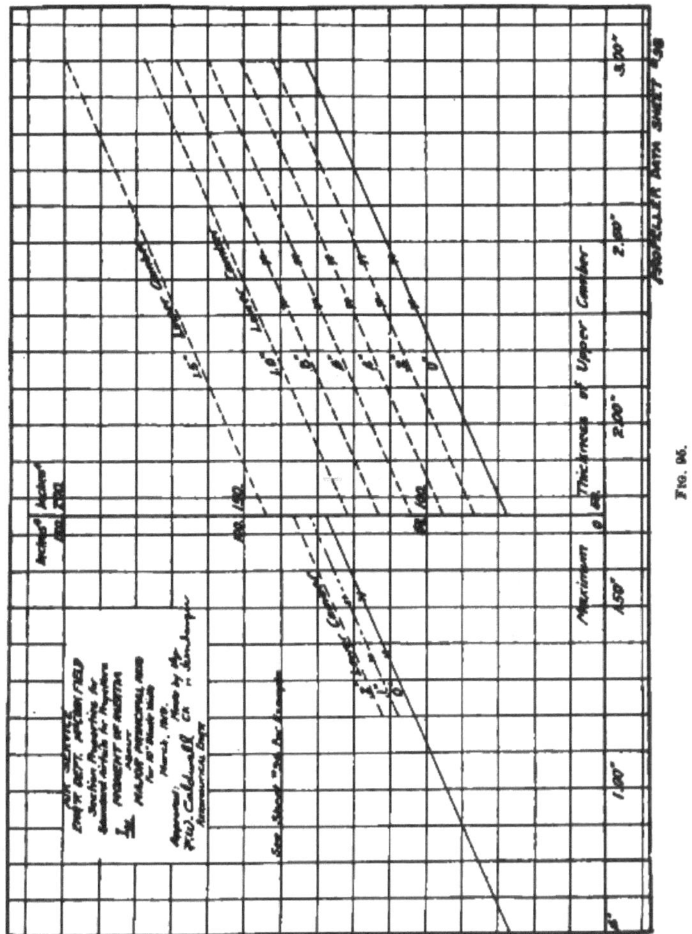


Fig. 95.

Draw loading bar on load sheet, figure 98, and locate the points of application for the various forces. Then draw these forces into scale. Enter the coordinates for the centrifugal force in the plan and elevation view and for the torque and thrust in the cross-sectional views.

PROPELLER X-6305
COORDINATES OF EXTREME FIBERS
CROSS SECTIONS LOOKING AWAY FROM HUB
ELEVATION
PLAN

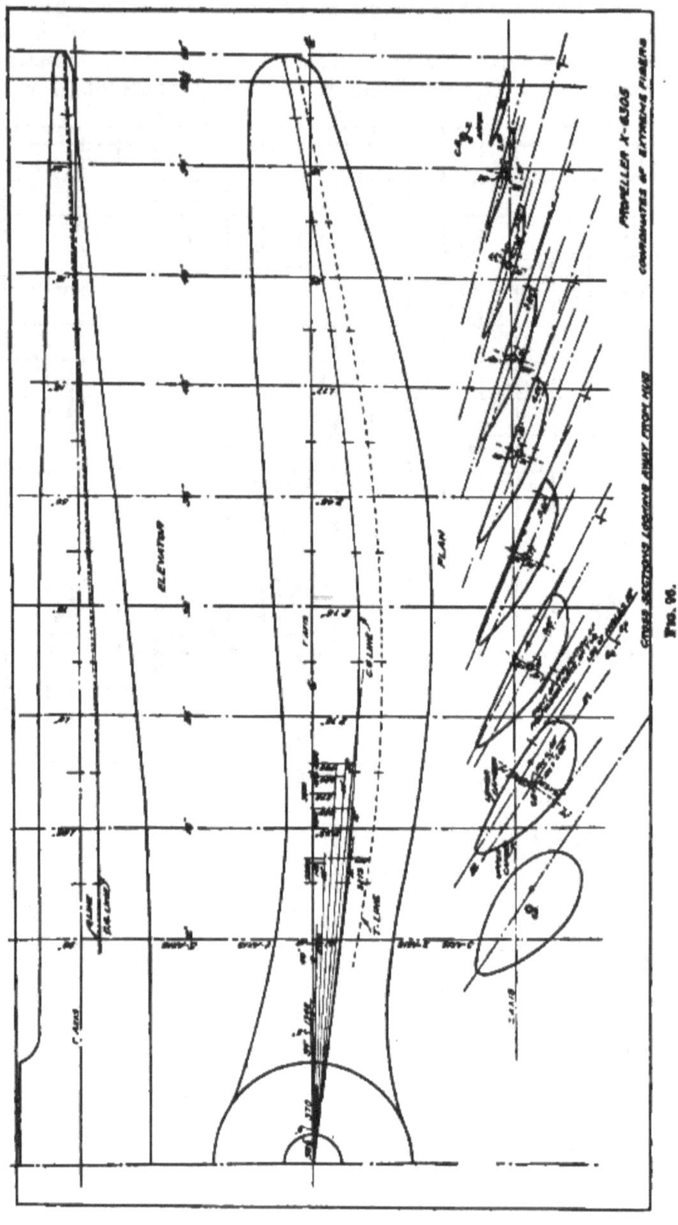

FIG. 98.

CURVES GIVING INCREMENTS OF LOAD
PER 6" ELEMENT OF PROPELLER
ΔF ΔF
500 5000
400 4000
300 3000
200 2000
100 1000
0 0
STATION

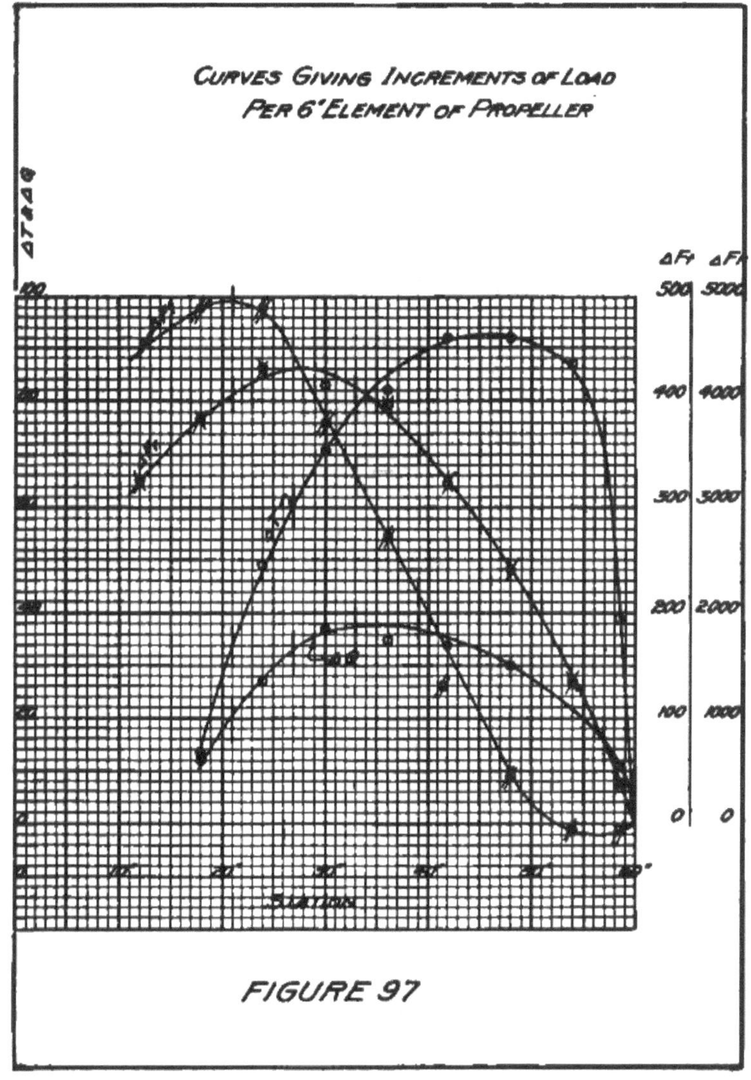

FIGURE 97

ELEVATION
PLAN
SECTIONS TAKEN LOOKING TOWARD RIGHT
PROPELLER X-6305
LOAD SHEET

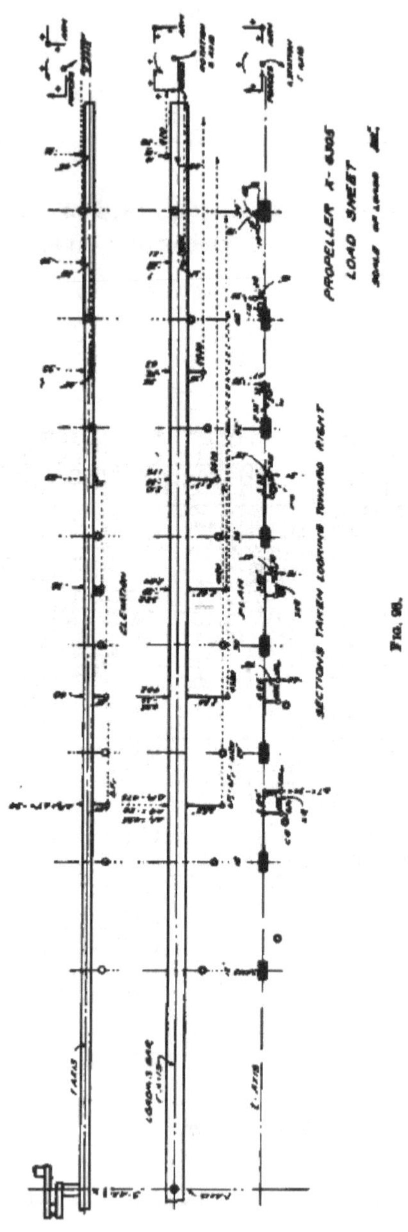

Fig. 98.

*Moments.*—Consideration must now be given to signs of forces, moments, and arms of moments. Figure 99 shows the direction of forces that have been considered positive.

Figure 100 shows the rotations that are considered positive. When the sight is as directed, positive rotations are anticlockwise. This figure also shows the positive arms when the forces are positive.

TABLE IV.—*Propeller 34291.*

DATA FOR LOADING.

| Item. | Unit. | Section considered. | | | | | | | | |
|---|---|---|---|---|---|---|---|---|---|---|
| | | 12-inch. | 18-inch. | 24-inch. | 30-inch. | 36-inch. | 42-inch. | 48-inch. | 54-inch. | 58½-inch. |
| $V$.... | Ft./sec...... | 184.3 | 276.5 | 368.6 | 461 | 553 | 645 | 737.2 | 830 | 899 |
| $A$.... | Sq. in....... | 23.88 | 18.40 | 15.45 | 11.98 | 9.50 | 6.7 | 4.28 | 2.14 | 1.06 |
| | | | | | | | | | | .1425 |
| $\Delta W$.... | Lbs......... | 3.153 | 2.43 | 2.04 | 1.58 | 1.254 | .885 | .565 | .282 | 2 |
| $V^2(.373)+r$. | Ratio....... | 1,050 | 1,585 | 2,120 | 2,650 | 3,180 | 3,650 | 4,230 | 4,760 | 5,150 |
| $\Delta F$.... | Lbs......... | 3,310 | 3,860 | 4,330 | 4,190 | 3,990 | 3,230 | 2,390 | 1,340 | 370 |
| $\Delta F_r$.... | .....do...... | 3,273 | 3,827 | 4,304 | 4,170 | 3,980 | 3,230 | 2,390 | 1,340 | 370 |
| $\Delta F_t$.... | .....do...... | 453 | 490 | 490 | 380 | 272 | 140 | 47 | −5 | −9 |
| $\Delta T$.... | .....do...... | ........ | 13 | 49 | 71 | 82 | 92 | 92 | 87 | 39 |
| $\Delta Q$.... | .....do...... | ........ | 12 | 27 | 37 | 38 | 34 | 30 | 26 | 11 |

Stresses in this propeller are worked out for a plane speed of 125 m. p. h. and a propeller speed of 1,760 r. p. m.

TABLE V.—*Propeller 34291.*

FINAL LOADS AND COORDINATES.

| Item. | Section considered. | | | | | | |
|---|---|---|---|---|---|---|---|
| | 18-inch. | 24-inch. | 30-inch. | 36-inch. | 42-inch. | 48-inch. | 54-inch. |
| $\Delta F_r$.................. | 4,120 | 4,320 | 4,120 | 3,570 | 2,630 | 1,900 | 670 |
| $\Delta F_t$.................. | 495 | 435 | 322 | 210 | 95 | 7 | −10 |
| $\Delta T$.................. | 34 | 60 | 78 | 89 | 93 | 90 | 75 |
| $\Delta Q$.................. | 20 | 33 | 38 | 35 | 32 | 28 | 18 |
| $t$ for $\Delta F$.................. | 2.58 | 2.80 | 2.67 | 2.17 | 1.33 | .31 | −.66 |
| $s$ for $\Delta F$.................. | −1.04 | −.87 | −.62 | −.33 | −.02 | .28 | .45 |
| $t$ for $\Delta T$.................. | 3.84 | 4.06 | 3.88 | 3.33 | 2.32 | 1.12 | −.08 |
| $s$ for $\Delta Q$.................. | −1.00 | −.78 | −.47 | −.22 | .07 | .33 | .52 |
| $t_c$ for C. G.................. | 2.33 | 2.75 | 2.75 | 2.45 | 1.77 | .84 | −.18 |
| $s_c$ for C. G.................. | −1.06 | −1.00 | −.75 | −.50 | −.15 | .12 | .38 |

NOTE.—Signs of coordinates are geometric. Positive is upward and toward observer. When these dimensions are used for moment arms the signs must be determined from the arm diagram for the axis about which rotation is considered.

These arms with these forces give positive rotation. A reversal of force or a change of arm to the opposite side of the axis will give negative rotation. When these two changes occur together, the rotation is again positive. The loads and moments are shown for each axis in Tables VI, VII, and VIII.

The $S_c$ axis is typical. For each section moments and direct loads are obtained first with reference to the $s$ axis of the loading bar, and then in the last step the moment about the $S_c$ axis is found from the load on the $S$ axis and the known moment about it.

In the $\Delta P_s$ line the load under any section and those to the right are the increments that come on this section and act along the $s$ axis. The total load on a section is given by $P_s$.

The increments of moments due to $\Delta P_r$ are given by $\Delta M$, where due consideration is given to the sign of the load and of the arm. The moment on the section from the $\Delta P_r$ forces to the right of it is given by $M$ and is the summation of the increments to the right of the section including the increment under it. The moments due to $\Delta P_t$ are found by first getting the $\Delta M$ or increment of moment on each section from the $\Delta P_t$ adjacent to the 54-inch section. Then obtain the $\Delta M$ on each section from the $\Delta P_t$ adjacent to the 48-inch section, etc. The sum of these increments gives the total $M$

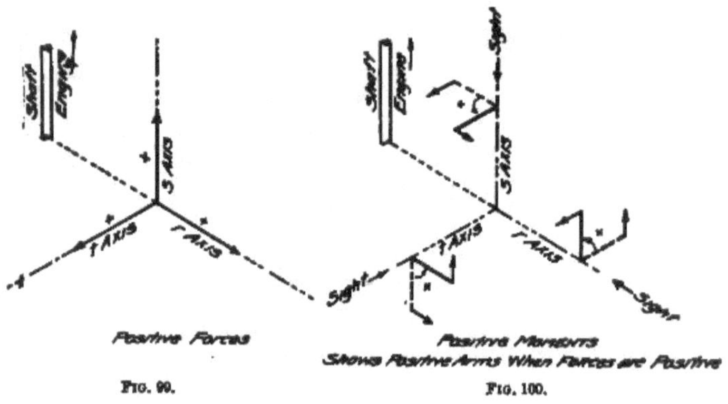


FIG. 99. FIG. 100.

on the section. An easy way to get the $\Delta M$ can be illustrated by taking the $\Delta P_t$ adjacent to the 54-inch section.

$$-3\times -28=84,\ 2\times 84=168,\ 84+168=252,\ 252+168=420, \text{ etc.}$$

The arm is 3 inches for the first section and is increased 6 inches for each successive station. The sign of the arm is minus.

The moment about the $S_o$ axis is obtained from that about the $s$ axis by the formula—

$$M_{so}=M_s+P_r\times t_o$$

The $S_o$ axis is parallel to the $S$ axis and passes through the center of gravity of the cross section.

$M_{so}$ is the moment about the $M_{so}$ axis.

$M_s$ is the moment about the $M_s$ axis.

$t_o$ is distance of C. G. from the $S$ axis.

$P_r$ is the total load normal to the section and acts along the $r$ axis.

The total load along the $S_o$ axis is $P_{so}$ and is equal to $P_s$.

Summing up the moments in each column will give $M_{so}$.

Do the same for the $t$ and $r$ axes.

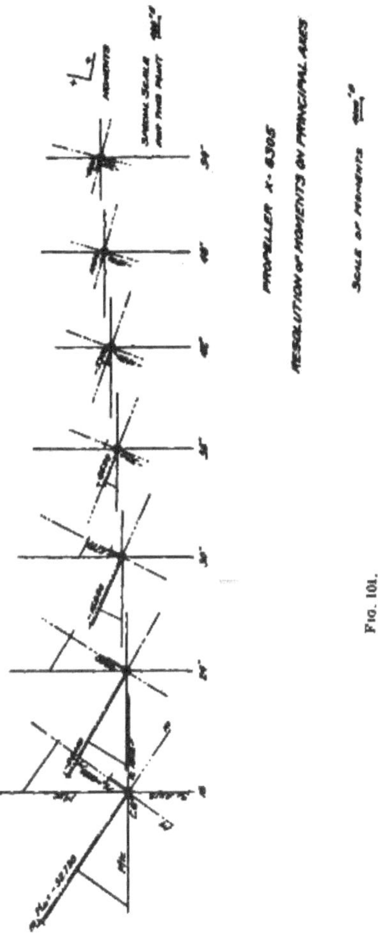


Fig. 101.

A piece of tracing paper is now laid on the blue print and the center of gravity, and the $s_0$ and $t_0$ axis traced on it and also the principal axes at all stations. The moments along the $s_0$ and $t_0$ axes are now

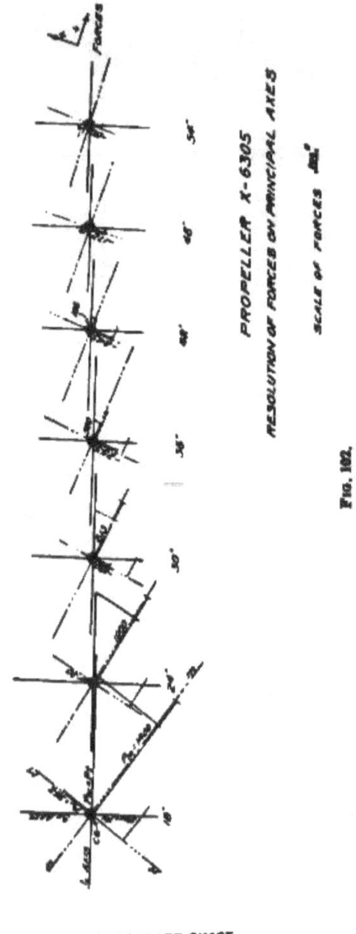


Fig. 102.

laid off to scale as shown in figure 101, entitled "Resolution of Moments on Principal Axes." The components of moments along principal axes are thus found. The moments are indicated by their axes of rotation.

**TABLE VI.**—*Propeller 34291.*

LOADS AND MOMENTS ON $r_e$ AXIS.

| Item. | Section considered. | | | | | | |
|---|---|---|---|---|---|---|---|
| | 18 inches. | 24 inches. | 30 inches. | 36 inches. | 42 inches. | 48 inches. | 54 inches. |
| $\Delta P_r$ | 4,120 | 4,320 | 4,120 | 3,570 | 2,830 | 1,900 | 670 |
| $P_r$ | 21,530 | 17,410 | 13,090 | 8,970 | 5,400 | 2,570 | 670 |
| $\Delta F_c$ | 495 | 435 | 322 | 210 | 95 | 7 | −10 |
| $S$ | −1.04 | −.87 | −.62 | −.33 | −.02 | .28 | .45 |
| $\Delta M$ | −515 | −378 | −200 | −70 | −2 | +2 | −5 |
| $M$ | −1,168 | −653 | −275 | −75 | −5 | −3 | −5 |
| $\Delta T$ | −34 | −60 | −78 | −89 | −93 | −90 | −75 |
| $t_e$ | −3.84 | −4.06 | −3.88 | −3.33 | −2.32 | −1.12 | .08 |
| $\Delta M$ | 131 | 244 | 303 | 296 | 216 | 101 | −6 |
| $M$ | 1,285 | 1,154 | 910 | 607 | 311 | 95 | −6 |
| $\Delta Q$ | −20 | −33 | −38 | −35 | −32 | −28 | −18 |
| $s$ | −1.00 | −.78 | −.47 | −.22 | .07 | .33 | .52 |
| $\Delta M$ | 20 | 26 | 18 | 8 | −2 | −9 | −9 |
| $M$ | 52 | 32 | 6 | −12 | −20 | −18 | −9 |
| $P_c$ | 1,350 | 875 | 473 | 189 | 14 | −49 | −29 |
| $S_e$ | 1.06 | 1.00 | .75 | .50 | .15 | −.12 | −.38 |
| $M$ | 1,432 | 875 | 354 | 94 | 3 | 7 | 11 |
| $P_s$ | −519 | −485 | −425 | −347 | −258 | −165 | −75 |
| $t_e$ | 2.33 | 2.75 | 2.75 | 2.45 | 1.77 | .84 | −.18 |
| $M$ | −1,209 | −1,333 | −1,170 | −850 | −456 | −139 | 14 |
| + $M$ grand total | 2,769 | 2,061 | 1,270 | 701 | 314 | 102 | 25 |
| − $M$ grand total | −2,377 | −1,986 | −1,445 | −937 | −481 | −160 | −20 |
| $M_{re}$ | 392 | 75 | −175 | −236 | −167 | −58 | 5 |

**TABLE VII.**—*Propeller 34291.*

LOADS AND MOMENTS ON $S_e$ AXIS.

| Item. | Section considered. | | | | | | |
|---|---|---|---|---|---|---|---|
| | 18 inches. | 24 inches. | 30 inches. | 36 inches. | 42 inches. | 48 inches. | 54 inches. |
| $\Delta P_s$ | −34 | −60 | −78 | −89 | −93 | −90 | −75 |
| $P_s$ | −519 | −485 | −425 | −347 | −258 | −165 | −75 |
| $\Delta P_r$ | 4,120 | 4,320 | 4,120 | 3,570 | 2,830 | 1,900 | 670 |
| $l$ | 2.58 | 2.80 | 2.67 | 2.17 | 1.33 | .31 | −.66 |
| $\Delta M$ | 10,620 | 12,100 | 11,000 | 7,750 | 3,770 | 590 | −440 |
| $M$ | 45,390 | 34,770 | 22,670 | 11,670 | 3,920 | 150 | −440 |
| $\Delta P_t$ | 475 | 402 | 284 | 175 | 63 | −21 | −28 |
| $\Delta M$ from $\Delta P_t$ adj. 54 inches. | 1,092 | 924 | 766 | 528 | 420 | 252 | 84 |
| 48 inches. | 693 | 567 | 441 | 315 | 189 | 63 | ........ |
| 42 inches. | −1,701 | −1,323 | −945 | −567 | −189 | ........ | ........ |
| 36 inches. | −3,675 | −2,625 | −1,575 | −525 | ........ | ........ | ........ |
| 30 inches. | −4,260 | −2,556 | −852 | ........ | ........ | ........ | ........ |
| 24 inches. | −3,618 | −1,206 | ........ | ........ | ........ | ........ | ........ |
| 18 inches. | −1,425 | ........ | ........ | ........ | ........ | ........ | ........ |
| − $M$ | −14,679 | −7,710 | −3,372 | −1,092 | −189 | ........ | ........ |
| + $M$ | 1,785 | 1,491 | 1,197 | 903 | 609 | 315 | 84 |
| $P_r$ | 4,120 | 4,320 | 4,120 | 3,570 | 2,830 | 1,900 | 670 |
| $t_e$ | −2.33 | −2.75 | −2.75 | −2.45 | −1.77 | −.84 | .18 |
| $M$ | −9,600 | −11,880 | −11,310 | −8,750 | −5,000 | −1,600 | 121 |
| + $M$ grand total | 47,175 | 36,261 | 23,867 | 12,573 | 4,529 | 465 | 205 |
| − $M$ grand total | −24,279 | −19,590 | −14,682 | −9,842 | −5,189 | −1,600 | −440 |
| $M_{se}$ | 22,896 | 16,671 | 9,185 | 2,731 | −660 | −1,135 | −235 |

TABLE VIII.—*Propeller 34291.*

LOADS AND MOMENTS ON $t_c$ AXIS.

| Item. | Section considered. | | | | | | |
|---|---|---|---|---|---|---|---|
| | 18 inches. | 24 inches. | 30 inches. | 36 inches. | 42 inches. | 48 inches. | 54 inches. |
| $\Delta P_t$ | 475 | 402 | 284 | 175 | 68 | −21 | −28 |
| $P_t$ | 1,350 | 875 | 473 | 189 | 14 | −49 | −20 |
| $\Delta P_r$ | 4,120 | 4,320 | 4,130 | 3,570 | 2,830 | 1,900 | 670 |
| $s$ | 1.04 | .87 | .62 | .33 | .02 | −.28 | −.45 |
| $\Delta M$ | 4,200 | 3,760 | 2,560 | 1,190 | 60 | −530 | −300 |
| $M$ | 10,980 | 6,730 | 2,970 | 420 | −770 | −830 | −300 |
| $\Delta P_c$ | −34 | −60 | −78 | −89 | −93 | −90 | −75 |
| $\Delta M$ from $\Delta P_c$ adj. 54 inches. | −2,925 | −2,475 | −2,025 | −1,575 | −1,125 | −675 | −225 |
| 48 inches. | −2,970 | −2,430 | −1,890 | −1,350 | −810 | −270 | ........ |
| 42 inches. | −2,511 | −1,953 | −1,395 | −837 | −279 | ........ | ........ |
| 36 inches. | −1,869 | −1,335 | −801 | −267 | ........ | ........ | ........ |
| 30 inches. | −1,170 | −702 | −234 | ........ | ........ | ........ | ........ |
| 24 inches. | −540 | −180 | ........ | ........ | ........ | ........ | ........ |
| 18 inches. | −102 | ........ | ........ | ........ | ........ | ........ | ........ |
| $M$ | −12,087 | −9,075 | −6,345 | −4,029 | −2,214 | −945 | −225 |
| $P_c$ | 21,530 | 17,410 | 13,090 | 8,970 | 5,400 | 2,570 | 670 |
| $S_c$ | −1.06 | −1.00 | −.75 | −.50 | −.15 | .12 | .38 |
| $M$ | −22,800 | −17,410 | −9,820 | −4,485 | −810 | 308 | 254 |
| $+M$ grand total | 10,930 | 6,730 | 2,970 | 420 | ........ | 308 | 254 |
| $-M$ grand total | −34,887 | −26,485 | −16,165 | −8,514 | −3,794 | −1,775 | −525 |
| $M_{tc}$ | −23,957 | −19,755 | −13,195 | −8,094 | −3,794 | −1,467 | −271 |

On another piece of tracing paper lay off the axes as before and resolve the direct force along the principal axes. This is shown in figure 102, entitled "Resolution of Forces on Principal Axes."

*Stress.*—The stress table, as in Table IX, may now be filled out down to the second main division. The neutral axis is then found by means of the formula:

$$\text{Tan} -I\, a_N = (M_M + I_M) + (M_m + I_m)$$

This formula is a rearrangement of the expression:

$$\frac{I_m}{I_M} \times \frac{M_M}{M_m}$$

Or

$$\text{Tan}^{-1} a_N = \frac{I_m}{I_M} \tan^{-1} a_M$$

Where

$$a_M = \text{angle made by resultant moment with minor axis} = \tan^{-1} \frac{M_M}{M_m}$$

The angle between the principal axis and neutral axis is laid off by means of the above antitangent, a 10-inch leg being used. Manner of obtaining these angles is given in the left-hand end of sheet. Instead of drawing in the neutral axis and then drawing tangents that are parallel to the perimeter the tangents only are shown, the direction being found from the little scratch marks at the end of each

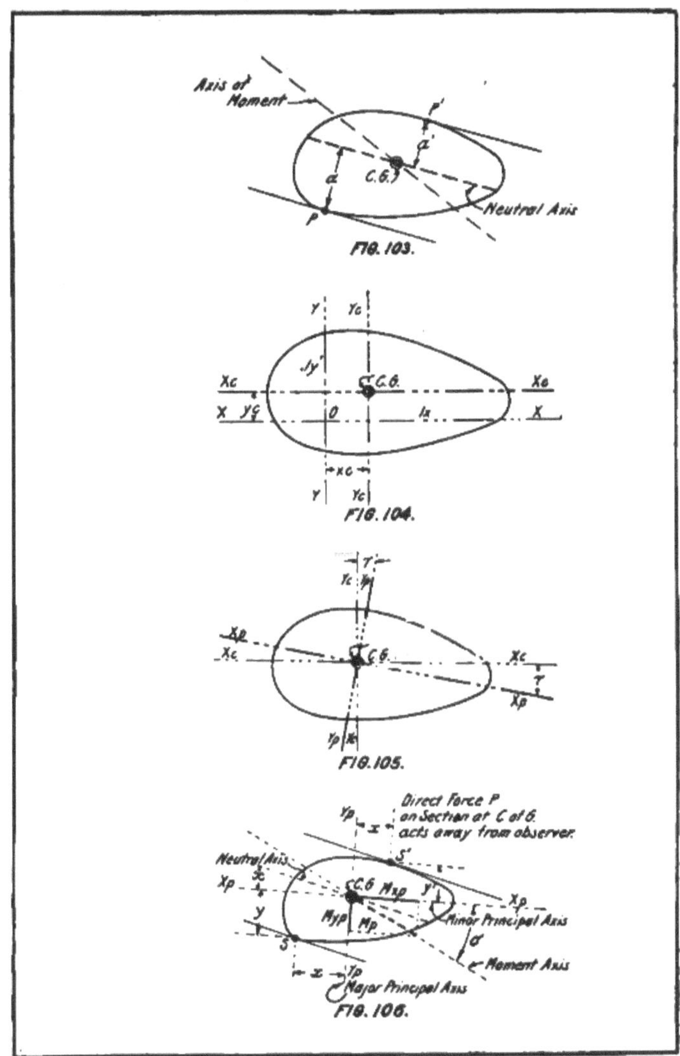


FIG. 103.

FIG. 104.

FIG. 105.

FIG. 106.

section. These points of tangency give the extreme distance from the neutral axis, and the coordinates of these points are measured with reference to the principal axes $M_m$ and $M_n$. These coordinates

161988—21——13

are given in the third section of the stress table. Large $M$ denotes measurement along the major axis and small $m$ along the minor axis. Large $U$ refers to a point of the upper camber, and large $L$ is a point in the lower camber. The arm diagram at the end shows the positive arm for a positive force on the section and will give a positive moment. These forces on the section have the same sign as the loads. In fact they are really the external forces broken up into a different arrangement when considered as actions. The reactions, of course, would have just the opposite signs, but in our present case we are considering the action on the section. It should be remembered in connection with these end views that we are looking in the direction opposite to that when looking at the other two views of the drawing, that is, in these cross sections we are looking away from the hub, while in the other views we are looking or work toward the hub.

The fourth division of Table IX gives the total loads along the axes. In the case of principal axes these are shears; in the case of the $r$ axis these are direct forces on the section.

TABLE IX.—*Propeller 34291.*

STRESS TABLE.

| Item. | Section considered. | | | | | | |
|---|---|---|---|---|---|---|---|
| | 18-inch. | 24-inch. | 30-inch. | 36-inch. | 42-inch. | 48-inch. | 54-inch. |
| FIRST DIVISION. | | | | | | | |
| $A$ | 18.40 | 15.45 | 11.96 | 9.50 | 6.70 | 4.28 | 2.14 |
| $I_m$ | 9.36 | 4.67 | 2.04 | 1.03 | .43 | .156 | .034 |
| $I_M$ | 84.5 | 84.3 | 69.5 | 53.9 | 32.6 | 15.1 | 4.38 |
| $M_m$ | −32,730 | −25,400 | −16,850 | −8,550 | −3,340 | −1.050 | −200 |
| $M_M$ | 4,950 | 4,950 | 2,750 | −320 | −1,960 | −1,540 | −310 |
| $M_{rs}$ | 392 | 75 | −175 | −236 | −167 | −58 | 5 |
| SECOND DIVISION. | | | | | | | |
| $M_m+I_m$ | −3,500 | −5,550 | −7,770 | −8,300 | −7,770 | −6,730 | −5,880 |
| $M_M+I_M$ | 58.6 | 58.7 | 39.6 | 5.93 | 59.8 | 102 | 70.7 |
| $\tan^{-1} a_M=(M_M+I_M)+(M_m+I_m)$ | −.0167 | −.0106 | −.0051 | .00072 | .0077 | .0152 | .0121 |
| THIRD DIVISION. | | | | | | | |
| $M_U$ | 1.58 | 1.18 | .93 | .73 | .58 | .41 | .29 |
| $M_L$ | −1.31 | −1.01 | −.79 | −.66 | −.49 | −.37 | −.25 |
| $m_U$ | .66 | .47 | .77 | .60 | .71 | .89 | .37 |
| $m_L$ | 1.83 | 2.43 | 3.51 | 4.26 | 3.95 | 3.37 | 2.53 |
| FOURTH DIVISION. | | | | | | | |
| $P_m$ | 1,400 | 1,000 | 613 | 300 | 106 | 0 | −3 |
| $P_M$ | 362 | 30 | −180 | −253 | −243 | −172 | −80 |
| $P_r$ | 21,530 | 17,410 | 13,090 | 8,970 | 5,400 | 2,570 | 670 |
| FIFTH DIVISION. | | | | | | | |
| $P_m+A$ | 76 | 65 | 51 | 32 | 16 | 0 | .6 |
| $P_M+A$ | 20 | 2 | −15 | −27 | −36 | −40 | −37 |
| SIXTH DIVISION. | | | | | | | |
| $P_r+A$ | 1,170 | 1,127 | 1,091 | 945 | 806 | 600 | 313 |
| $(M_m+I_m)\times M_U$ | −5,510 | −6,550 | −7,230 | −6,060 | −4,500 | −2,760 | −1,705 |
| $(M_m+I_m)\times M_L$ | 4,570 | 5,610 | 6,140 | 5,480 | 3,810 | 2,490 | 1,470 |
| $(M_m+I_m)\times m_U$ | 39 | 28 | 33 | −4 | −42 | −91 | −26 |
| $(M_m+I_m)\times m_L$ | 108 | 144 | 140 | −28 | −236 | −344 | −179 |
| SEVENTH DIVISION. | | | | | | | |
| $f_U$ | −4,301 | −5,395 | −6,106 | −5.119 | −3,736 | −2,251 | −1,418 |
| $f_L$ | 5,848 | 6,881 | 7,371 | 6,397 | 4,380 | 2,746 | 1,604 |

The fifth division gives the average shear along the two principal axes. The average direct load on cross section is given in the sixth division.

The unit stress from the moment about the $I_m$ axis is then found using moment about this axis obtained from the upper part of the table and the fiber distances obtained from the third section. This stress is combined with the stress found at this point from the moments about the $I_m$ axis and is given in the fourth item of this division of the table. To these stresses from these two sources is added the average unit stress on the section, and the total stress on the extreme fiber of the upper camber or the maximum stress on this side of the axis is obtained. The maximum fiber stress of the opposite kind is similarly obtained. Minus sign indicates compression and plus tension.

In the seventh division of the table are given the unit fiber stresses. No conclusion should be drawn from the unit fiber stresses here shown, as no allowance has been made for the deflection of the propeller. This reduces the moment from centrifugal force and consequently also the unit stresses. The figures in this stress table are to be considered as a first approximation. The final position of the center of gravity line is obtained by trial, and the operations for obtaining the unit stress are repeated.

The method of finding the section properties for any cross section is given below.

*Method of determining the point of maximum stress in any section.*—Figure 103 represents a moment acting normal to a given section and about the center of gravity and is indicated by its axis of rotation.

The neutral axis corresponding to this moment is also indicated.

The maximum stress is at point $p$, whose fiber is at the maximum distance $a$ from the neutral axis.

The point of maximum stress of the opposite kind will be $p^1$, if $a^1$ is the maximum distance of a fiber on the opposite side of the axis.

To obtain the neutral axis, use is made of the principal axes of the section which were gotten in the following manner:

The principal axes through any point of a section are a pair of rectangular axes about one of which the moment of inertia is a minimum (minor axis), and about the other it is a maximum (major axis).

Rectangular trial axes through any convenient point of the section are first chosen and the following properties are obtained:

(1) Area ($A$).

(2) Location of the center of gravity ($x_o$, $y_o$), figure 104, $x_o = \frac{M_y}{A}$.

(3) Moment of inertia of section about the $X$ $X$ axis ($I_x$).

(4) Moment of inertia of the section about the $Y$ $Y$ axis ($I_y$).

(5) Products of inertia of section about the pair of axes $X\ X$ and $Y\ Y$ ($K$). ($K = \Sigma xy.dA$.)

Next find the properties of the section about parallel axis through the center of gravity. These are obtained from the above by the following formulas:

$$I_{xc} = I_x - Ay^2_c \quad (1)$$

($x_c$ and $y_c$ referred to 0 as origin)

$$I_{yc} = I_y - Ax^2_c \quad (2)$$

$$K_c = K - Ax_c y_c \quad (3)$$

Then determine the angle between the $X_c$ and the $X_p$ axis. This is:

$$2\gamma = \tan^{-1}\frac{2K_c}{I_{yc} - I_{xc}} \text{ or } \gamma = \frac{1}{2}\tan^{-1}\frac{2K_c}{I_{yc} - I_{xc}}. \text{ (See fig. 105.)} \quad (4)$$

Minus values swinging clockwise from the $X_c$ axis and plus opposite.

The values of the moments of inertia about the principal axes are determined from the following formulas:

$$I_{xp} = I_{yc}\sin^2\gamma + I_{xc}\cos^2\gamma - 2K_c\cos\gamma\sin\gamma \quad (5)$$

$$I_{yp} = I_{yc}\cos^2\gamma + I_{xc}\sin^2\gamma + 2K_c\cos\gamma\sin\gamma \quad (6)$$

$$K_p = 0 \quad (7)$$

The neutral axis can now be determined with reference to the principal $X_p$ axis for a given moment axis, as follows. (See fig. 106.)

$$\lambda = \tan^{-1}\left(\frac{I_{xp}}{I_{yp}}\tan_\sigma\right) \quad (8)$$

$$\text{Where } \sigma = \tan^{-1}\frac{M_y}{M_x} \quad (8a)$$

The functions of the angles should be given their proper sign.

$M_{xp}$ and $M_{yp}$ are defined under equation 10, below.

Equations 1 to 8, inclusive, have been adopted from derivation given in standard works on mechanics.

The maximum fiber distance each side of the neutral axis is found by drawing parallels to this axis that are tangent to the perimeter, as in figure 106.

The coordinates of these points to the principal axes are measured. The theoretically correct stresses are given by the formulas below. The applied loads are behind the plane.

$$f_s = \frac{M_{xp}}{I_{xp}}y + \frac{M_{yp}}{I_{yp}}x - \frac{P}{A} \text{ for point } s. \quad (9)$$

$$f_s{}^1 = -\frac{M_{xp}}{I_{xp}}y' - \frac{M_{yp}}{I_{yp}}x' + \frac{P}{A} \text{ for point } s'. \quad (10)$$

FIG. 107.—AMSLER'S INTEGRATOR.

Where $M_{xp}$ is the component of the resultant moment around $X_p$ axis.
$M_{yp}$ is the component of the resultant moment around $Y_p$ axis.
$P$ is the direct force on the section.
$A$ is the area.

These two formulas are special for this example as regards their signs. If points $s$ or $s^1$ are in other quadrants, the signs will change accordingly. In this case also the direct force was taken as a tension.

As will be noted, moments are indicated by their axes of rotation, as textbooks on mechanics suggest, and not by their plane of turning.

The use of an integrator in connection with these calculations is explained below.

*Obtaining section properties with an integrating machine (description of machine and its use).*—The area of section, the location of its center of gravity, and the moment of inertia about any chosen axis are obtained by means of Amsler's integrator, an integrating machine which is shown in figure 107, page 199. Specifications for this machine are given in Keuffel & Esser Co.'s catalogue, page 319, in the thirty-fifth edition.

If such a machine should not be available, these properties may be obtained with a planimeter, as explained on page 206.

The operations are as follows for obtaining the center of gravity and moment of inertia with reference to any axis of a figure; also the area (see fig. 107):

(*a*) Set steel track parallel to the axis and back from it a distance determined by the two gage bars $G$.

(*b*) Place the machine $S$ on the track and insert counterweight $W$.

(*c*) Place the tracing point $P_1$ at any convenient starting point $O$, and take the reading of each of the three wheels $A$, $M$, and $J$ and place them in their respective columns on the line marked "Zero reading" on the form shown on page 208.

(*d*) Trace around the perimeter of the figure in a clockwise direction on back to the point $O$ and again read the wheels and place in line marked "First traverse."

(*e*) Subtract the "Zero reading" from the "First traverse" and get the net travel of wheels.

Net travel of wheel, $A = a$.

$$a \times 0.02 = \text{area in square inches.} \tag{11}$$

Net travel of wheel, $M = m$.

$$m \times 0.04 \text{ moment in inches.}^{3} \tag{12}$$

(or inch-pounds if 1 square inch of area is considered equal to 1 pound).

Net travel of wheel, $J = i$.

$$0.32\, a - \frac{i}{10} = \text{moment of inertia in inches.}^{4} \tag{13}$$

NOTE. -If tracing is done with point $P_2$ in place of $P_1$, the following factors are used:

Area $= a \times 0.01$ square inch.
Moment $= m \times 0.01$ inch.[3]
Inertia $= 0.04a - \frac{i}{80}$ square inch $x$ square inch.

(*f*) The center of gravity is obtained from:

$$\frac{m}{a} = y \text{ in inches.} \tag{14}$$

If $m$ is minus, $y$ is on the side of the axis away from the track.

(*g*) For another axis, such as one at right angles to $X$ $X$ of the

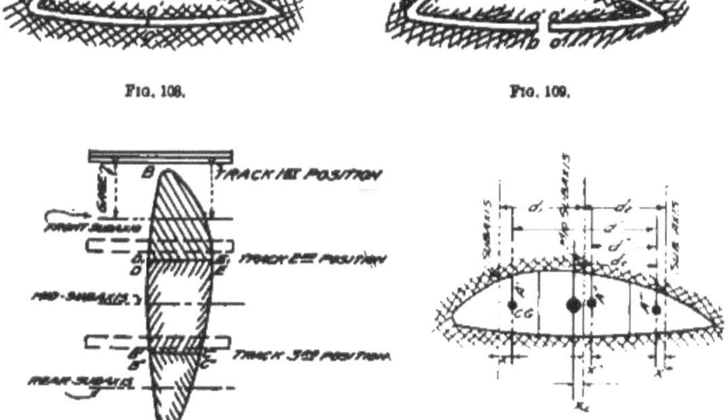

FIG. 108. FIG. 109.

FIG. 110. FIG. 111.

photograph, the same operations are repeated with the track parallel to and at the gage distance from this new axis.

*A hollow figure.*—For a hollow section, such as in figure 108, the tracing point is crossed over from $O$ to $O'$ after making the exterior traverse. The interior traverse is made anticlockwise around to $O'$ again and then back to $O$. An exaggeration of the notion here used is shown in figure 109.

*Figure beyond machine's reach.*—The point $P$ can only sweep 6.5 inches upward from the chosen axis, or 7.2 inches downward. When the figure is higher than 13.7 inches, it should be broken into two or more subfigures, each within the reach of the machine, as in figure 110.

These figures are treated as before, each having its own subaxis about which the moment and moment of inertia are found. The area is obtained along with these and the center of gravity can be then calculated. The area of the whole figure is (fig. 111):

$$A = A' + A'' + A''' \tag{15}$$

The center of gravity for the whole figure is: ($x'$, $x''$, and $x'''$ having already been found and $d_1$ and $d_2$ being known).

Taking moments about the center of gravity of $A'''$, figure 111:

$$d_0 = \frac{A'd' + A''d''}{A} \tag{16}$$

knowing which $x_0$ can be obtained.

The total moment of inertia of the whole figure about the axis through the center of gravity of the whole figure is:

$$I_{y_0} = I_y' = I_y'' + I_y''' + A'[(x_0')^2 - (x')^2] + A''[(x_0'')^2 - (x'')^2] + A''' [(x_0''')^2 - (x''')^2]. \tag{17}$$

The above formula is obtained from the following considerations:

(*a*) Inertia of subfigure about axis through point $O$ is (see fig. 112):

$$I_{y_0}' = I_{y_c}' + A'(x_0')^2.$$

(*b*) But $I_{y_c}' = I_y' - A'(x')^2$ because

$$I_y' = I_{y_c}' + A'(x')^2.$$

Therefore

(*c*) $I_{y_0}' = I_y' - A'(x')^2 + A'(x')^2 = I_y' + A'[(x_0')^2 - (x')^2]$.

For the three subfigures $I_{y_0} = I_{y_0}' + I_{y_0}'' + I_{y_0}'''$, which gives the formula of equation (17).

On page 151 is given calculations for two subfigures.

Where the figure is too long for traversing along the track, it may be broken up into subfigures, as in figure 113. In this case it is only necessary to add the corresponding items for each subfigure to get this item for the whole figure, thus:

$$A = A' + A'' + A'''$$
$$M = M' + M'' + M'''$$
$$I = I' + I'' + I'''$$

*Limits of accuracy of machine.*—It was found that for areas with a small moment of inertia about a chosen axis, e. g., 1 $\text{inch}^4$, the result from the machine might be a minus quantity. Experimentation with small rectangular sections proved that even with careful use of the tracing point no greater accuracy than 3 $\text{inches}^4$ might be expected.

To overcome this, the section was enlarged so as to give an $I$ of about 100, from which an accuracy of about 3 per cent might be obtained. This was further improved upon by making three traverses and taking the average as the correct quantity.

A reading is taken after the first traverse so as to provide a check for the average.

Where the figure might be too cumbersome when enlarged so as to have an $I$ of 100, a smaller $I$ may be used, such as 50, in which case six traverses would be made. In a hollow steel propeller the enlargement used for each section was:

| Section. | Enlargement. | Multiple of original $I$. | Enlarged $I$.[1] |
|---|---|---|---|
| 18 inches | Twice | $2^4 = 16$ | 177 |
| 30 inches | 3.5 times | $3.5^4 = 150.66$ | 154 |
| 42 inches | 3.5 times | $3.5^4 = 150.66$ | 40 |
| 54 inches | 7.5 | $7.5^4 = 3,164$ | 56 |

[1] Six traverses made.

After obtaining the properties for the enlarged section these are reduced for the original section by use of the following multiples, where $n$ is the ratio of the enlarged section to the original one.

For area .......... $\frac{1}{n^2}$

For coordinates of the center of gravity .......... $\frac{1}{n}$

For $I$ .......... $\frac{1}{n^4}$

For $K$ (products of inertia) .......... $\frac{1}{n^4}$

*Use of machine to obtain products of inertia.*—In developing the formula for finding the $I$ about any axis $X'\ X'$ from the known $I$ about any pair of rectangular axes through the point $O$ there appears the quantity $\Sigma xy \cdot dx \cdot dy$ or $\Sigma xy \cdot dA$, which has been called the products of inertia with respect to the axes $X\ X$ and $Y\ Y$, where $x$ and $y$ are the coordinates of $dA$ to these axes and is represented by the letter $K$. (See upper right-hand quadrant, fig. 114.)

A practical method to obtain this summation is to come along the $X\ X$ axis and at each 1 or 2 inch station get the *rate* of the "products" per running inch, thus:

Say that the strip taken in figure 114 is $\frac{1}{4}$ inch long and $h''$ high at point $x$, $y$.

The product for this strip would be: $\frac{1}{4}\ h \cdot x \cdot y$.

The rate per inch would be 4 times this, or $4(\frac{1}{4}\ h \cdot x \cdot y) = h\ x\ y$.

Note that whatever the length of strip taken the *rate* would have been the same, viz, that given by the height of the strip at the point $x$, $y$, multiplied by the product of the coordinates.

Plot these rates for the various stations assumed and draw a curve through the points, as on the left-hand side of figure 115.

Now, the products of inertia for a small segment of the section astride any station along the axis is obtained by multiplying the ordinate to the curve at this point (which is the rate per running inch) by the length of the segment. Thus at the 6-inch station in figure 115 the product will be represented by the area of the strip on that station. These strips could be taken contiguous and the sum of the products for the whole figure could be obtained by adding together all the strips under the curve for the whole figure. The greater the number of strips the closer will the result be to the

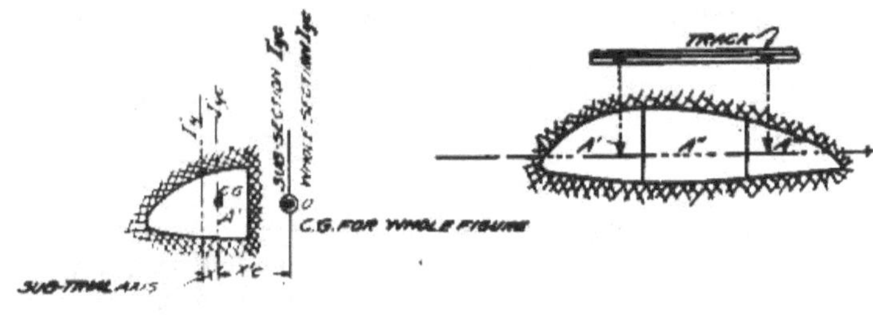


FIG. 112. FIG. 113.

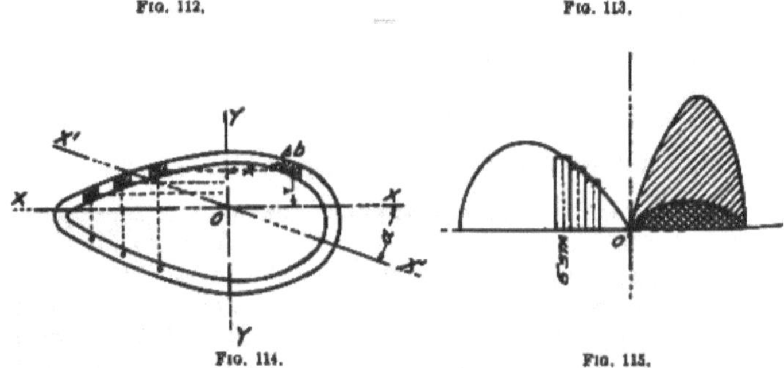


FIG. 114. FIG. 115.

actual sum. It can be seen that the area between the curve and the axis is the *exact* sum, or the products for the whole figure. The right half of figure 115 gives product for right half of figure 114.

Figure 116 is a representation of the curves found for a hollow section. Curve 1′ is for the part of the section in the upper right-hand quadrant marked 1. This curve has positive values. Curve 2′ has negative values. The difference between these two areas is the sum of the products for the two quadrants and will be the lighter-shaded portion. Similarly, the light-shaded portion on the left is the difference between the positive quantities given by curve 4′ and the negative quantities for curve 3′.

Start the planimeter at $O$ and run along curve 3′, going clockwise, then return to $O$ by means of 4′, then continue on to 1′, which will cause an anticlockwise traverse and return to $O$ by means of 2′. The net reading of the area will give the products for the whole figure

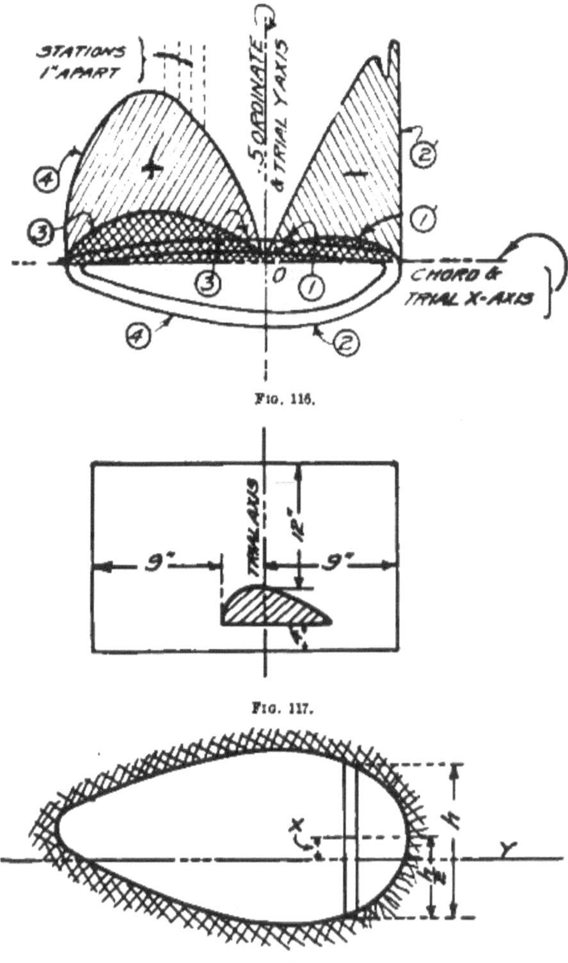


FIG. 116.

FIG. 117.

FIG. 118.

provided that the rates have been plotted to full scale, i. e., if the rate is 8 cubic inches per inch, the height to the curve would be made 8 inches (each inch of the height being understood to represent a cubic inch).

Where the curve would be too large, the scale may be reduced by dividing the rates of $K$ by some factor $n$. Then the final area obtained by the machine must be multiplied by this factor $n$.

*General notes on use of integrator.*—(*a*) The size of sheet outlined in figure 117 provides enough paper to keep the integrating wheels on the sheet and to get the sheet under the track and provides for $K$ curve.

(*b*) Care must be exercised to see that the paper has no tendency to creep when the tracing point is run around it. When thin paper is used thumb tacks should be placed near the figure with the paper stretched as much as possible.

(*c*) When the figure is large, it is often a question whether the thousandth dials have made one or more complete revolutions or whether their readings may be minus in place of plus. A convenient way to determine this is to make a rough traverse of the figure, starting the wheel in question at zero and watching its turning while the tracing point is roughly passing around the perimeter. A memorandum of the reading expected to the nearest hundredth is made at the bottom of the column of the wheel in question, as is shown in parentheses in the table on page 208. This is done for each wheel.

*Moments and inertias by means of a planimeter.*—Where an integrating machine is not available, a planimeter can be utilized for determining the moments and the $I$ of a section about a chosen axis if rate curves for these quantities are made as described above.

The rate for any element about $Y\ Y$, as shown in figure 118, is:

$$\text{For } M \text{ rate per inch} = \frac{\Delta b \cdot h \cdot x}{\Delta b} = h\ x$$

$$\text{For } I \text{ rate per inch} = \frac{\frac{1}{12}x\Delta b\ h^3 + \Delta b\ h\ x^2}{\Delta b} = \frac{h^3}{12} + h\ x^2$$

The rate areas for the moment should be plotted subtractive, like the products of inertia, figure 115. The rate areas for the $I$'s are all plus, and each area should be plotted in its respective quadrant, and the machine should run around the whole figure in a clockwise direction, thus adding together the areas of the four quadrants if the figure is hollow. For a solid figure curves will only appear in two quadrants.

Page 208 gives a form for tabulating the readings of the planimeter, while on page 208 the calculations for a section are carried out. These calculations were carried out for the 30-inch section of propeller, drawing X–8399. A detail of this section is given in figure 119.

24.2°
A
B
FIG. 119-A
TRACK
DETAIL OF 30" STATION
FROM DRAWING X-8399

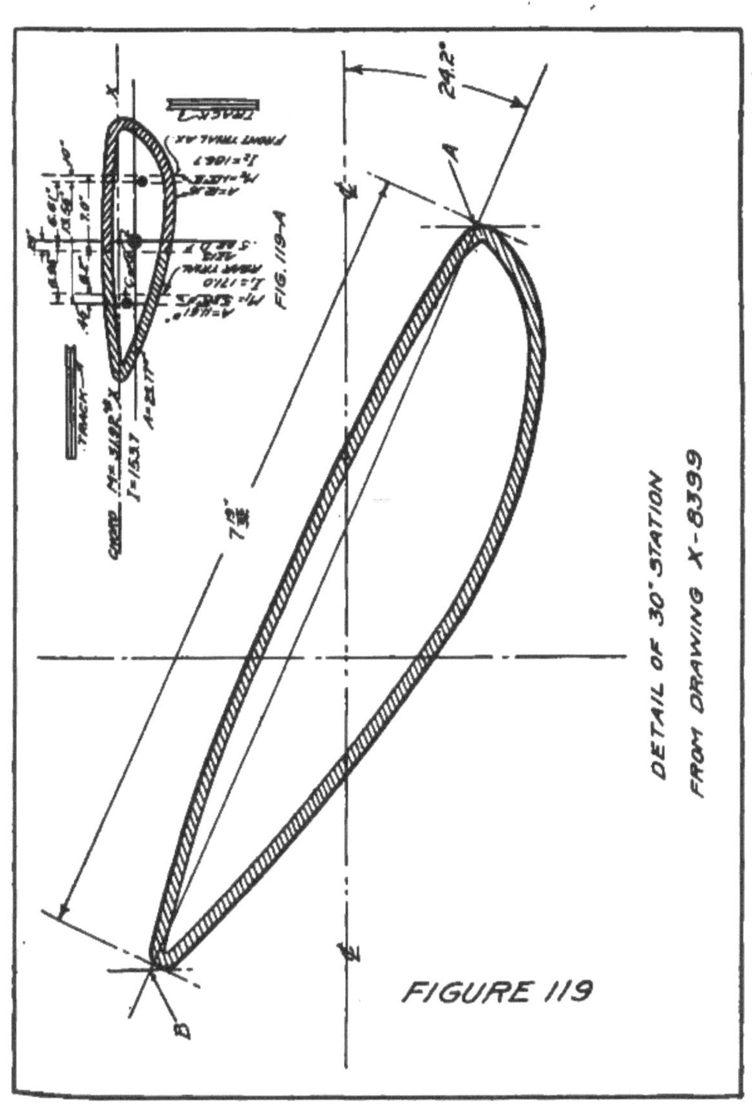

FIGURE 119

TABLE No. 5.—*30-inch section. Date for section properties using integrator.*

[Section 3.5 times that on X–8399. Zero ordinate at leading edge.]

$I_x$ Taken about chord. Track near flat face.

| Traverses. | $A$ | $a$ area $\times 0.02$ inches². | $M$ | $m$ moment $\times 0.04$ inches. | $J$ | $i$ inertia $\times 0.32 - \frac{i}{10}$ inches⁴. |
|---|---|---|---|---|---|---|
| 3 | 4314 | $\frac{1}{3}(3541\times 0.02)=23.61$.... | −7984 | $\frac{1}{3}(-2394\times 0.04)=-31.92$ | 7684 | $\frac{1}{3}\left(3541\times 0.32-\frac{6720}{10}\right)=153.7$ |
| 1 | 2351 | $1178\times 0.02=23.56$....... | −9582 | $-796\times 0.04=-31.84$.... | 3183 | $1178\times 0.32-\frac{2240}{10}=153.0$ |
| 0 | 1173 | ........................ | 0878 | ........................ | 0943 | |

Minus moment shows C. of G. on side of axis opposite track.

Partial $I_y$ Taken about subaxis 7 inches in front of 0.5 ord. Track near 0 ord.

| Traverses. | $A$ | $a$ | $M$ | $m$ | $J$ | $i$ |
|---|---|---|---|---|---|---|
| 3 | 1850 | $\frac{1}{3}(1825\times 0.02)=12.16$.... | 4050 | $\frac{1}{3}(-88\times 0.04)=-1.17$... | 0915 | $\frac{1}{3}\left(1825\times 0.32-\frac{239}{10}\right)=186.7$ |
| 1 | 0629 | $604\times 0.02=12.08$....... | 4117 | $-21\times 0.04=-0.84$...... | 0748 | $604\times 0.32-\frac{72}{10}=186.0$ |
| 0 | 0025 | ........................ | 4138 | ........................ | 0676 | |
| | | | (−100) | | (10) | |

Partial $I_y$ Taken about subaxis 6.2 inches behind 0.5 ord. Track near 0.5 ord.

| Traverses. | $A$ | $a$ | $M$ | $m$ | $J$ | $i$ |
|---|---|---|---|---|---|---|
| 3 | 2568 | $\frac{1}{3}(1741\times 0.02)=11.61$.... | −9669 | $\frac{1}{3}(-396\times 0.04)=-5.28$.. | 0553 | $\frac{1}{3}\left(1741\times 0.32-\frac{440}{10}\right)=171.0$ |
| 1 | 1413 | $586\times 0.02=11.02$....... | −9281 | $-134\times 0.04=-5.36$..... | 0269 | $586\times 0.32-\frac{156}{10}=171.9$ |
| 0 | 0827 | ........................ | 0065 | ........................ | 0113 | |

K Taken about chord and 0.4 ord. Curve heights give full rates of K.

| For areas within maximum reach of arm. | | | Remaining area. | |
|---|---|---|---|---|
| 3 | 8845 | $\frac{1}{3}(1947\times 0.02)=12.98$ | 447 | $\frac{1}{3}(-1458\times 0.02)=-9.72$ |
| 1 | 7547 | $649\times 0.02\times 12.98$ | 1419 | $-486\times 0.2=-9.72$ |
| 0 | 6898 | | 1905 | |
| | (760) | | | |

SAMPLE CALCULATIONS FOR SECTION.

*Properties of propeller, X–8399, 30-inch station.*

To secure accurate mechanical results, the section from which the readings were taken was increased 3.5 times that shown in figure 119.

(1) *Center of gravity.*—

$$y'=\frac{-31.92}{23.77}=-1.35''$$

Applying the reduction factor 3.5 to obtain the property for the actual section, we have—

$$y=\frac{-1.35}{3.5}=-0.38''$$

To obtain $x$—

$$\frac{-1.17}{12.16}=-0.10''; \quad \frac{-5.28}{11.61}=-0.46''; \quad \frac{12.16\times 13.56}{23.77}=6.95''$$

$$x'=(6.95''-6.66'')=0.29''$$

$$x=\frac{0.29}{3.5}=0.08$$

(2) *Area.*—

$$\text{Area}=\frac{23.77}{(3.5)^2}=1.94 \text{ in.}^2$$

(3) *Inertia about axis through C. G. and parallel to original axes.*—

$$\begin{aligned} I_{xo} &= 153.70-23.77(1.35)^2=153.70-43.25 \\ &= 110.45 \text{ in.}^4 \\ I_{yo} &= 171.0+11.61[(6.95)^2-(0.46)^2]+186.7+12.16[(6.61)^2-(0.10)^2] \\ &= 171+559.54+186.7+529.88 \\ &= 1447.12 \text{ in.}^4 \\ K_o &= (12.98-9.72)-23.77(-2.367)(-1.35) \\ &= 3.26-76.00 \\ &= -72.74 \text{ in.}^4 \end{aligned}$$

(4) *Inertia about principal axes.*—

$$\text{Angle } \gamma=\frac{1}{2}\tan^{-1}\frac{2(-72.74)}{1447-110}=-\frac{1}{2}\tan^{-1} 0.1085$$

$$=-\frac{1}{2}(6.2^\circ)=-3.1^\circ$$

$$\text{Sin}-3.1^\circ=-0.055$$
$$\text{Cos}-3.1^\circ=0.998$$

From equation (5), page 198—

$$\begin{aligned} I'_{xp} &= 1447(-0.055)^2+110(0.998)^2-2(-72.74)(0.998)(-0.55) \\ &= 4.38+109.8-8.0 \\ &= 106.18 \text{ in.}^4 \end{aligned}$$

$$I_{xp}=\frac{106.18}{3.5^4}=0.70 \text{ in.}^4$$

From equation (6), page 198—

$$\begin{aligned} I'_{yp} &= 1447(0.998)^2+110(-0.055)^2+2(-72.74)(0.998)(-0.055) \\ &= 1442+0.33+8.0 \\ &= 1450.33 \text{ in.}^4 \end{aligned}$$

$$I'_{yp}=\frac{1450.33}{3.5^4}=9.70 \text{ in.}^4$$

Angle $\gamma$ = (Minimum principal axis referred to vertical axis).
= Blade angle $-\gamma$
= $(-24.2° - 3.1°)$
= $-27.3°$

PROPELLER 34291.

*Nomenclature.*

$A$ *Area* of airfoil cross section.
$a_N$ *Angle* between *neutral* axis and *minor* principal axis.
$a_p$ *Angle* between *principal* axis and chord of airfoil.
$b$ *Blade* width of airfoil.
$b_e$ Distance of *center* of gravity from leading edge measured on *blade* width.
C. G. *Center* of *gravity* of airfoil.
$\Delta$ Denotes *increment*, as $\Delta T$, or increment of *thrust*.
$\Delta F$ Resultant centrifugal *force* produced by an element of the propeller.
$F_r$ Component of centrifugal *force* along *r axis*.
$F_t$ Component of centrifugal *force* along *t axis*.
$f_L$ Maximum normal *fiber* stress on *lower* camber face in any air-foil cross section.
$f_U$ Maximum normal *fiber* stress on *upper* camber face in any air-foil cross section.
$g$ Acceleration due to *gravity*.
$h_e$ *Height* of *center* of gravity above chord.
$h_L$ Maximum height of *lower* camber (measured from chord).
$h_U$ Maximum height of *upper* camber (measured from chord).
$I_M$ Moment of *inertia* about *major* axis.
$I_m$ Moment of *inertia* about *minor* axis.
$M$ *Moment* about any axis. $\Delta M$ is increment of moment.
$MM$ *On drawing* denotes *major* axis.
$M_L$ Ordinate of a point of the *lower* camber measured along *major* axis.
$M_U$ Ordinate of a point of the *upper* camber measured along *major* axis.
$M_M$ *Moment* about *major* axis.
$M_m$ *Moment* about *minor* axis.
$M_r$ *Moment* about *r axis*.
$M_{re}$ *Moment* about $r_e$ *axis*.
$M_s$ *Moment* about *s axis*.
$M_{se}$ *Moment* about $s_e$ *axis*.
$M_t$ *Moment* about *t axis*.
$M_{te}$ *Moment* about $t_e$ *axis*.
$m\,m$ *On drawing* denotes *minor* axis.
$m_L$ Ordinate of a point of the *lower* camber measured along the *minor* axis.
$m_U$ Ordinate of a point of the *upper* camber measured along the *minor* axis.
$P_M$ Total *lead* along *major* axis.
$P_m$ Total *lead* along *minor* axis.
$P_r$ Total *lead* along *r axis*; $\Delta P_r$ is increment $P_r$.
$P_s$ Total *lead* along *s axis*; $\Delta P_s$ is increment $P_s$.
$P_t$ Total *lead* along *t axis*; $\Delta P_t$ is increment $P_t$.
$Q$ Total *torque* on a section. $\Delta Q$ increment of torque coming onto propeller from element adjacent to a given station.
$r$ Distance measured along the *r axis* from any station in it as origin. When it is the *radius* the shaft is origin.
$r$ axis *Radial* axis coming out from the hub.

$r_o$ axis Through X of airfoil and parallel to *r axis*.

$s$ Distance measured along the *s axis* from the *r* axis.

$s$ axis Passes through *r* axis and parallel to *shaft* which rotates propeller. It may be located at any station.

$s_o$ Distance of X of airfoil section from *r* axis measured along *s axis*.

$s_o$ axis Through C. G. of airfoil and parallel to *s axis*.

$T$ Total *thrust* on a section. $\Delta T$ is increment of thrust coming onto propeller from element adjacent to a given station.

$t$ Distance measured along the *t axis* from *r* axis.

$t$ axis Passes through both the *r* and the *s* axis and is perpendicular to each. It is parallel to the *tangent* to the rotating disk at the end of the *r* axis.

$t_o$ Distance of X of airfoil section from *r* axis mentioned along the *t axis*.

$t_o$ axis Is through C. G. of airfoil and parallel to *t axis*.

$V$ Rotational *velocity* of any point in the propeller.

$W$ Weight.

$\Delta W$ Is *weight* of element of propeller adjacent to a given station.

## TORSION TESTS FOR DEFLECTIONS.

*Description of apparatus and its use.*—The term "twisting" will be used here in place of "torsion," so as to avoid confusion with the term "torque" used elsewhere in this manual.

The unit stresses arising from the twisting in a propeller designed along the usual lines will generally be small. However, the angular deflection, due to this twisting, may be large enough to be taken into account in the design.

Because of the small quantities involved, and also on account of the difficulties involved in the theory, it will be more satisfactory to determine these deflections experimentally.

A simple device has been built through which it is expected to obtain experience enough to design a more elaborate machine for standard tests. A drawing of this apparatus is shown in figure 120. Figure 121 shows the apparatus ready for use with a propeller in place. It consists of a wooden or concrete clamp gripping the propeller, and from the extended arm of this a load is applied by means of a cord which supports the primary loading bar. The load thus brought onto the clamp is counteracted by the tension in the upper cord, which hangs to the left side of the upper supporting beam. This beam is suspended from the ceiling, as shown. Figure 122 shows how the loads act in this arrangement.

The clamp suspension need not be over the center of the section. Figure 123 shows an assumed offset "$a$" of the suspension support. The formula shows that the moment is always the force multiplied by the distance between the forces of the couple:

$$Wa + W(1-a) = W1$$

Figure 122 shows the twisting moment to be:

$$\text{Twisting moment} = \frac{W}{2} \times 1$$

161988—21——14

In the present case 1 is 10 inches, so that:

$$\text{Twisting moment} = 10\ W.$$

The first propeller tested was the Smith steel propeller No. 6, made to propeller drawing No. X–8399. It is a hollow steel tube

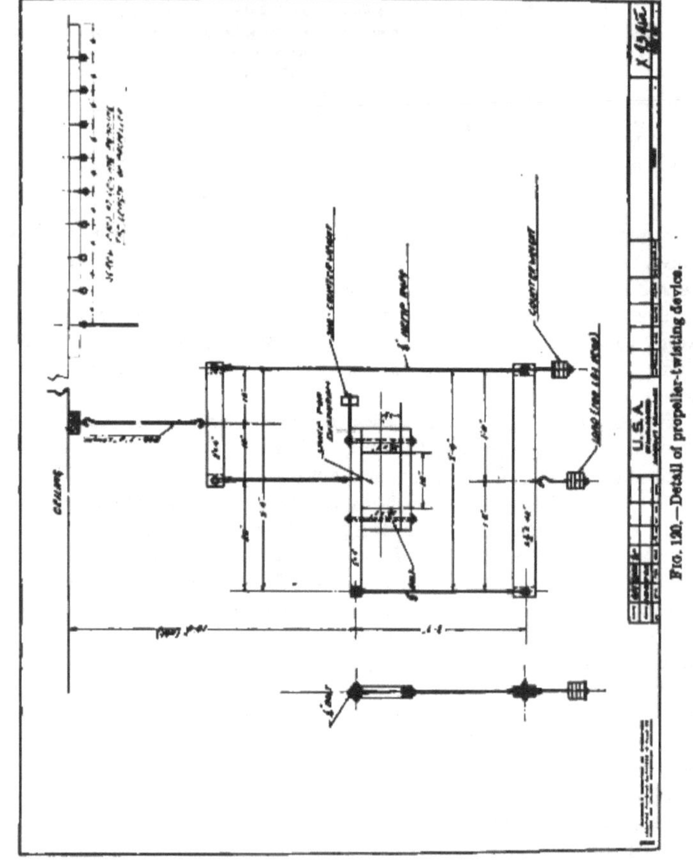


Fig. 120.—Detail of propeller-twisting device.

pressed into the shape of an air foil and is internally braced with a thin steel lattice bar.

A cement diaphragm cast into the clamp proved to be very satisfactory for this smooth steel surface. For this propeller it was only necessary to apply the load at the very end. A maximum twisting moment of 1,000 inch-pounds was applied and readings at every

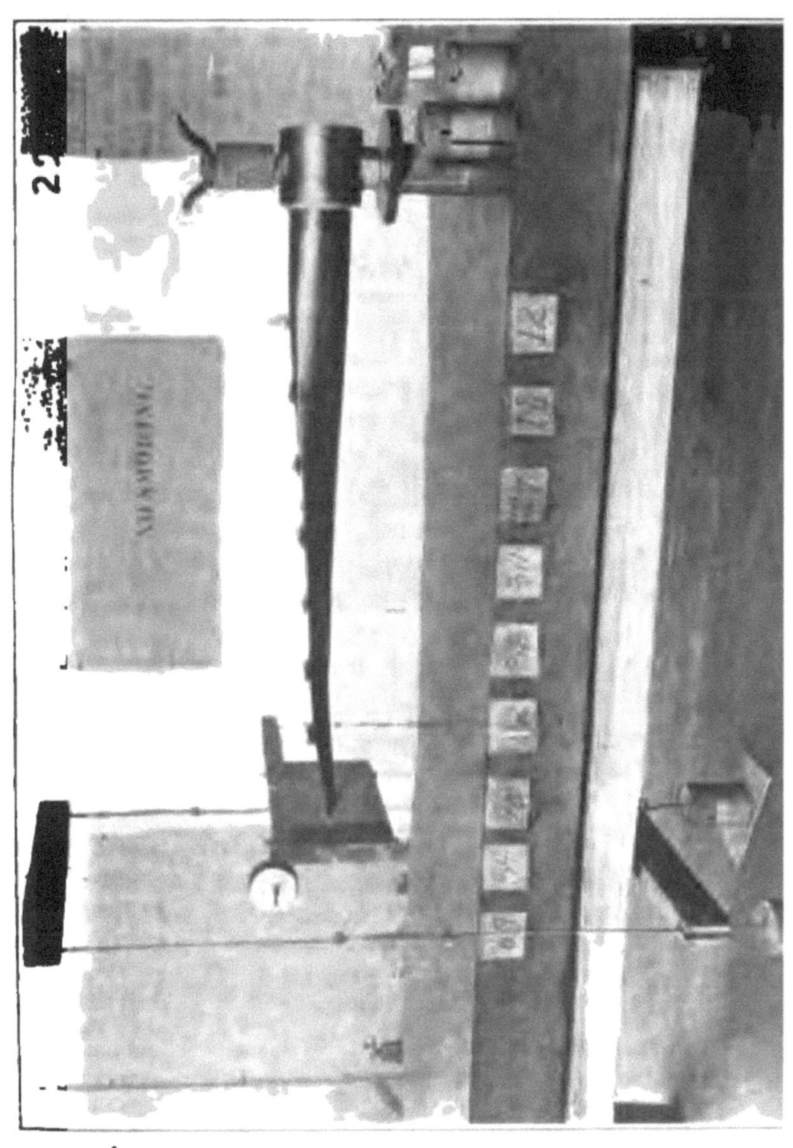

6-inch station were taken, zero readings being first obtained. The moment was then reduced by 100 inch-pounds and a second set of readings was taken. In this way the moment was successively

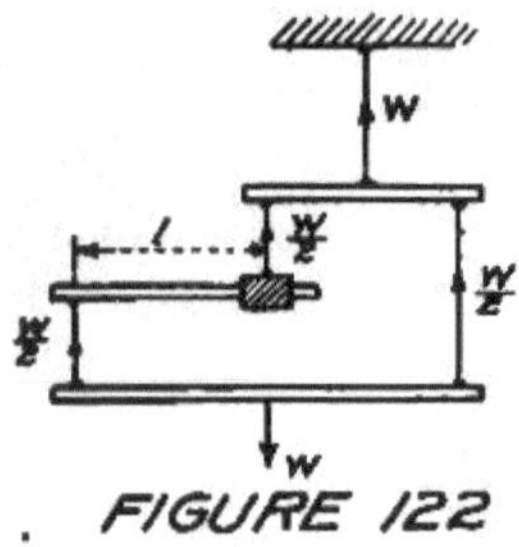


FIGURE 122

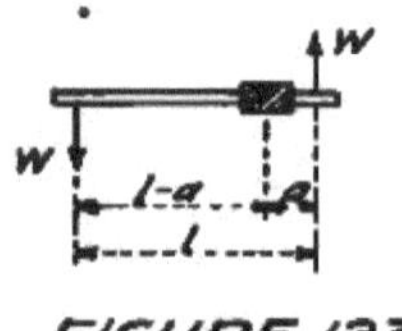


FIGURE 123

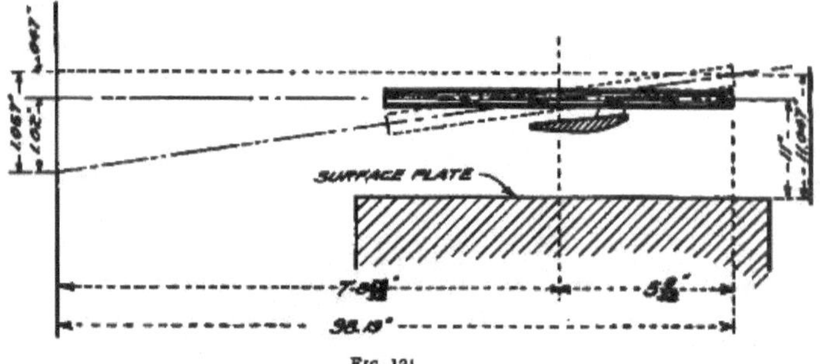


Fig. 124.

reduced down to the zero load, and readings were taken at all stations where deflections were apparent.

To measure the deflections, paper tubes about ⅜ inch in diameter by 4 inches long were set on the propeller in putty beds at the stations required. These beds were glued to the propeller by means of shellac. (See Fig. 121.)

A line of sight was obtained by punching a small pinhole through each closed end of the tubes. The screen on which the line of sight played was a strip of drawing paper tacked onto the wall and located about 8 feet from the center line of the propeller. The observer would sight through a tube and locate the point of a lead pencil operated by an assistant, and when the point appeared in the middle of the aperture a mark was made on the screen. The height above the table of the end of the tube toward the observer was measured. When twistings were applied the line of sight was again located on the screen and the new position of the end of the tube above the table was found. The angular deflection is then calculated, as shown in figure 124, which is for the 54-inch station when a twisting moment of 1,000 inch-pounds is applied at this point and stresses the propeller for its full length. As noted, the quantities involved are not large and any small inaccuracies in measurement for the sight end of the line will be negligible.

$$\text{Angle} = \tan^{-1} \frac{1.067}{98.19} = \tan^{-1} .0108 = 0° \; 37'$$

The sheet of curves for this test, figure 125, gives the deflections found at each station for the various twisting moments acting from the 54-inch station to the hub. From it the following table is obtained, which gives the total angular deflection at the various stations for the twisting moments as calculated in the stress analysis of this propeller. The increments are found as follows:

Suppose that 700 pounds twisting moment exists between stations 48 and 42. By subtracting the angular readings of station 42 from station 48 we find the angle through which the cross sections twist for the given moment acting on this length. This process is begun at the hub where the section is fixed and the increment of angle for the moment acting between the hub and the first station is found. Then the increment corresponding to the moment between the first station and the second station is found, and this is added to the deflection of the first station, thus obtaining the total deflection for the second station. This is continued throughout the length of the propeller.

*Smith steel propeller No. 6. Total deflections from calculated twisting moments.*

TEST No. 1.

| Item. | Distance out from shaft. | | | | | | | |
|---|---|---|---|---|---|---|---|---|
| | 12 inches. | 18 inches. | 24 inches. | 30 inches. | 36 inches. | 42 inches. | 48 inches. | 54 inches. |
| Moment on section | 1,017 | 934 | 839 | 737 | 607 | 451 | 310 | 149 |
| Increment of deflection | ........ | 0° 2′ | ½′ | 1′ | 3′ | 1½′ | 1′ | 1′ |
| Total deflection of section | ........ | 0° 2′ | 0° 2½′ | 0° 3½′ | 0° 6½′ | 6′ | 9′ | 0° 10′ |

Moments are in inch-pounds; deflections are in minutes.

## Propeller Tipping.

The most ordinary service demands that the tips of a propeller be made of or protected by material that has a very great resistance

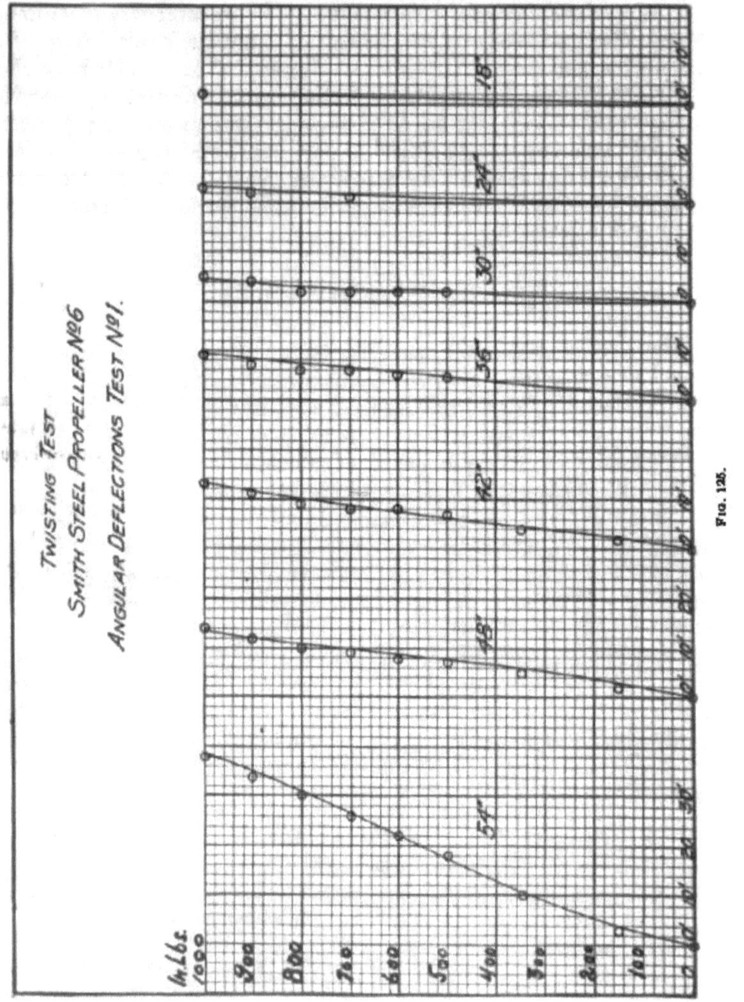


Fig. 125.

to the abrasive action of sand, small gravel, grass, sea spray, and rain. A 30-minute flight in a rainstorm will damage an unprotected wooden propeller beyond repair.

It is estimated that training planes are on the ground 20 to 25 per cent of the time their engines are running. This is much more ground service than any other type of machine is subjected to. The reasons for this are: Longer runs in taking off and in landing; many short practice flights to develop skill in landing; excessive taxing, due to the fact that beginners are required to land far out on the field away from hangars and other obstructions; and frequent emergency landings in rough fields, weeds, tall grass, etc., arising out of practice in cross-country flying. In taking off, landing, and taxying the indraft of the propeller picks up sand, gravel, sticks, grass, and other loose objects. These are struck by the rapidly revolving blades while passing through the plane of rotation.

The following calculation shows the severe usage imposed upon a rotating propeller when it comes into contact with these hard objects:

Consider a propeller having a 10-foot diameter and turning at the rate of 1,750 r. p. m. Such a one is used with the Liberty motor on the U. S. D-4 plane. The peripheral velocity of this blade is 917 feet per second, or 637 m. p. h. An ounce of material at the tip sets up a tensile stress, under the action of centrifugal force, of 326 pounds. Should the tip of this propeller *strike squarely* a pebble weighing 1 *ounce, 815 foot-pounds of energy* would be developed at the point of contact. Practically all of this energy must be absorbed by the propeller tip. Hence the necessity for an extremely durable propeller tip is obvious.

### POINTS TO BE CONSIDERED IN SPECIFYING A TIPPING MATERIAL.

*Weight.*—Lightness is always an important factor in any material entering into the construction of an airplane. Emphasis is laid on this quality in particular for propeller tipping because of the enormous centrifugal stresses due to the high r. p. m.

Uniformity of weight and thickness and homogeneity of texture are essential to the proper balancing of the propeller.

*Physical properties.*—The necessary physical properties vary with the kind of material and the proposed method of attachment. The resistance of the material to abrasion and its imperviousness to moisture are prime considerations. Sheet copper, brass, and steel have proved very satisfactory from these standpoints.

*Application.*—The relative merits of tipping materials are largely dependent upon their adaptability to efficient methods of application and upon the facility with which repairs can be made to damaged tips.

### DESCRIPTION AND RELATIVE MERITS OF STANDARD MATERIALS AND METHODS OF TIPPING.

*Sheet copper, brass, and steel.*—Specifications Nos. 11043A and 11040 cover, respectively, the requirements for sheet brass and copper for use in tipping propellers. Specifications for steel to be used in tipping propellers have not been issued by the Bureau of Aircraft Production. However, steel has been used with very good success on the Olmsted propeller. For instructions covering the application of metal tips see Specification 29500D, paragraphs 37 to 40, inclusive.

It is the practice to fasten these materials to the tips with nails, rivets, or screws soldered into place. Great care must be exercised in placing these fasteners so that consecutive ones do not come in line with the grain of the wood. This is to prevent splitting. At the same time the rivets must be placed so as not to destroy the balance of the propeller and at a sufficient number of points to hold the tips in place against the high centrifugal stresses. It is obvious that a template can not be used for locating and drilling the screw holes. This makes the application of *each* tip a difficult and *special* operation. The acid used in soldering the heads of the fasteners to the tip is always a possible source of danger. It is liable to get between the metal and the wood in sufficient quantity to deteriorate the metal or the surrounding wood. Should this occur, the rivets, and possibly the whole tip, would become loose. Danger from this source is greater when the propellers are stored for a long period before being put into service.

Considerable difficulty has been encountered with sheet-metal tipping due to failure caused by fatigue in the metal from blade fluttering. Parts of the tips have let go from this cause, resulting in accidents of a more or less serious nature.

Illustrations of failures in metal tips, due to fluttering, are given in figures 83 and 85.

An illustration of the protection afforded by a brass tip to a wooden blade against the abrasive action of spray or rain is given in figure 84.

Aside from the objections as pointed out, sheet-metal tips are very satisfactory in service.

*Pigskin.*—Leather tanned from pigskin in accordance with Specification No. 24109A has been used with some success as a tipping material. The hides are oak-barked tanned. The use of a non-drying oil or acid during the process of tanning or rinsing is prohibited, as the presence of oil in the leather will destroy its gluing qualities.

The method of applying the leather to the propeller tips is described in Specification No. 24110A. Considerable skill is required to properly apply these tips, and the method is wasteful of leather.

*Linen and cotton fabric tips.*—Specifications 16004A and 16003A give the requirements for cotton and linen tipping fabric, respectively. The method of applying these materials is standardized and is covered in Specification No. 29500D, paragraphs 33 to 36, inclusive.

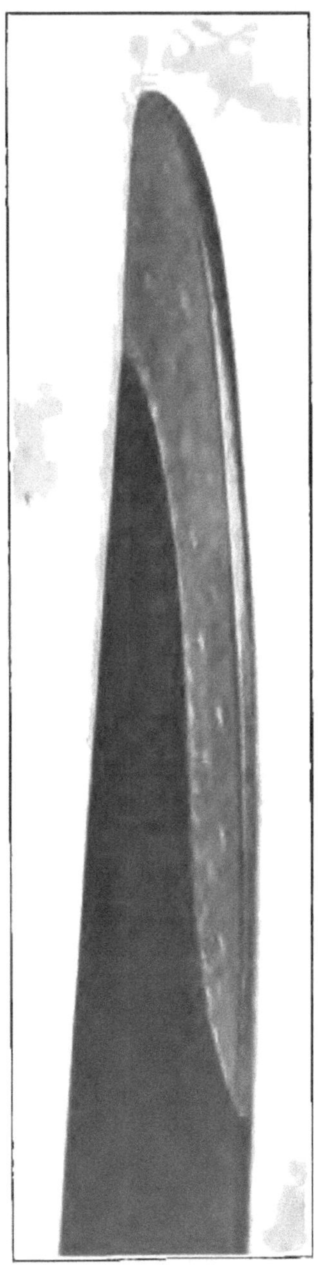

FIG. 126.—ELECTROLYTICALLY DEPOSITED COPPER TIP.

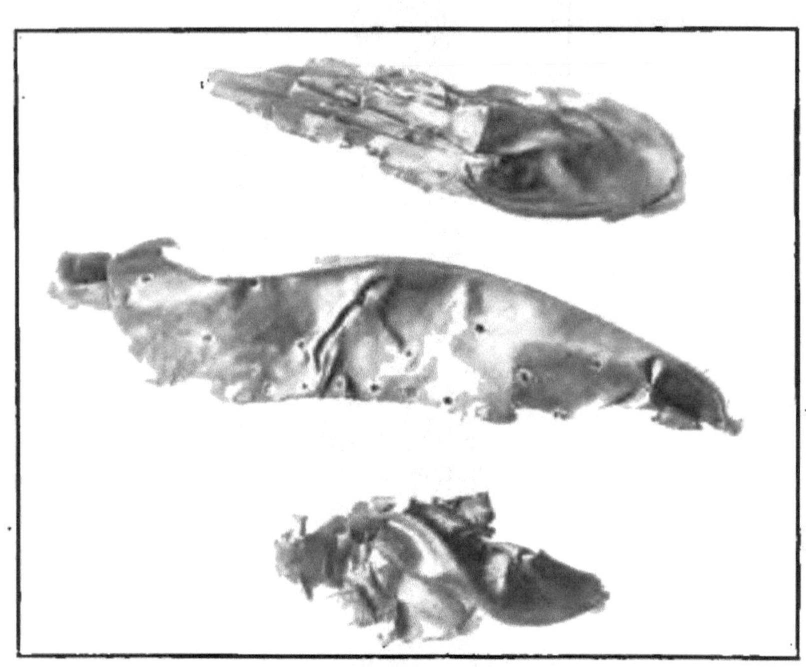

FIG. 127.—ELECTROLYTICALLY DEPOSITED COPPER TIPS AFTER D. W. T. NO. 19.

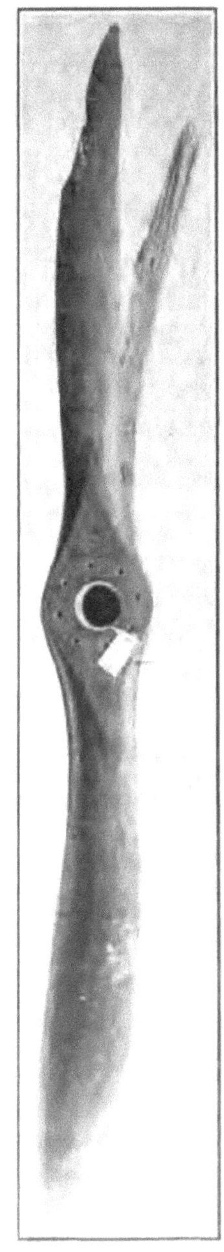

FIG. 128.—PARAGON OAK PROPELLER AFTER ELECTROLYTICALLY DEPOSITED COPPER TIPS FAILED IN D. W. T. NO. 19.

These materials protect a propeller against ordinary wear to an appreciable extent. They are easily and cheaply applied.

*Comparisons.*—From the standpoint of service the sheet-metal tips give by far the best results on small diameter propellers of low r. p. m. and have the longest life under severe usage. They are, however, the most difficult to apply. A damaged sheet-metal tip can not be repaired except by replacing the entire covering. A repair job is difficult, as the old rivet or screw holes can not be used the second time and the addition of new holes reduces the strength of the propeller blade. Also the new rivets are liable to be placed close to the old holes, thus allowing the tip to work loose.

Pigskin-leather tips afford slightly better protection than linen tips but are much more difficult to apply. Their application requires as much skill as that of the sheet-metal tips.

Linen tips are by far the easiest to attach. They afford an appreciable amount of protection to the wood and are easily and quickly repaired when damaged. They can also be used on the high-speed propellers with perfect safety.

For these reasons this type of tip has been the most commonly and universally used. For comparative tests of these types of tips on their resistance to abrasive action of water spray see Water-spray abrasion tests, Chapter V.

*Experimental types of tips.*—Having described the tips which are in common use, we will now take up those which have been experimented with in an attempt to develop a more durable and satisfactory protection.

A number of materials and methods have been experimented with, some of which show good results.

*Electrolytically deposited copper tips.*—These tips were applied as follows: First, brass screws were inserted into the wood in the area to be electrolytically plated. These screws were to anchor the tips firmly to the wood. Next the area to be covered was painted with a substance conductive to electricity. The copper was then deposited by the ordinary electroplating process. A white oak paragon propeller designed for a *JN-4* plane was used for the experiment. The appearance of the finished tip is shown in figure 126. This propeller was given a destructive whirling test during which the tips were thrown off under the action of centrifugal force. Figure 127 shows the tips after this test and figure 128 shows the condition of the blades after the tips were thrown off.

Electrolytically deposited copper is very soft and of low tensile strength. This accounts for the tips being thrown off. These tips showed such unpromising results that further experiments were discontinued.

*Fabric attached with blood albumen glue.*—The use of blood albumen glue in the manufacture of plywood suggested the possibility of using this glue to fasten fabric-tipping material to propeller blades. This glue is waterproof, a decided advantage for this purpose. Another feature of this glue which appeared to be an advantage for propeller tipping is the method of applying. The glue is made up in a liquid of about the consistency of thick cream. This is then applied to one side of the fabric-tipping material and allowed to dry, when the glue is again brought into the liquid state by the application of heat, which process also produces a chemical change, leaving the glue insoluble in water.

Specifications for applying this glue to fabric and for tipping propellers with the glue-coated fabric are given below.

*Propeller section experimental specification for coating fabric with blood albumen glue.*—This specification is based on experiments made by the propeller section of McCook Field.

*General.*—This specification covers the quality of the material, workmanship and final product.

## MATERIAL.

(1) *Fabric.*—The fabric shall be either cotton or linen. Linen shall conform to Signal Corps Specification No. 16001A.

Cotton shall conform to Signal Corps Specification No. 16002.

(2) *Glue.*—The glue shall be blood albumen glue prepared in accordance with Propeller Section Experimental Specification.

*Finished product.*—(1) The fabric shall be glue coated on one side only. The glue coat must be free from lumps and of a uniform thickness. The weight or thickness of the glue coat shall be such as will require 1 gallon of the glue mixture for 10 square yards.

(2) Uneven or patchy coating will not be allowed. The glue coat must cover the entire width of the cloth.

(3) When exposed to ordinary atmospheric conditions the glue coat shall not peel off the fabric.

## PROPELLER SECTION EXPERIMENTAL SPECIFICATION FOR SHEATHING PROPELLERS WITH BLOOD ALBUMEN GLUE-COATED FABRIC.

This specification is based on experiments made by the propeller section of McCook Field.

*General.*—This specification covers the method of applying specially prepared blood albumen glue-coated fabric. The operations are very simple and easily done. The tools required are a sharp knife or a pair of scissors and a heavy electric iron.

*Material.*—The sheathing fabric shall be coated with black blood albumen glue in accordance with propeller section experimental specification.

*Process.*—The surface to be sheathed shall be clean and free from dust and other foreign substances. Immediately before the fabric is applied, the surface of the wood shall be *slightly* dampened by rubbing with a moist cloth or sponge. Care must be taken not to apply too much water. The wood must be dampened, *not wet*. An excess of water at this stage of the operation will soften the glue too much and cause it to penetrate entirely through the fabric. Very little of it will be pressed into the pores of the wood, consequently it will not adhere properly.

After the surface to be covered is dampened, the fabric is laid into the position with the glue side next to the wood. The glue is now set with pressure and heat applied by a hot iron. A 12-pound tailor's electric iron has proven satisfactory for this use. Particular attention must be paid to the heat of the iron. Too much heat will ruin the glue, causing it to blister and lose its adhesive properties. Too little heat will not "set" it sufficiently to cause it to become waterproof. The temperature at which the glue sets properly is about 176° F. It has been found that the iron should be maintained at a temperature that will barely produce a sizzle when a wetted finger is applied to its surface.

The ironing operation must be thorough. It takes about five minutes at 176° F. for the glue to set properly. The setting process consists of a chemical change in the glue. After this takes place the glue is insoluble in water. The best results have been obtained by holding the iron stationary on the cambered surface and rocking it back and forth very slowly, at the same time applying pressure. The rocking can not be carried out on the flat side of the blade; instead, the iron is used in the regular manner but moved very slowly.

The ironing operation should be of sufficient duration to keep the glue at a temperature of approximately 176° F. continuously for five minutes.

Considerable pressure should be applied. It is impossible to exert too much by the hand operation as described. It is necessary to press the glue, softened by the heat, into the pores of the wood and fibers of the cloth.

After the operation is completed on the back side of the blade, as described above, the fabric is trimmed away along the leading edge and around the tip end so that it comes flush with the working face of the blade. The working face is now dampened and the fabric stretched around the trailing edge, and onto the working face, where it is secured with the hot iron. The operation is completed by making a ¾ to 1 inch lap onto the cambered side of the blade along the leading edge at the tip. Where it is necessary to lap the fabric back over sharp radii, as at the tips, the width of the lap may be reduced to prevent folds or the fabric may be V notched. Before attempting to set the lap joints, the first layer of fabric should be slightly dampened with a sponge in the same manner as described for sponging the wood.

After allowing to condition for two hours, the tip is ready for the finish.

*Finishing.*—The uncovered portions of the blade shall be finished in the regular manner by the aluminum leaf process. The first of the finished coats applied to the fabric sheathing shall be a liquid substance having water-resisting qualities. It shall thoroughly penetrate the fibers of the cloth.

(1) Apply two coats of orange shellac. The first of these coats should be thinned with alcohol to secure good penetration of the fabric. Sand lightly between each coat and allow each to dry for two hours before applying the next.

(2) Apply a coat of sizing varnish and aluminum leaf.

(3) Apply one coat of gray enamel.

(4) Apply one coat of spar varnish.

Several specimens were made up in accordance with these specifications. These specimens were placed in water at a temperature of about 65° F. for a period of 16 days. At the end of this time the glue joints were in good condition. One full-sized propeller has been tipped with this material. No tests have been made on this propeller. However, there is no doubt but that the glue is sufficient to hold the tipping in place. Due to the insolubility of the glue, a greater protection will probably be afforded a tip running in a rainstorm than is the case with the ordinary linen tip attached with hide glue.

The ease with which repairs can be attached is a point in favor of this material.

As to its resistance against abrasion, no more protection is hoped for than is usual with ordinary fabric tips.

*Fabric attached with casein glue.*—These tips are attached with casein glue in much the same manner as is employed in attaching fabric tips with hide glue. The advantages of this method are the waterproof qualities of the glue and the fact that the glue can be worked cold.

Tests were made on fabric impregnated with casein glue to determine what deteriorative effect would develop in the fabric from the action of lime in the glue. The results showed that any effects were negligible for this purpose.

A propeller tipped by this process was given a water spray test. See abrasive tests, Chapter V. The results were very good and compare favorably with tests on other fabric tips.

*Rubberized fabric tips.*—These tips were built up of fabric and raw rubber and vulcanized by heating before applying to the propeller blade. The process was similar to the manufacture of an automobile tire.

First a tip form was built. This was an exact duplicate of the propeller blade on which the finished propeller tips were to be used. This form was used to build the tip around and as a mold during the vulcanizing. High-grade cotton fabric, of automobile tire stock, weighing about 10 ounces per square yard, was friction coated on one side with raw rubber. This rubber-coated fabric was then built to the desired tip shape around the form. The uncoated side of the fabric was placed next to the mold. A triple thickness was applied along the leading edge. Figure 129 is a cross section through the tip near the extreme end taken at *A A*, figure 130.

The fabric was placed around the form on the bias so as to allow the finished tip to stretch easily in any direction. This makes it adjustable to any slight differences in shape which might exist between the fabric tips and propeller blades built to the same design.

After being built to the proper shape on the forms, the tips were wrapped tightly with tape to hold the fabric in position during the

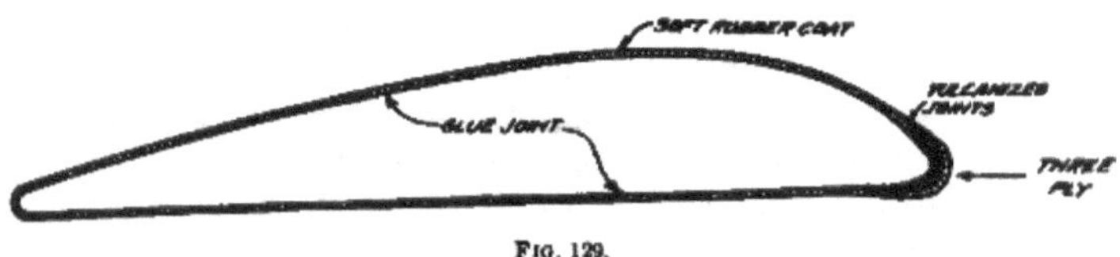

FIG. 129.

curing process. They were cured in live steam under a pressure of 60 pounds, which corresponds to a temperature of about 292° F.

After this process the tips were turned wrong side out and the cloth surface was buffed to remove all rubber particles from the fabric side and prepare it for gluing. This completed the process of manufacture. The finished tips are shown in figure 130.

The process of attaching the tips to the propeller blades was very simple. Hide glue was prepared in the usual way and applied to both the inside of the tips and to the propeller blades. The tips were then slipped into place. The glue was kept in a fluid state by heat from a warm iron. The iron was also used to work out the excess glue.

After allowing the glue to set for 24 hours the tips were given two coats of orange shellac. This completed their application. These tips were given a standard water-spray test, the results of which are given in Chapter V.

Compared with the usual linen or cotton fabric tips, these tips give slightly better results under the action of abrasion. However, this

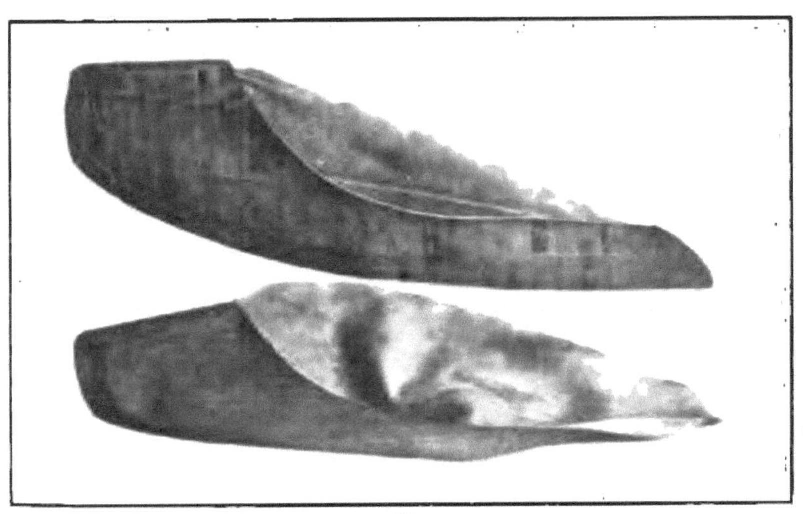

FIG. 130.—RUBBERIZED FABRIC TIPS BEFORE APPLYING TO PROPELLER.

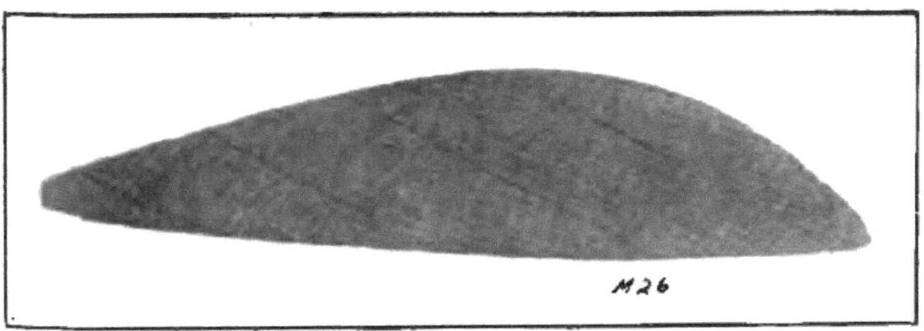


FIG. 131.—SECTION CUT FROM BRUNSWICK-BALKE RUBBER-COVERED PROPELLER, SHOWING OPEN GLUE JOINTS. RUBBER COVERING REMOVED.

advantage is small. An improvement might be effected by using a harder rubber. Other material, namely, rawhide, shows so much better results that further experiments with the rubber tips was discontinued.

*Rawhide tips.*—The hardness and toughness of rawhide and its use in the manufacture of gear wheels suggested its possibilities as a tipping material. Accordingly, samples of rawhide were made up from calfskin, cowhide, pigskin, and horsehide taken from the buttocks where cordivan leather is obtained.

Sample tips were made up and subjected to a water-soaking test which showed good results. This test demonstrated the superiority of casein glue over hide glue for attaching.

The method of attaching these tips is given in the tentative specification below.

One propeller tipped in accordance with this method, using calfskin, was given a water-spray test. The results of this test showed that this tip has much greater resistance to abrasive action than any others tested excepting those of metal and of hard rubber. (See fig. 88.)

## PROPELLER SECTION EXPERIMENTAL SPECIFICATION FOR TIPPING AIRPLANE PROPELLERS WITH RAWHIDE.

### MATERIALS.

*Rawhide.*—The rawhides shall be obtained from cattle; they shall be free from holes, cuts, and other imperfections.

*Glue.*—"Certus" casein glue, or its equivalent, shall be used.

*Shellac.*—Signal Corps Specification No. 14012.

*Varnish.*—Signal Corps Specification No. 14001C.

### PROCESS.

*Preparation of surface.*—The surface to be tipped shall be free from grease and dirt.

Since the casein glue discolors mahogany and some other woods it is necessary to fill and shellac the blade adjacent to the tip area to prevent the glue from coming into contact with the wood.

*Preparation and application of skin.*—The dry rawhide shall be soaked in cold water for about two hours to render it soft and pliable. After soaking, all excess or free water shall be thoroughly wiped off. The skin is then ready to be applied to the propeller.

Much of the success of the finished work depends upon the amount of water absorbed by the leather during the soaking. Too much soaking saturates the skin so that later in the tipping process this water is given up to the glue, making it too thin.

The skin should be soaked only long enough to make it pliable.

Glue is applied to the cambered side of the blade first. The skin is stretched and formed over this surface and the excess glue squeezed out toward the edges by working from the center outward with a brad stick.

The skin is now trimmed off about ⅛ inch from the entering edge and around the tip end. This edge shall now be trimmed or skived to a "feather edge" and thoroughly fastened to the pressure face of the blade.

The application of the skin to the pressure face is identical with the method just described. The process is finished by lapping the skin over the leading edge and tip ends. Glue is applied to the lap.

Care must be exercised in stretching the skin around the sharp curvature at the tips to avoid folds.

The tips are now allowed to dry thoroughly, which will require about 24 hours.. After drying the excess skin is carefully trimmed away, leaving a lap of ¾ to 1 inch wide. The joint is now pared off with a cabinet scraper and fine sandpaper. The tip is now ready for the finish.

*Finish.*—(1) Apply two coats of orange shellac. Allow each coat to dry for three hours. Sand lightly after each coat.

(2) Apply two coats of battleship gray enamel. Allow each coat to dry for 24 hours. The last coat should be a flowing coat.

*Hard rubber tips.*—Several propellers, covered entirely with a coating of hard black rubber approximately ⅛ inch thick and vulcanized in place, were built by the Brunswick-Balke-Collender Co., of Chicago.

The composition of the hard rubber used in this method has not been divulged by the manufacturer; neither is much known pertaining to their method of manufacture. However, it is supposed that the vulcanizing process has been necessary in connection with attaching the rubber covering. This process would undoubtedly be ruinous to the glued joints between the wooden laminations on account of the high heat necessary to vulcanize the rubber. Also the moisture content of the wood would necessarily be reduced to a very low figure.

A section cut from one of these propellers, after it had been subjected to the water-spray abrasive test, is shown in figure 131. The open glue joints are very apparent here. These open joints extended from the tip to within 12 inches of the center.

Since these propellers stood up very well when subjected to the destructive whirling test, it is believed that the rubber covering furnished sufficient strength to hold the laminations in place, in spite of the impaired glued joints.

This method of protecting propellers looks promising. This material showed very good results under the action of the water spray. For details of these tests, see Water-spray Abrasive Test, Chapter V.

## MICARTA PROPELLERS.

*Definition.*—At the date of this writing Micarta propellers are still in the experimental stage. Consequently little is known of them except among those who have been working with them. They are most frequently heard of as "Bakelite" propellers, this term being derived from the name of the binder used in the manufacture of "micarta," which latter name is applied to the composite material of which the propellers are made. Since bakelite is used as a binder in other composition materials which can not be satisfactorily employed as propeller materials, and which also go by the name of "bakelite", the more specific term "micarta" will be used herein.

Micarta, therefore, refers to the product resulting from the proper application of heat and pressure to bakelized paper, or bakelized cloth, such as duck, cotton sheeting, cambric, etc., or combinations of paper and cloth.

Micarta propellers are a molded product, made up of layers or laminations of duck or paper previously bakelized and baked under pressure to produce a homogeneous and uniformly compact material structure. The baking temperature is in the neighborhood of 200° C., and the pressure about 800 to 1,000 pounds per square inch, the latter being a little higher than necessary. They are completely formed when they come from the mold, it being only necessary to remove the overflow burs of excess bakelite and dress the corners and edges.

*Historical.*—There were a number of reasons that prompted the experiments with micarta as a propeller material, and these may be divided into two general classes. The first were fundamentally urgent or primary reasons, which were responsible for the inauguration of the research in this field. The second were consequential or secondary reasons, which grew out of the experience gained during the course of the investigations.

A great deal of trouble was being experienced with wooden propellers undergoing considerable changes due to changes of climatic conditions encountered in the course of transportation and storage. Since micarta is little affected by climatic changes, it was thought that many of the troubles experienced with wooden propellers in this respect might be eliminated with its use.

The possible scarcity of good propeller woods in the future prompted the early consideration of substitutes. The enormous supply of the raw materials for the production of micarta firmly secured its future. This would conserve other propeller materials for uses to which micarta could not be put.

A material as homogeneous, as reliable in its structure, and giving so uniform a product was quite desirable. With woods this is not the case. The most extreme care must be exercised in its selection, attended with large waste, to insure even a comparatively uniform and reliable product.

It appeared that after once being established the manufacture of micarta propellers would lend itself to speed production much more readily than of wooden propellers. It would further insure a more uniform and reliable product with the elimination to a high degree of the uncertain natural and human elements.

It was thought possible to make up a propeller with some sort of light core, such as cork or light wood, or even hollow. If this could be successfully accomplished, it would mean a great reduction in the weight, because the working material would be distributed to the outer fibers, where it would be more effective in strength production and the region of the neutral axis would be filled with a very light material or be made entirely hollow.

As a material, micarta is very durable and capable of withstanding a lot of punishment.

Micarta being almost exclusively a Westinghouse Electric & Manufacturing Co. product, and since this company was the most experienced in its production and the most completely equipped for handling such an enormous molding proposition, the lot fell to them to develop the materials and processes of manufacture.

After numerous experiments it proved most feasible to first direct attention to the construction of *solid* micarta propellers. The promising possibilities developing out of the experiments in this direction, and the fact that their real values could only be determined by trials, seemed to fully justify a continuation of this work.

As can be seen by looking at the end of the grain, that of micarta may be defined as planimetric and that of wood as lineal. To point out more clearly the difference, it may be said that the first is somewhat analogous to a stack of cards which can only be split in planes parallel to those of the cards; and the latter is similar to a bundle of fine wires which can be split in any plane that is parallel to the axis of the bundle. On the basis of this feature of micarta, the use of thinner and more efficient blade sections than could be safely allowed with wood appeared very advisable.

On account of its durability against erosive and abrasive actions micarta would probably give good service on sandy fields, where the tipping of wooden propellers was a real problem.

Because steel reenforcing can be molded into the leading edges of the blades it was hoped to be able to counteract the objectionable increase of blade angle that takes place when a propeller is heavily loaded aerodynamatically. This phenomenon will be discussed later.

The high density of micarta (1.38) as compared with wood was the first and outstanding objection to a solid propeller of this substance. But by eliminating the metal hub entirely and driving the micarta directly by the engine shaft key, or at the most supplying it with a very light hub or bushing, this, with thinner blades, gave promise of producing a solid micarta propeller whose weight would easily be favorably comparable with that of wooden propellers.

A number of different bakelized materials were experimented with by the Westinghouse Co., but bakelized duck, or paper, or a combination of duck and paper were decided upon as the only advisable materials for the best results with the existing means of manufacture.

Experiments with cork for a core proved that this construction was undesirable. In order to get enough back pressure from the cork to form good micarta on the outside, the cork had to be compressed to a density so great that the advantage sought in its use—lightness—was entirely lost. It also required so great a bulk of cork to supply the compression that it resulted in buckling the applied duck covering. It *added* weight without contributing to the strength. A soft strong wood, such as poplar or light pine, proved much better than cork, and the hardwoods promised to be even better. But wooden cores would have to be formed in the same manner as wooden propellers.

In all cases the material was hard to apply to the wood on such an irregular form as a propeller, and the uncertainty of distribution, consequent compression, and balance, and the failure to obtain the lightness sought, made the proposition impracticable.

Attention was then turned to the possibility of making a hollow propeller. Many schemes were proposed, but the most favorable one was to employ a core of low-melting alloy which would become molten at or even much below the baking temperature of the micarta and through which medium pressure could be applied to the interior of the baking propeller. This, it was thought, would force the material to completely fill out the interior of the closed mold and give the same a proper pressure per square inch to the forming micarta at all points. The alloy core could then be poured out when the baking was completed.

Accordingly a hollow propeller was designed by the propeller section, A. E. D., B. A. P., for the Curtiss *OX–5* engine and the *JN-4* training plane. This propeller was to absorb 85 horsepower at 1,450 r. p. m. when *flying* at about 70 m. p. h. It had a pitch of 5 feet and a diameter of 8 feet. The Westinghouse Co. immediately proceeded with the design and construction of a mold which, according to the best knowledge of the art at the time, would do the work. The A. E. D. drawing number for this propeller is "S-21."

In the meantime experiments were conducted on the forming of hollow bodies by the low-melting alloy-core process, and some very successful results were obtained. The scheme seemed quite possible of application to the making of a propeller.

However, the equipment and skill required for this work compared with what would be gained placed the process in the academic rather than in the commercial class. So it was decided to first try out propellers of solid micarta.

Since the mold was by this time practically completed and could not be altered, it was necessary to use it as it was with the outside dimensions and form of the propeller unchanged. This of course resulted in an excessively heavy propeller, as the design was for a hollow one. Nevertheless, very much good information was gleaned from this mold and the propellers it produced for a new design to be considered later.

*Mold for 8-21.*—The mold in which the propellers of this design were made was a manganese-bronze casting of four primary or forming parts—top and bottom plungers and the two sides. These were attached to and held in place by heavy, channeled, steel retaining plates. The sides were also bound together at each end by a heavy I link.

Through passages cored in these bronze primary members, steam or water was passed for the purpose of respectively heating or cooling the mold. This feature of cored passages proved very unsuitable, for the steam would pass through even the small pores of the casting and into the mold with the micarta, causing blistering.

The excess metal allowed for "finish" on the castings had to be worked by hand from those faces forming the propeller. This was slow and tedious. From the other surfaces it was machined. There was no previous experience in exactly this line to guide those entrusted with the work.

The mold was mounted on a 450-ton, slow-acting die press with a 75-ton pullback. It gave a pressure of about 1,000 pounds per square inch—higher than necessary—on the micarta while baking and cooling. Experience with this press soon showed that it was not long enough nor of the type suitable for the accommodation of such a long mold. Nevertheless, the best was made of what was available.

In extracting a propeller from the mold the top plunger was raised and the propeller pulled out by the "mold pin" extending through the hub. This method was very injurious to the blades, generally distorting them and frequently damaging the edges.

*Laminations.*—A micarta propeller is made up of laminations in the same manner as a wooden propeller except that the layers are only as thick as the duck of which the propeller is constructed. The big difference, however, is that the duck laminations have to be cut to the exact size and form the first time, no "rough cutting" being possible if a good propeller is to be expected. The laminations have to fit the mold when the latter is closed. Consequently accurate patterns or templates have to be made and the materials cut accordingly. The ideal way would be to have a template for each sheet of duck corresponding to its particular location in the propeller. But this is practically impossible and unnecessary. A propeller with a 4-inch hub (axially) would require over 200 templates. With propellers with exceptionally large sweepbacks it was found that a template for every tenth inch of material was very good. This leads to the group or step method, wherein each template serves as a pattern for a number of laminations, enough, say, to form a tenth of an inch of micarta. There is sufficient flow in the duck when the bakelite is softened by the heat to allow it to make up for the slight difference between what the laminations really are and what they should be theoretically.

These groups are laid in the mold in their proper order one at a time after previously turning end for end every other lamination of the group in order to equalize any unsymmetry that may have resulted from unavoidable errors in the laying out and sawing process. As each group is placed in the mold its laminations are "fanned" slightly (as one would fan a bunch of cards in order to easily and quickly count them) so as to conform to the contour of the mold.

*Manufacture of 8-21.*—In preparing the duck (or paper) for molding into a propeller it was first impregnated to the right ratio with bakelite varnish. The solvent was then evaporated from it, leaving it properly "cured" and dry enough to be easily handled and cut into the correct forms. For cutting, the sheets of duck were piled together with sufficient layers to make up the proper number of laminations per group. They were then laid out according to the templates, marked and sawed out with a band saw. The templates have been, for this experimental work, of varnished 18 or 20 mil paper.

The groups of laminations thus sawed were then loosened, and, as previously mentioned, alternated and stacked in the proper order for laying into the mold. It was found that the best results were obtained by covering both faces of each blade with a sheet of duck or cambric and not exposing the edges of the laminations. One set of covers were placed in the mold, the groups of laminations were stacked in, and then the other set of blade covers were put in place.

The mold was then closed and the steam turned on. With full pressure without, the baking was continued for about two hours. The water was then turned on and the mold and propeller cooled to a temperature low enough to make sure that the micarta would not blow when the pressure was released. This cooling required about 45 minutes.

An attempt was made to bake the hub section a little longer than the blades because of its much greater bulk. This could only be approximately accomplished by turning the steam into the sides of the mold about 10 to 15 minutes before it was turned into the top and bottom plungers.

After removal from the mold a number was stamped on each of the propellers, the bakelite burrs were filed off, and the edges dressed up and rounded. This work was most successfully and rapidly done with a large, fine-cut mill file.

When finally dressed the propellers were put on the balancing ways to determine the center shift for balancing. This part of the operation is completely discussed under the separate head of "Propeller balancing."

Each was then first centered on a large boring mill with respect to the mold-pin hole, and then shifted so that the center of the completed bore would go through the center of gravity. The keyways were then slotted or broached into the micarta.

The propellers next went to the rebaking mold, where they were clamped in place one at a time and given a slow curing at about 115° C. for 24 hours. This re-forming mold consisted of a steel slab with a tapered center pin near its middle, which exactly fit the bore of the hub. Mounted so as to engage with the thrust face of each blade was a steel plate warped to the proper angles all along the blade. The blades were pulled down to these plates to keep them from warping during the course of the rebaking and correct any existing errors. The blade angles were thus made symmetrical with respect to the axis of the hub.

Then came the final balancing. This slow curing or rebaking, as it is called, was liable to change the balance slightly, and there was always a chance that the shift of centers was not exact. If the unbalance was not too great, the difference could be made up with paint or varnish, but if it was off very much it was necessary to resort to plugging. The manner of doing this is also described under the head of "Propeller balancing."

*Propeller balancing.*—In order to give an idea of the method employed in the balancing of micarta propellers, there follows a reproduction of the instructions and explanations issued by the research division of the Westinghouse Co. These were issued February 14, 1918, and, while there have since been some alterations made and limitations established regarding the size and location of the plug, the scheme of solution for balance is still the same, except in so far as the balance can be secured with paint.

### SHIFTING CENTERS.

#### CASE I.

The discussion under Case I will deal with a propeller that is unbalanced when its longitudinal axis is in a horizontal position.

Place the propeller, supported by a balancing mandrel on knife edges, in a *horizontal* position, as shown in figure 132; that is, with its longitudinal axis *B B* horizontally located. Its transverse axis *A A* is then in a vertical position.

The axes *A A* and *B B* divide the propeller into four *quadrants*, which are arbitrarily and for convenience numbered clockwise, when viewed on the thrust (or down-stream) side, as shown in figure 132, starting with the upper left-hand quadrant. In this position the propeller blades are numbered, from left to right, blade 1 and blade 2.[1]

Now, allow the propeller to freely swing into its position of stable equilibrium or rest. Assume that it takes the position shown in figure 133. The *angle of horizontal unbalance* is the angle of the position of stable equilibrium of the propeller with a horizontal position. In other words, it is the smallest angle through which the propeller can turn to get from a horizontal position to its stable position. The complement of this angle is the *angle of vertical unbalance.*

The case of locating the transverse axis on the micarta propellers now being made has prompted the use of this axis as a base line from which to measure the angles of unbalance.

The old position of *A A* of figure 133 was vertical, as shown in figure 132, but now it makes an angle $\theta$ with the vertical. Therefore, $\theta$ is the smallest angle of turning from a horizontal position to a stable position and is consequently the *angle of horizontal unbalance.*

The center of gravity (*c. g.*) is below the center of support ($C$) a certain distance ($r$) on the plumb line ($OP$) when an unbalanced propeller, suspended as above, is allowed to assume its stable position. ($r$) is proportionally exaggerated in figure 133 for the sake of clearness. By the "Method of quadrants" the position of *c. g.* with respect to $C$ (center of support) can easily and unquestionably be specified as "distance $r$, angle $\theta$, fourth quadrant," or more condensed as "4th $q-\theta-r$ inches."

Example.—If $\theta=30°$ and $r=0.045$ inches, then the location of *c. g.* is: $Q-30°-0.045''$.

A perfectly balanced propeller will have its *c. g.* on its center line of support. There are two ways of accomplishing this:

1. By shifting *c. g.* to $C$.
2. By shifting $C$ to *c. g.*

A combination of the two might be called a third way.

1. To shift *c. g.* to $C$ would mean to do either or both of two things which would give a resultant *c. g.* coinciding with $C$. First, "plug" on $OP$ on opposite side of $C$ from *c. g.*; second, remove material on the same side of $C$ as is *c. g.*

[1] Blade 1 is from the left-hand end and blade 2 is from the right-hand end of the mold as viewed from the control side of the press.

For a number of reasons[1] the removal of material is out of the question at present with micarta.

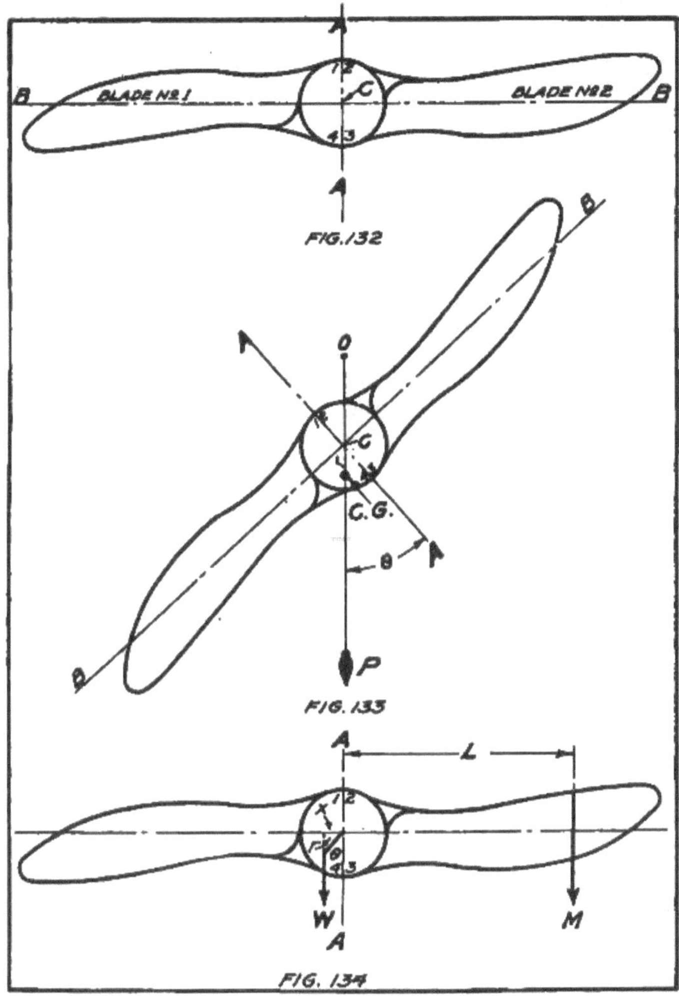


Plugging mars the appearance of the propeller and adds to the total weight. In some cases it may require a plug weighing in the neighborhood of a pound.

[1] Too much material would have to be removed; hard, finished surface would be spoiled; micarta very hard to work for this purpose and hard on cutting tools.

2. By shifting $C$ to $c.\ g.$ advantage can be taken of the total weight of the propeller and consequently a very small shift may suffice to accomplish the desired result and at the same time not increase the weight, as would plugging, nor involve the objectionable features of removing material.

The first attempt at balancing (and by all means the last one necessary, if possible) should then be that of shifting $C$ to $c.\ g.$ Knowing $\theta$ and the quadrant, it only remains to find $r$.

At some distance $L$ (fig. 134) out along the blade add weight $M$ until the transverse axis $A\ A$ is again brought into a vertical position. The greater $L$ is, the more accurately it can be measured; that is, the smaller the percentage of error of measurement becomes. Of course, as $L$ is increased, the necessary weight $M$ for balancing is decreased, but weight can with little effort be measured to a very small percentage of error and much closer than length. Therefore, make $L$ the greatest convenient distance.

The moment $M\ L$ is now balanced by the total weight ($W$) of the propeller acting through $c.\ g.$ with a moment arm $x$.

Then $Wx = M\ L$.

But $x = r \sin \theta$.

Therefore $Wr \sin \theta = M\ L$.

And $r = \dfrac{M\ L}{W \sin \theta}$.

The most convenient unit in which to measure $M$ is grams. Therefore, $W$, which is more easily obtained in pounds, must be converted to grams by multiplying by 453.6, since there are 453.6 grams in a pound. If $L$ is measured in inches, then $r$ will be in inches.

$$r \text{ (inches)} = \frac{M \text{ (grams)} \times L \text{ (inches)}}{454 \times W \text{ (pounds)} \times \sin \theta}. \qquad (1)$$

CASE II.

It may happen that the horizontal position is a *stable* position, but still the propeller would not be in *neutral* equilibrium and, if freed with its longitudinal axis ($B\ B$) in a vertical position, would at once swing to a horizontal position. This would indicate that $c.\ g.$ was somewhere on the transverse axis ($A\ A$) below $C$ (fig. 135).

In this case $\theta$, $M$, and $L$ are zero, and the above formula (1) can not be used to determine the value of $r$.[1]

Drive a tack or brad into the light side of the propeller hub on the axis $A\ A$ and from this suspend enough weight ($M$) to bring the axis $B\ B$ into a vertical position, as shown in figure 136. $L$ is again the distance of the point of suspension of the balancing weight ($M$)

[1] Conditions of *small* values of $\theta$ should also be classed under Case II.

from the center of support ($C$), $r$ the distance of $c.\ g.$ from $C$, and $W$ the weight of the propeller. In this case the moment arm of $W$ is $r$, since it is acting directly against $M$ and not at an angle.

Therefore, $Wr = M\,L$

And $r = \frac{M\,L}{W}$

with $M$ in grams, $W$ in pounds, and $L$ in inches, to get $r$ in inches the formula needs to convert $W$ to grams, and hence takes the form:

$$r = \frac{M\,L}{454\,W}. \tag{2}$$

### PLUGGING.

#### CASE I.

If no errors have been made in determining and specifying the value and position of $r$, and if the workman in boring the new hole has shifted its center according to specifications, the propeller will be balanced when supported about its *new* center and, no matter in what position it may be placed, will always be in *neutral* equilibrium.

If, however, the new center has not been correctly shifted, it will be necessary, since the new hole is to be of proper size for the engine shaft, to complete or correct the balance by plugging.

Note that now the balancing will be done about the new center of support $C$ at a distance $r$ angle $\theta$ from $C$ and $A\ A$. The new axes are $A'\ A'$ and $B'\ B'$, corresponding to the old axes $A\ A$ and $B\ B$. The new condition is represented to a greatly exaggerated scale in figure 137. In future illustrations $A'\ A'$ and $B'\ B'$ will be in the positions heretofore occupied by $A\ A$ and $B\ B$, because the proportional shift of axes can hardly be represented on such a small scale. However, the idea must not be gained from this that in actual balancing these changes of conditions can be ignored.

In balancing by plugging two things must be known: First, the location of the plug; second, the necessary weight of the plug.

In order to simplify future equations, let us now eliminate, combine, and simplify some terms and expressions whose values we know and can at once put in condensed formulæ.

Assume that we know the weight of unbalance at the point where the plug is going to be placed. But in boring a hole for the plug material is being removed that was assisting in balancing, so that the plug will have to weigh enough more than the unbalanced weight to make up for that lost by boring.

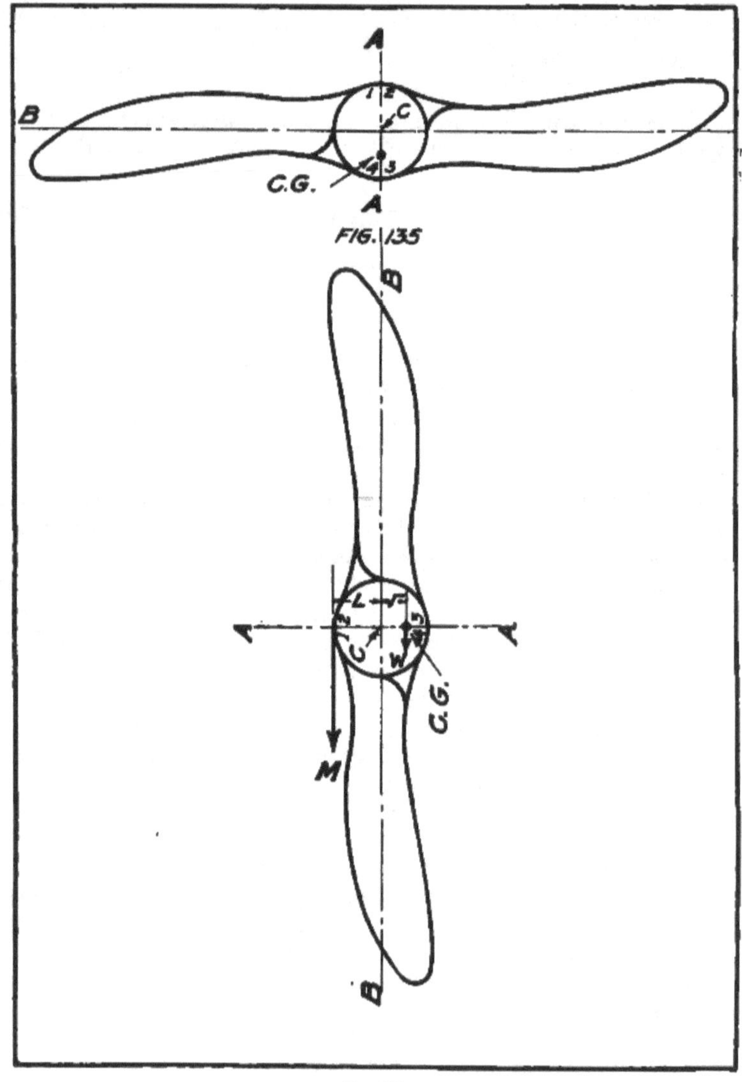


Fig. 136.

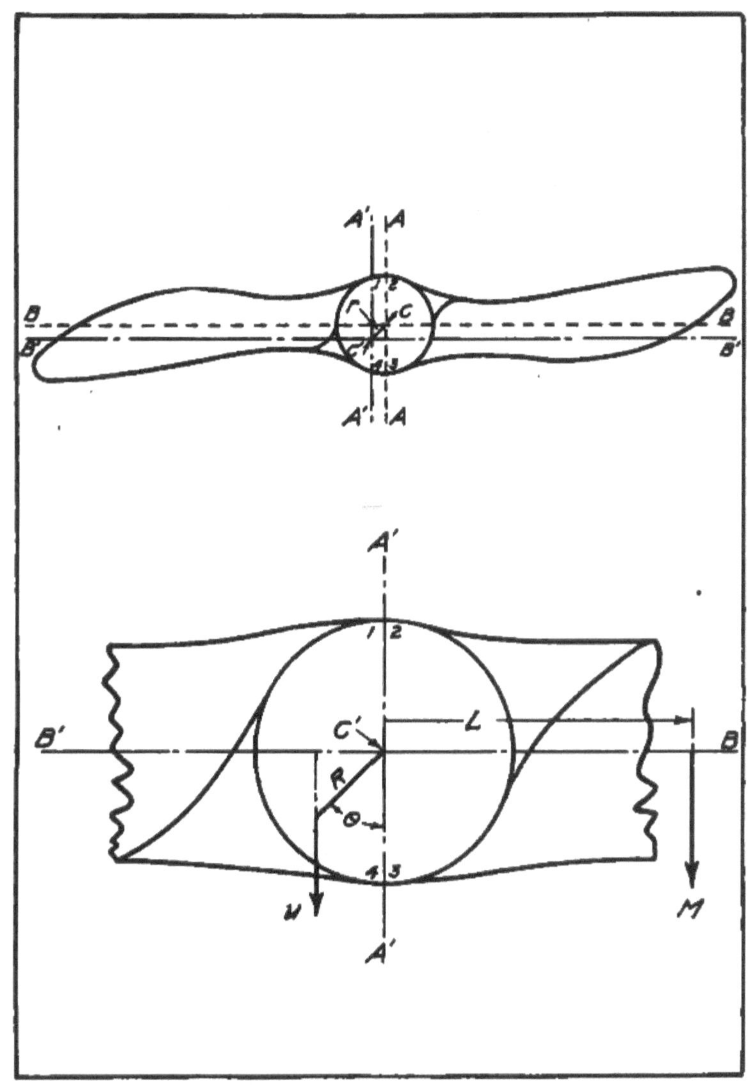


FIGS. 137-138.

Let the following symbols hold throughout the discussion:

$R$ = Distance of plug center from center of hub.

$u$ = Weight of unbalance at a distance $R$ from center of support, as determined on balancing knives.

$b$ = Weight of propeller material bored out to accommodate plug.

$w$ = Weight of plug.

$D$ = Plug diameter.

$l$ = Length of plug.

$\delta$ = Density of plug metal.

$\rho$ = Density of propeller material.

Since the weight of a body is the product of its density and its volume, then the weight of the material bored out is—

$$b = l\rho\frac{\pi d^2}{4}$$

and the weight of the plug is—

$$w = l\delta\frac{\pi d^2}{4}$$

But

$$w = u + b$$

or

$$\delta l\frac{\pi d^2}{4} = u + \rho l\frac{\pi d^2}{4}$$

Solving for the necessary diameter of the plug hole ($d$):

$$u = \frac{\pi l d^2}{4}(\delta - \rho)$$

$$d^2 = \frac{4u}{\pi l(\delta - \rho)}$$

and

$$d = 2\sqrt{\frac{u}{\pi l(\delta - \rho)}} \qquad (3)$$

In the *c. g. s.* (centimeter-gram-second) system the density of a material is the same as the specific gravity, so that formula (3) holds as it now stands in that system; that is, $d$ and $l$ are in centimeters, $u$ is in grams, and $\delta$ and $\rho$ are the weights per cubic centimeter of the plugging material and propeller material, respectively.

While it is most convenient for us to measure $u$ in grams, still it is more convenient and causes less confusion in the shop if $l$ and especially $d$ are measured in inches. Therefore let $d$ and $l$ be measured in inches and expressed in centimeters when inserted in the formula:

$d$ (inches) = (2.54 $d$) centimeters.
$l$ (inches) = (2.54 $l$) centimeters.

Substituting in formula (3)

$$2.54\ d=2\sqrt{\frac{u}{\pi 2.54 l(\delta-\rho)}}$$

$$d=2\sqrt{\frac{u}{\frac{2.54\ \pi\ l\ (\delta-\rho)}{2.54}}}\text{ inches when } l \text{ is in inches.}$$

Therefore

$$d=0.287\sqrt{\frac{u}{l\ (\delta-\rho)}} \tag{4}$$

Formula (4) can be further simplified if the following points are settled as permanent conditions in consideration of the propellers now being made:

1. The smaller the diameter of the plug hole the better the appearance; so make $l$ as long as possible; that is, run it the full thickness of the hub, which is 4 inches.
2. Use lead for plugging; then $\delta$ becomes 11.3.
3. For micarta $\rho=1.38$.

Formula (4) then is reduced to

$$d=0.287\sqrt{\frac{u}{l\ (11.3-1.38)}}$$

or

$$d=0.0884\sqrt{\frac{u}{l}} \tag{5}$$

and by running the hole clear through the hub—

$$d=0.0884\sqrt{\frac{u}{4}}$$

or

$$d=0.0442\sqrt{u}. \tag{6}$$

When the hole diameter, with the plug extending the full thickness of the hub, is too small for practical purposes, it may be necessary to shorten $l$. In this case formula (5) should be used with the value of $l$ *in inches* inserted. If it is possible to plug clear through the hub, then $l$ will be 4 inches and formula (6) should be used.

For algebraic convenience let $G$ stand for constant 0.0442 in formula (6):

Then

$$d=G\sqrt{u}. \tag{7}$$

The next step is to determine the value of $u$.

161988—21——16

The angle of horizontal unbalance $\theta$ (fig. 138), is determined in the same manner as for the purpose of shifting centers.

Let $R$ represent the distance that the center of the plug will have to be from the center of the hub. To find the value of $u$ for this value of $R$, bring the axis $B'$ $B'$ into a horizontal position by adding weight $M$ at a distance $L$ from the hub center, as was previously done. According to the system of moments:

$$u\,R \sin\theta = ML$$

and

$$u = \frac{ML}{R \sin\theta} \tag{8}$$

Substituting this value of $u$ into equation (7), we get:

$$d = G\sqrt{\frac{ML}{R \sin\theta}} \tag{9}$$

In equation (9) $ML$ and $\theta$ have been measured and $G$ is a known constant. By obtaining either the value of $d$ or $R$ the question of plugging can be quickly settled by the use of this formula.

It is very evident that the further the plug is from $C'$ the less it will need to weigh and the smaller its diameter will be; that is, the greater the value of $R$ the less will be the value of $d$. But for a number of reasons [1] it is desirable to keep the plug within the hub-face circle $A'$ $B'$ $A'$ $B'$, with the radius $H$ (fig. 139). Also the outside of the plug, when in the region near $A'$, should still have sufficient material between it and the side of the hub for the sake of safety. Therefore the distance $n$ has been taken to allow for this, and the *limit circle* (dotted in fig. 139) with the radius $H-m=k$ has been drawn.

We know that $R$ must be of such a value that the edge of the plug will come to the circumference of the limit circle, but not beyond.

That is:

$$R + \frac{D}{2} = k$$

or (10)

$$R = k - \frac{d}{2}$$

But from equation (9):

$$d = G\sqrt{\frac{ML}{R \sin\theta}}$$

---

[1] These reasons are:

1. The hub will suffer less from boring than will the blades.
2. The plug can always be so located that only one will be needed.
3. The above formulæ and those to follow can be applied without alterations, and the scheme of balancing can be put on a productive shop basis.
4. The location of the plug will not result in a tendency to warp or twist the blades or generate abnormal and objectionable stresses when subjected to centrifugal forces.

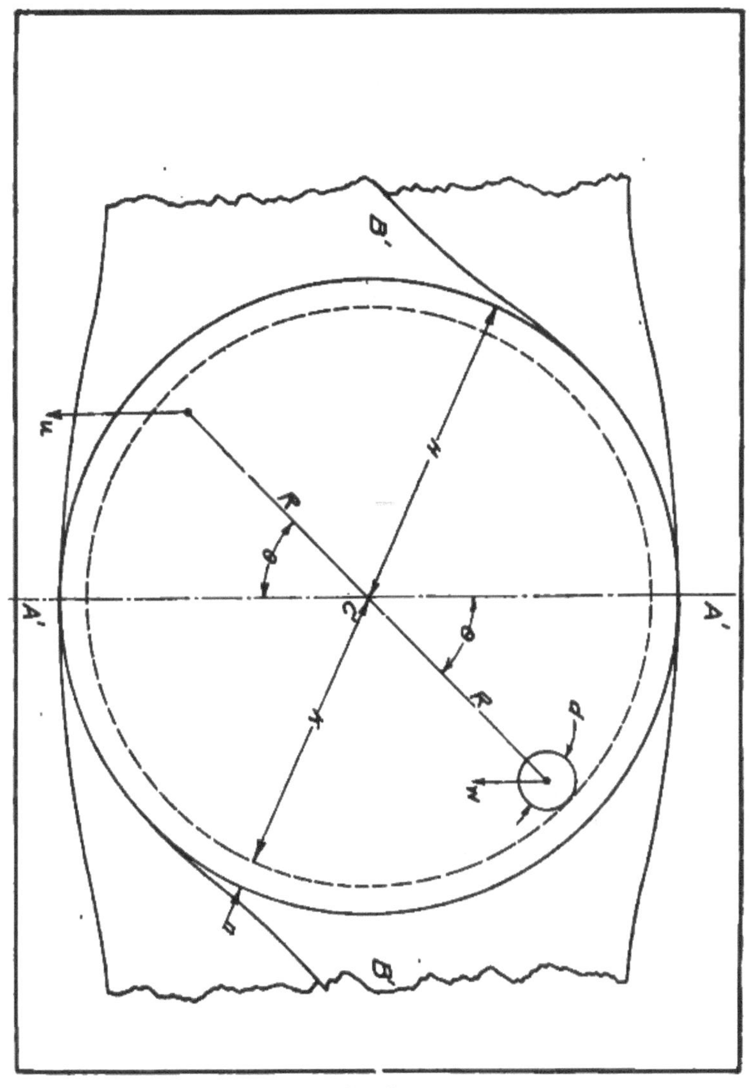


Fig. 130.

Substituting in equation (10) the expression for $d$ given in equation (9), we get:

$$R = k - \frac{G}{2}\sqrt{\frac{ML}{R \sin \theta}} \tag{11}$$

$$= k - \frac{G}{2R^{1/2}}\sqrt{\frac{ML}{\sin \theta}} \tag{12}$$

Multiply (12) through by $R^{-1/2}$):

$$R^{3/2} = KR^{1/2} - \frac{G}{2}\sqrt{\frac{ML}{\sin \theta}} \tag{13}$$

or, transposing:

$$KR^{1/2} - R^{3/2} = \frac{G}{2}\sqrt{\frac{ML}{\sin \theta}} \tag{14}$$

We now have an equation expressing the relation of $R$ (the maximum moment arm our balancing plug can have and still be totally within the *limit circle*) to other quantities which we either know as constants or whose values we can easily measure. Thus $K$ is known, because it is the radius of the *limit circle*, which we arbitrarily (or otherwise) set. $G$ is a constant, depending for its value upon the plugging material used, the propeller material, and the units of measurement of $M$, $L$, and $R$. $M$, $L$, and $\theta$ are known by measurement.

However, equation (14) is a cubic equation whose solution is too complicated for shop use and would require too much time if it had to be carried out individually for each propeller. But it is very easy to assign values to $R$ within the useful range and plot a curve with $R$ for one axis and $\frac{G}{2}\sqrt{\frac{ML}{\sin \theta}}$ for the other.

This has been done for values of $R$ between $R = 1$ inch and $R = 2.75$ inches. We know these limits set a sufficiently large range, because 2.75 inches would carry $R$ to the *limit circle*, making $R = K$. But this would mean that no plug whatever would be needed because its diameter would be zero, since:

$$d = 2(K - R) \quad \text{(Refer to equation 10.)} \tag{15}$$

If $R$ should even come as low as 1 inch in value, the propeller would be so badly unbalanced that it would have to be discarded, for the center of the plug hole would come to the edge of the bore, and the plug diameter would be 2 $(2.75 - 1.00) = 3.5$ inches, overlapping the bore for the shaft. It is very improbable that such a condition would ever arise.

The accompanying "*propeller plugging curve*" (fig. 140) was drawn to take care of the following conditions:

1. Plugging bakelite micarta propellers with a density ($\rho$) of 1.38.

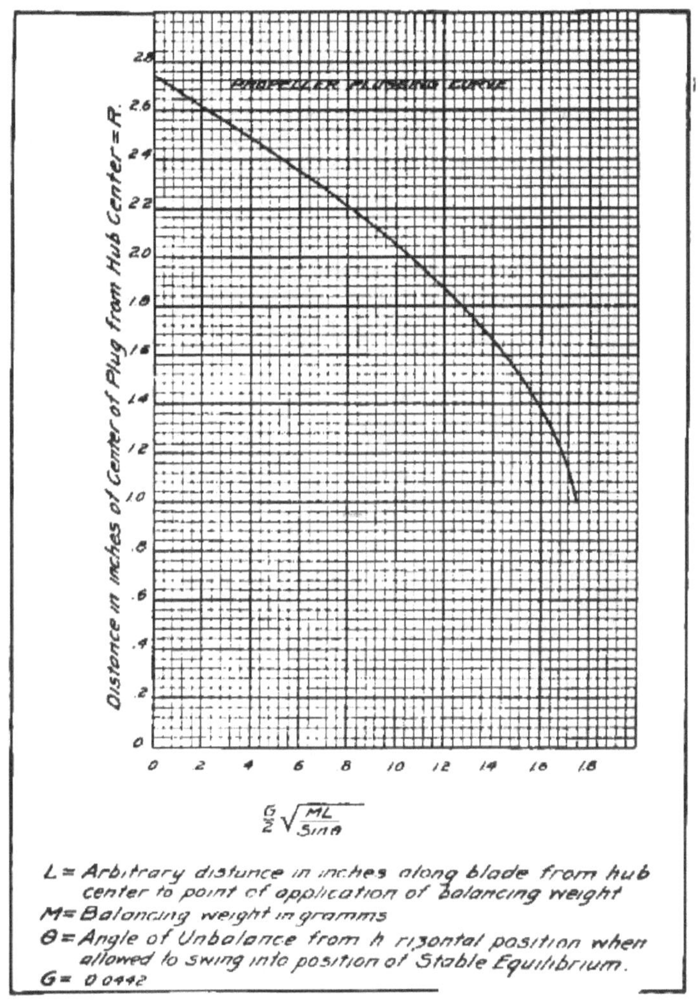


FIG 140.

2. Using lead, with a density ($\delta$) of 11.3, as the plugging material.

3. Using a plug with a length (1) of 4 inches; that is, running completely through the hub—parallel to the axis of rotation—of the

micarta propeller now being made. (United States Signal Corps drawing 8-21.)

4. Having the radius ($K$) of the limit circle set at 2.75 inches.
5. Measuring $M$ in grams.
6. Measuring $L$ in inches.

CASE II.

When this condition exists (good horizontal balance but bad vertical balance, see Case II, under "Shifting centers"), values for $M$ and $L$ should be determined in the same manner as described for Case II in "Shifting centers."

In equation (14), $K R^{1/2} - R^{3/2} = \frac{G}{2}\sqrt{\frac{ML}{\sin\theta}} \cdot \frac{ML}{\sin\theta}$ is the necessary balancing moment for Case I. In Case II the necessary balancing moment is simply $M L$, as previously shown. Therefore to use our plugging curve, $\frac{G}{2}\sqrt{ML}$ should be used in place of $\frac{G}{2}\sqrt{\frac{ML}{\sin\theta}}$. In other words, let $KR^{1/2} - R^{3/2} = Y$ which is the fundamental equation of our curve.

For Case I,

$$Y = \frac{G}{2}\sqrt{\frac{ML}{\sin\theta}}$$

For Case II,

$$Y = \frac{G}{2}\sqrt{ML}.$$

*Hubbing.*—First the micarta propellers were bored to accommodate the standard *OX-5* hub. Then the scheme of doing away with the hub entirely and cutting a keyway directly in the micarta to receive a key from an engine shaft was conceived. The depth of this keyway was made ⅜-inch and tested out on the belt-driven test rig at the Westinghouse plant. It was run for 10 hours at 1,800 r.p.m. and then run up to 2,346 r.p.m. and 806 horsepower. Neither the propeller nor keyway showed any signs of failure. However, when this type of drive was subjected to service tests and engine explosions, the keyways failed in nearly all cases after only very short runs. This showed up very pronouncedly that the keyway tests with a smooth running electric motor could not be taken as representative of what would happen when driven with a gas engine. Keyways only $\frac{5}{32}$ inch and $\frac{3}{32}$ inch deep withstood the same tests on the test rig as did the ⅜-inch keyways.

A complete analysis showed that the micarta at the keyway was greatly overloaded when subjected to the maximum pressure of the explosion. This lead to the development of a special hub or bushing with four keys driving in the micarta and without any flanges, except a small shoulder, or bolts. So far this has proven satisfactory. These bushings were pressed into the propeller hub immediately after boring and broaching.

*Reinforcing.*—It is well known that the center of pressure of an air foil of the type used in propellers is nearer the leading edge than is the center of gravity of the section. In a homogeneous section the center of gravity coincides with the center of resistance. Now, if in micarta, for instance, steel wires were molded into the leading edge of the section, the steel being stronger than micarta, the homogeneity would be changed and the center of resistance would be moved toward the wire. Since the center of pressure is nearer the leading edge than the center of gravity by putting steel wire in the leading edge, it was proposed to bring the center of resistance nearer or even ahead of the center of pressure, and thereby eliminate the tendency

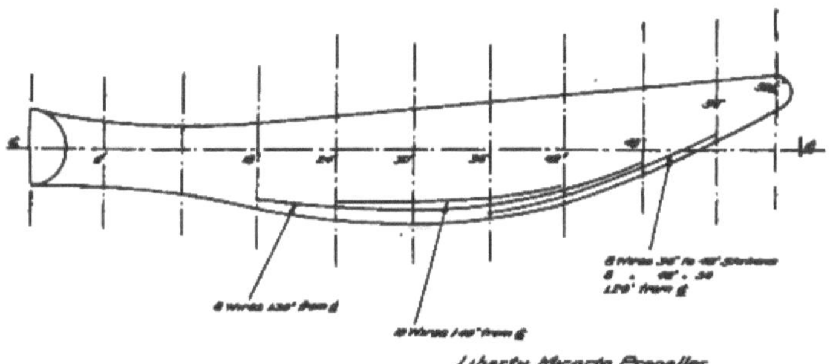


FIG. 141.

of the blade angle to increase when heavily loaded. The greatest loading of a propeller is when the plane is on the ground or going slowly, and this is the time of all times that the blade angles should not be increased, for already a part of, if not all of, the normal pitch angle is added to the angle of attack to form the operating angle of attack. It was even hoped that with the flexibility of micarta an actual automatic reduction of the angle could be effected.

The wire first molded into the leading edge of micarta propeller was steel piano wire about 0.02 inch in diameter, hand woven with soft cotton yarn into a braid form. No machines were found that could handle this stiff wire and weave the braid. To do it by hand was laborious and very slow. A more speedy method of binding wires together at intervals of 6 inches was employed for a while. However, this made a knotty, flat strand and was very hard to handle and put in place. A much better and faster method, that of "preforming" the braid, was later adopted.

Tests of small beams with these reinforcing wires molded into outer fibers showed up very good for the reinforcing and method of application. When properly made, the wires in these beams would always be pulled in two without slipping when sufficiently loaded in a testing machine.

Figures 141 to 148, inclusive, show the distribution of the steel reinforcing wires in the micarta Liberty propeller (which propeller is discussed later) and serve to illustrate the practical application of the above principle to the manufacture of micarta propellers.

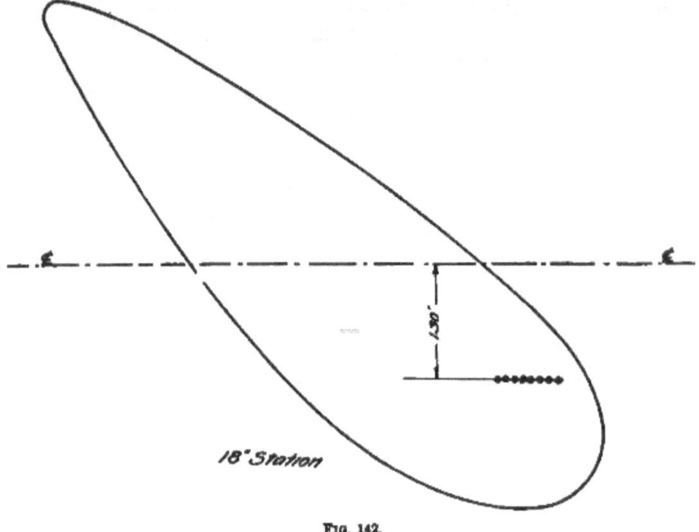


FIG. 142.

*Tests.*—A large number of whirling tests and a few flight tests were made on micarta propellers.

The second micarta propeller that was made was given a destructive whirling test at the Westinghouse plant in East Pittsburgh. It was run for one-half hour at each of the speeds 1,000, 1,200, 1,400, and 1,600 r. p. m., and then for 10 hours at 1,800 r. p. m. At this speed it took 300 horsepower. After this endurance run a destructive run was made. The speed was boosted to 2,015 r. p. m., the corresponding horsepower being 441. After 19 minutes at this speed about one-third of one of the blades broke off where a crack had appeared during the 10-hour run. Symmetrically opposite on the other blade a crack had appeared at the same time. Both of these cracks were due to flaws in the thin trailing edge where the lamination templates were in error.

Propellers Nos. Z–4 and Z–5 were made the same day, and except for a little difference in the amount of material in each they were duplicates. No. Z–4 turned out a little light. This one was given a similar test to that of No. Z–2 and failed at 2,208 r. p. m and 711 horsepower. This failure was due to the same error in templates as that of No. Z–2. No. Z–5 was not given a destructive test, but after 10 hours at 1,450 r. p. m. was prepared for shipment to McCook Field for flight tests. It was first tried out on a *JN–4* on December 18, 1917. The first flight lasted 10 minutes and the second for 4 minutes. No speed or climbing tests were made at this time on account of the risks encountered in starting and landing on a snow-

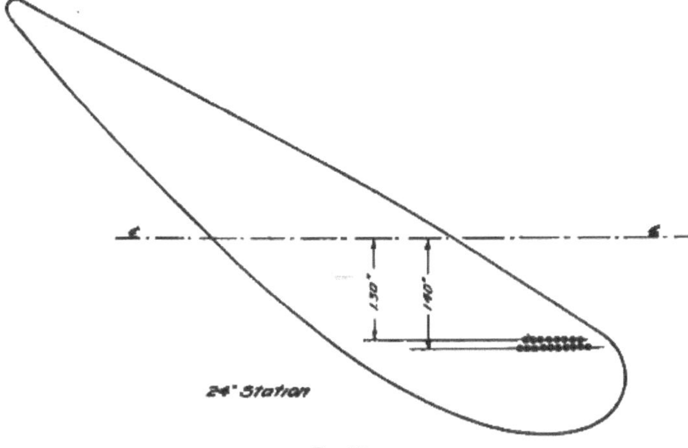


FIG. 143.

covered field. Further testing of this propeller was indefinitely postponed.

At this stage the scheme of reinforcing the leading edges of the blades was proposed. Hand-woven steel braid was molded into the leading edge of No. Z–6 and the propeller given a destructive whirling test. At the lower speeds there seemed to be between the test results for this propeller and the others no perceptible difference that could be confidently attributed to the presence of the wire.

The next idea to be tested out was whether or not the heavy steel hub, with its bolts and flanges, could be entirely eliminated. No. Z–7 was taper bored and a ⅜-inch keyway was cut directly into the micarta. Previous to cutting the duck for this propeller the templates were corrected to remove the cause of the flaws heretofore occurring. On test this propeller successfully withstood over 800 horsepower and 2,400 r. p. m. The keyway showed no signs of fail-are, and the results of the test indicated that direct drive in the micarta could be safely employed. It was impossible to break this propeller on a speed test with the power available, and no succeeding propellers ever failed out on the blades.

There was some question as to the advisability of boring out the micarta in the hub section for the purpose of plugging. In order to determine what the margin of safety was, eight 1-inch holes equally spaced on a 4½-inch circle about the hub center, were bored axially through the hub of No. Z–8. This propeller did not fail until the speed of 2,200 r. p. m. was reached.

Seven propellers were next given flight tests at McCook and Wilbur Wright fields. During these tests it was found that the single keyway as provided in these propellers could not withstand the force of

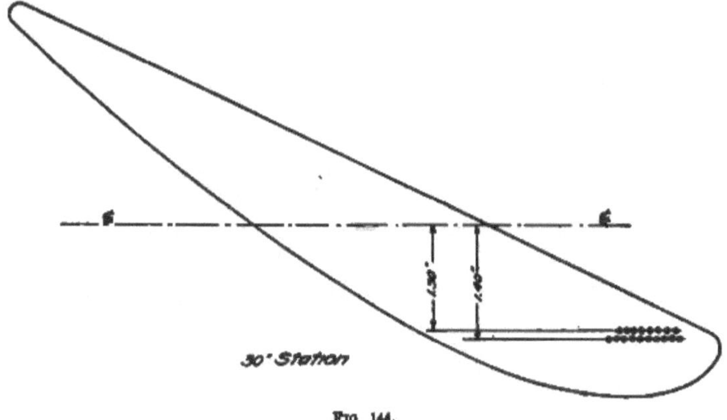


Fig. 144.

the engine explosions. It was decided to design some form of light bushing with more keys extending into the micarta.

Twenty-five propellers were shipped to Kelly Field, but were not given service test on account of the fact that they allowed the engine to run at too high a speed and thus greatly shorten its life. Consequently it was decided to make up some propellers with special angles. For the purpose of comparison of the proposed angle changes, the following table is inserted:

| Station. | 24-inch. | 30-inch. | 36-inch. | 42-inch. | 46.5-inch. |
|---|---|---|---|---|---|
| | *Degrees.* | *Degrees.* | *Degrees.* | *Degrees.* | *Degrees.* |
| Specified | 20.2 | 17.6 | 16.3 | 13.5 | 12.5 |
| Change No. 1 | 20.4 | 18.0 | 16.9 | 14.3 | 13.5 |
| Change No. 2 | 20.5 | 18.2 | 16.2 | 14.7 | 14.0 |
| Change No. 3 | 20.6 | 18.4 | 16.5 | 15.1 | 14.5 |

For each of the three changes, three propellers were made, making a total of nine special angle propellers. It was attempted to make these changes by setting the angles of the reforming mold to the values desired. However, when the blades were twisted to these angles and rebaked they would, after being released, spring back to a value much lower than that desired. Consequently it was necessary to set the angles of the reforming mold much larger and allow the blades to spring back toward the value wanted. The result was both uncertain and unsatisfactory. It was proven that while small angle corrections could be made in the reforming mold, changes of angle vastly different from those for which the propeller was molded could not be accomplished with certainty. Nine propellers with increased angles were made up (not strictly according to the specified changes)

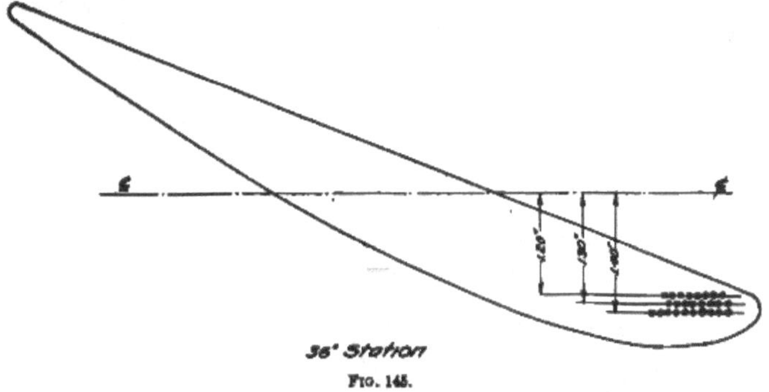

FIG. 145.

and sent to Ellington field. No word has been received of their whereabouts.

### MICARTA LIBERTY PROPELLER.

The A. E. D. drawing number for this propeller is X–5425. It was designed for the U. S. A. 12–A engine and the *DH–4* plane; to be made of micarta, solid, and with the following rating: 1,750 r. p. m., 420 horsepower, 10 feet diameter, 132 miles per hour, 6.64 feet effective pitch.

In this design was incorporated all the knowledge gained about micarta propellers, in regard to both manufacture and performance, from experience with 8–21. Improvements were made in the aerodynamic design and in the design of the mold and press to handle the job. The improvements may be summarized as follows:

1. The press was designed and built for this particular job and mold.

2. The stripping arrangement of the mold and press eliminates all possibility of straining and distorting the propeller when taking it from the mold.

FIG. 149.—PRESS AND MOLDS FOR LIBERTY MICARTA PROPELLER.

FIG. 150.—ARRANGEMENT FOR HEATING MOLD FOR LIBERTY MICARTA PROPELLER.

3. Sectional heating of the mold permits of baking each section consistent with its thickness.

4. Inserted steam and water pipes insure a dry mold with no leakage through the pores and into the propeller cavity.

5. The large diameter mold pin facilitates better baking of the hub section on account of its large amount of heat transfer than did the small pin of the old equipment.

6. The mold construction is an improvement throughout.

7. The blade sections were made comparatively thin for the purpose of gaining an aerodynamic efficiency.

8. The hub section was designed to be supplied with a special bushing driving into the micarta through four keys.

Two views of the Micarta Liberty propeller molding equipment are shown in figures 149 and 150. In figure 149 most of the mold and press can be seen. Of the three cylinders at the top of the press, the two outside ones are the cylinders which supply the pressure to the micarta while forming. The middle cylinder is the "pullback" that opens the mold. In this figure the mold is open and the top plunger that forms the thrust faces of the blades can be seen fastened to the top platen of the press. Below the level of the floor are two "stripper cylinders." These operate on a plate carrying heavy pins or posts that extend up through the bottom of the mold and are attached to the bottom plunger which forms the cambered faces of the blades. When the propeller is to be removed these stripper cylinders force the plate with its pins and the bottom plunger up between the sides of the mold. The propeller is thereby pushed out without bending or straining it. As shown in the figure, the bottom plunger has been forced up and can be seen resting on the stripper pins showing a little above the sides of the mold. Note the heating and cooling pipes leading into the sides of the mold. Also notice the equalizing shaft with rack and pinions employed as an extra precaution against one cylinder getting ahead of the other.

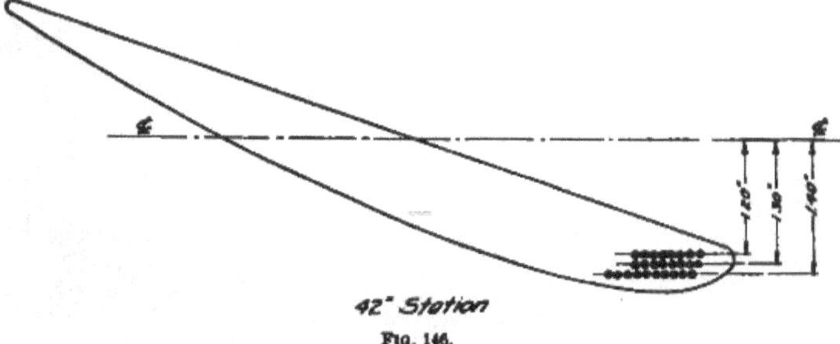
42" Station

Fig. 146.

Figure 150 shows only one end of the press after it has been closed, and the steam traps which prevent condensation from collecting in the mold tubes and cutting out the steam. Heating and cooling tubes can be seen at the end of the mold where they come out and loop back in.

A-1 (Westinghouse Electric & Manufacturing Co. number), the first propeller of this design to be made, was molded March 7, 1918.

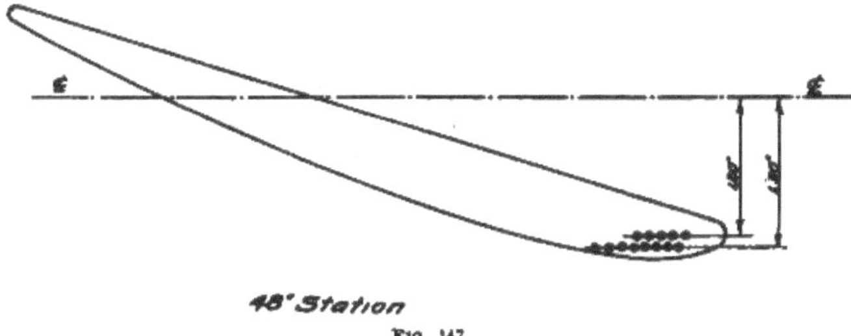

48" Station

Fig. 147.

It turned out an excellent propeller, apparently almost perfect, and the more noteworthy because it was the first attempt with the new equipment (mold, press, etc.) and with a propeller of this type and size.

The mold closed completely, and evenly, a thing never successfully accomplished with the old press and mold for the 8-21 propeller.

The only noticeable flaw was where one of the blade covers had pulled back from one of the tips. But this was not even of minor

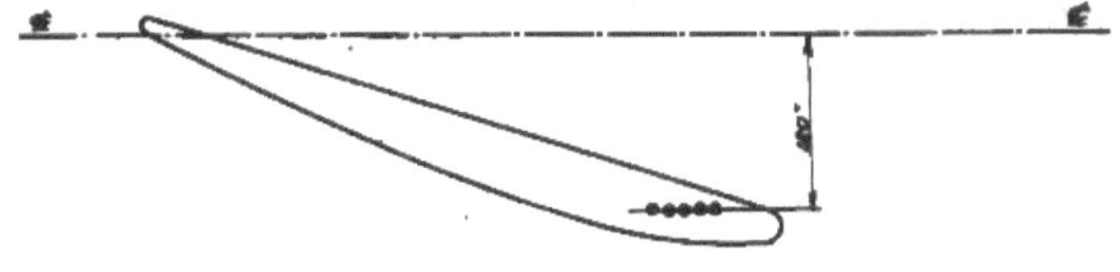

54" Station

Fig. 148.

seriousness and can easily be corrected. It seems advisable to use only "half" instead of "full" blade covers, the former being much more easily applied and molded and the latter being entirely unnecessary.

The weight of the complete propeller was 82.13 pounds, this including 13.75 pounds for the hub, and making the net weight of Micarta 68.38 pounds. This is very close to the calculated weight of 68.7 pounds.

In making the propeller the laminations were built up in the same manner as for the old 8-21 propeller, but for the Micarta Liberty propeller, they were much more easily located, and laid right in place. For this operation about one-half hour was required. The mold was closed with low hydraulic pressure and the heat turned on to the whole propeller. The heat was turned off the outer halves of the blades after they had baked for 1½ hours, but the hub section and the heavier portion of the propeller was allowed to continue baking until it had baked a total of 2½ hours, an hour longer than the thinner blades.

The whole propeller was then cooled for 1¾ hours. This cooling time can be reduced by modifying the cooling system, or also to a probably greater extent, if feasible, by using a different method.

That the hub section was very good was indicated by the smooth finish inside the hub about the mold pin. The large mold pin used greatly favors the thorough baking of the hub.

In an excellent manner the steel reinforcing molded into the leading edge of the propeller.

The propeller blades, even at the tips, were very stiff and not nearly so flexible as might have been expected from the appearance of the drawing and wooden pattern. The blade thicknesses were very close to exactly the values specified, and the blade angles were within acceptance limits of the specifications, even being very close just as they came from the primary mold. Also the propeller was very easy to "dress."

The propeller was bored for balance and to receive a special Micarta Liberty propeller hub, designed by Westinghouse engineers so as to duplicate the standard Liberty 12-A hub with respect to the internal dimensions, but to be applicable at the same time to the Micarta propeller

After hubbing, the propeller was further baked in a "reforming" or "secondary" mold for 20 hours at 115° C. This gave it a thorough slow curing.

To date only one of this design has been tested. This was given a destructive whirling test consisting of a power curve up to 600 horsepower, and 3 hours at this power at the Westinghouse Co., in East Pittsburgh, and then at McCook Field another power curve up to 600 horsepower, 10 hours at this power, an attempt at destruction wherein the propeller was given 1,325 horsepower at 2,100 r. p. m. for 5 minutes (all the test rig could stand), and then after this severe overloading it ran for 25.25 hours at 1,800 r. p. m. over 800 horsepower when it broke. It pulled in two at the hub.

Another propeller of this design with which it is proposed to make flight tests has been made. This propeller has been given a 10-hour destructive whirling test at 600 horsepower. It stood this test with no signs of weakening.

## Adjustable Pitch Propellers.

Immediately after the United States entered the war against Germany, the National Advisory Committee for Aeronautics called to the attention of inventors the necessity for an adjustable pitch propeller for airplanes. England and France had done some experimenting with this type of propeller without achieving satisfactory results. Both countries had given up the idea as a failure.

The first adjustable pitch propeller to make its appearance in this country was designed and built about the middle of 1917 in Los Angeles, Calif., by Robert I. Eustis and the writer. This propeller was designed for a Curtiss 75-horsepower engine. The mechanism for changing the pitch of the blades while in flight is essentially the same as that of the very latest designs. The hub was of metal made hollow. It was 18 inches long and 4 inches outside diameter. The wooden blades were set in metal ferrules of $3\frac{1}{8}$-inch inside diameter They were held in place by two steel dowels which were placed through the wood at 45° to the laminations and 90° to each other and about 2 inches apart. The end of the ferrules were tapered on the inside to a distance of a little over 1 inch from the end. After they had been pressed on the wood a circular steel wedge was driven into the end of the wood, causing it to spread outward until it conformed to the taper in the ferrule. Near the wedge end of the ferrule, at a given number of degrees from the center lamination line, a hole was drilled directly through the center of the ferrule and blade. Through this hole a steel sleeve was inserted through which the operating pin slides. The blades were then placed in the hub and held in position by a thrust collar which screwed to the end of the hub. Ball bearings ($\frac{7}{32}$-inch balls) were placed between the thrust collar and the ferrule to take the centrifugal load and to allow the blades to be easily adjusted while in flight. The operating mechanism is essentially a sliding member, which moves the end of the operating pins in a direction parallel to the axis of rotation. The operating pins slide in bushings through the propeller blade in such a way that the straight line motion parallel to the rotating axis of the propeller is converted to the angular motion necessary for pitch change. The hub and blade ferrules were cast of aluminum alloy having a tensile strength of 48,000 pounds per square inch. The blades were of birch and their diameter was 8 feet 2 inches. The weight of the complete propeller was approximately 45 pounds. The completed propeller is shown in figure 151.

Before attempting flight tests, block tests were made in a rather crude sort of way, as shown in figure 152. Enough data was obtained from these tests to justify flight tests.

161988—21——17

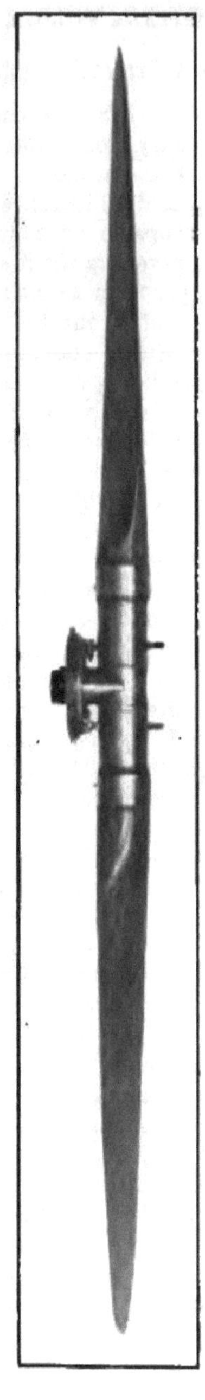

FIG. 151.—FIRST HART VARIABLE PITCH PROPELLER.

FIG. 152.—FIRST BLOCK TESTS OF HART VARIABLE PITCH PROPELLER.

Several tests were made without the pilot having control over the propeller while in flight as the variable-pitch mechanism is self-locking and the blades remain in any position at which they are set. Later a control was installed in the pilot's cockpit and connected with the variable-pitch mechanism of the propeller hub. This allowed the pilot to change the pitch of the blades while in flight. No locking device was used on the control and none seemed to be necessary.

From the results of the tests on this first propeller it developed that aluminum did not have sufficient strength to successfully withstand the high stresses set up in the hub.

A second hub was then built. Shelby seamless steel tubing was used for both the hub and the ferrules. The diameter and length of the hub remained the same as in the first, but it was made to fit a Curtiss *OX-5* engine. The construction of this hub differed from the first one inasmuch as it was necessary to weld the shaft plug to the hub and also to weld splines on the shaft plug for guides for the sliding member. The mechanism for changing the pitch was made just the same as used on the first propeller with the exception of the sliding pins, which were put a little closer to the wedge end of the ferrules. The blades were made of birch and the diameter over all was increased to 8 feet 4 inches. The range through which the pitch could be changed was 12°, the same as the first one.

When completed this propeller was taken to Rockwell Field, January 31, 1918, and installed in a Curtiss *JN-4* plane, equipped with a Curtiss *OX-5* motor. It was flown for over four hours by Lieut. Mairesse, of the French Flying Corps. These flights showed that the pitch of the blades could be changed during flight and also that the r. p. m. of the motor could be reduced by increasing the pitch of the blades or increased by decreasing the pitch of the blades. With a low pitch and a high r.p.m. it was found that a faster climb could be made. Also a better speed could be made in level flight by increased pitch and less r. p. m. than could be obtained with a fixed pitch propeller of all service design. Speed tests were also made over a marked course to ascertain the different airplane speeds obtained by the different pitch settings of the blades.

At the conclusion of these tests the propeller was shipped to McCook Field for further Government tests.

The British Government having learned of this propeller, and being desirous of testing it, cabled for one. One was immediately made up and delivered to Gen. Cormack, of the British Mission, in Washington, D. C. In the meantime the propeller that was tested at Rockwell Field became delayed in shipment to McCook Field. Upon Gen. Cormack's suggestion, the propeller made for the British Government was taken to McCook Field and tested until the one should arrive

FIG. 153.—HART VARIABLE PITCH PROPELLER FOR JN-4H PLANE—MAHOGANY BLADES.

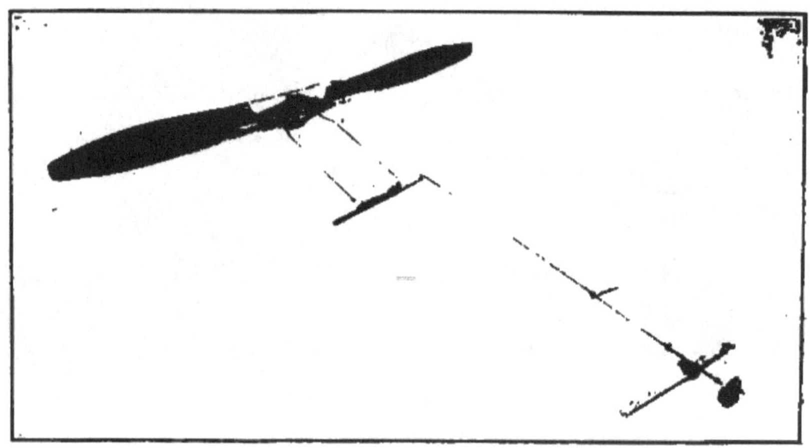

FIG. 154.—CONTROL MECHANISM USED IN JN-4H PLANE.

FIG. 135.—MAHOGANY BLADES AFTER D. W. T. OF HART VARIABLE-PITCH PROPELLER.

from Rockwell Field. The propeller section at McCook Field suggested that the propeller be given a destructive whirling test before trying it on an airplane. This was done on March 30, 1918, with the result that the wooden tips pulled out of the ferrules about ½ inch.

Aside from this the propeller stood the test perfectly. In view of this the propeller section recommended that new tips designed for a *JN-4H* plane be put in the same hub and flight tests be made. Before doing this the ferrules were given a greater taper at the wedge end and a thicker wedge was made. The new tips were made of mahogany. The completed propeller is shown in figure 153 as installed on the Curtiss *JN-4H* plane. The control mechanism is shown in figure 154.

The first flight was made on May 1, 1918. This flight called for setting the pitch of the propeller so that the motor would turn up 1,650 r. p. m. on the ground and then to fly to the maximum altitude without changing the pitch. The result was 18,000 feet in 60 minutes. The second flight called for climbing to maximum altitude by adjusting the pitch so as to keep the r. p. m. of the motor at 1,650 during climb. The result was 19,000 feet in 50 minutes and 20 seconds. The third flight called for setting the pitch to check with a fixed pitch propeller of the same design. This showed an altitude of 16,600 feet in 54 minutes and 45 seconds. The fixed pitch propeller on the same plane did 16,000 feet (ceiling) in 52 minutes. A complete report of these flights is given in report Serial No. 128, dated May 15, 1918, issued by the Propeller Section, Airplane Engineering Department, McCook Field, Dayton, Ohio.

After these tests were completed the propeller was shipped to the British Government. The one from Rockwell Field was shipped to the United States Navy to be used in tests.

The propeller section at McCook Field then began building another adjustable pitch propeller like the one that had been tested. It was to be made from a steel forging, and therefore took some time to finish. It was not ready for whirling tests until July 27, 1918. The angle of the operating pins was changed, which made the movement of the blades opposite to what they were on the former hub; also the operating pins were not allowed to go through the hub on the opposite side from the sliding member. This latter proved a mistake, as the pins had a bearing in only half of the blade when they were pulled out to their extreme position. The propeller was given a destructive whirling test, which lasted 5 hours and 16 minutes at 330 horsepower. As a result of this test both blades broke off in the ferrules, as shown in figure 155. The ferrules in this propeller were tapered toward the tips as well as toward the hub end of the blade. This allowed them to become loose and vibrate in case they pulled at all, which they evidently did as the wood was chafed around the ferrule. This outward taper was not used on the former ferrules.

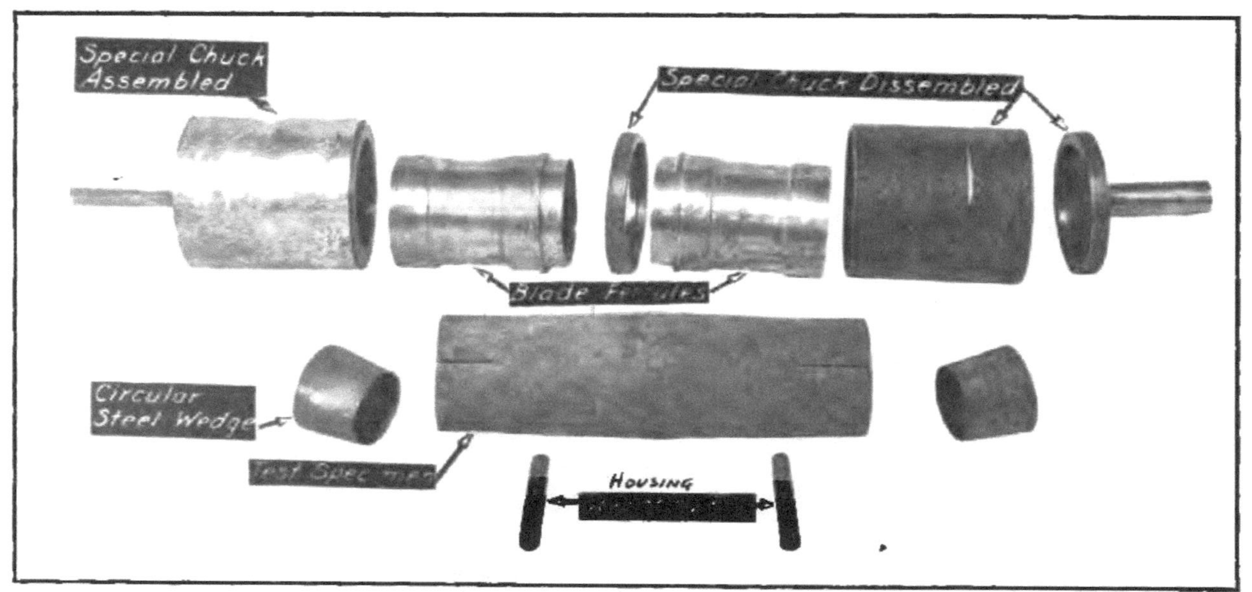


FIG. 156.—SPECIAL JAWS FOR HOLDING TEST SPECIMENS.

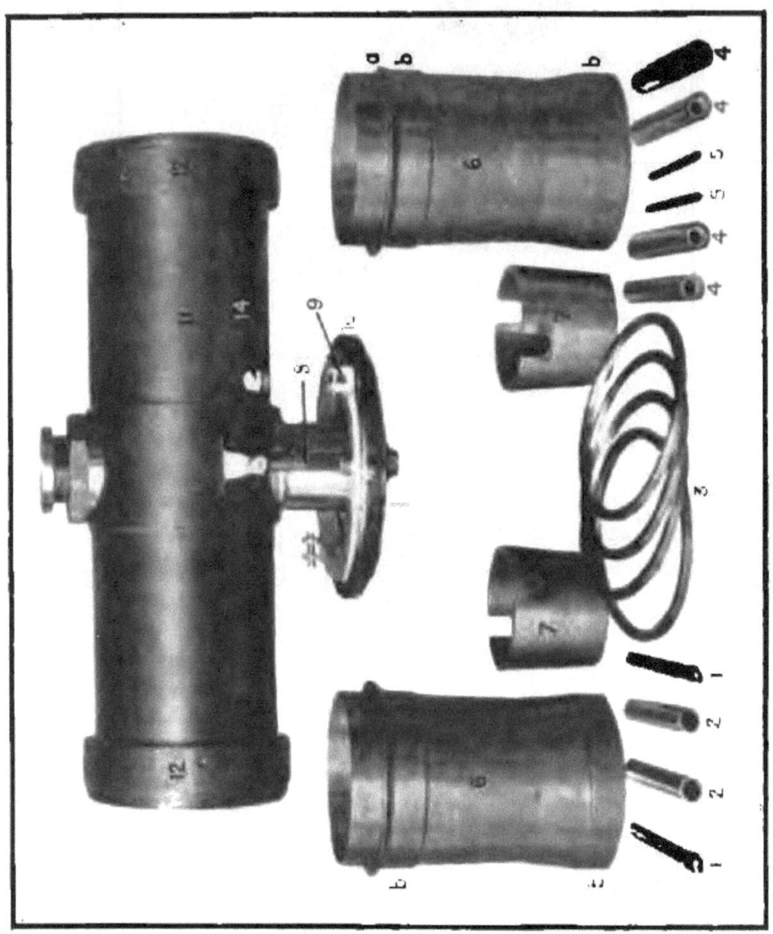

FIG. 157.—VARIABLE-PITCH PROPELLER HUB FOR LIBERTY-12 ENGINE.

268

FIG. 158.—ASSEMBLED VARIABLE-PITCH PROPELLER HUB FOR LIBERTY-12 ENGINE.

The propeller section then began a series of tests to ascertain if it was not possible to get a stronger wood for the blades than mahogany. A wood was desired which could be more successfully held in the ferrules. New ferrules were made up and a round block of wood made to fit these ferrules. They were then doweled and wedged the same as would be done in actual propeller construction. Then by means of a simply constructed chuck the sample ferrules were held in the jaws of a standard tension-testing machine and a load applied. The tendency was to pull the wood from the ferrules. As a result of these tests, birch seemed to be the most suitable wood for the blades. Figure 156 illustrates the special jaws for holding the test specimens during the tests. Mahogany developed a tensile strength of 452 pounds to the square inch, while birch developed 919 pounds to the square inch.

After the conclusion of these tests the writer was asked to enter the employ of the Government for the purpose of superintending the building of two adjustable pitch propellers for further tests, one for the Liberty 12 motor and one for the Hispano 150 motor. The outside diameter of the hub was increased to 5½ inches for the Hispano propeller and to 6 inches for the Liberty propeller. The Hispano hub and ferrules were made from Shelby seamless steel tubing. It was impossible to get this same tubing in large enough diameter for the Liberty hub, so a piece of steel was purchased and hogged out to the desired inside diameter. This latter steel did not have the tensile strength desired, but it was afterwards found to stand up well under all the necessary tests. The Hispano hub had the welded shaft plug and welded splines. The Liberty hub was redesigned so as to do away with the welding. A separate shaft plug was machined to fit the shaft plug hole in the hub. This plug was fitted with four keys and was turned to a press fit in the hub. A shoulder about 3 inches from the motor end of the plug fitted against a counterbore in the hub. On the other end it is threaded. After being pressed in the hub a nut is screwed on these threads and held in place by a set screw. Two views of the complete hub are shown assembled in figures 157 and 158. In figure 157 the hub is partially assembled, and in figure 158 it is wholly assembled. In place of the welded splines on the shaft plug two brackets were screwed onto the hub. From these a brass tube was carried to a support offset on the end of the shaft plug. The brass tube replaced the spline slides of the earlier designs and served as a guide upon which the sliding ball race operates. The two views of the assembled hubs give a very comprehensive idea of the working principles of the mechanism.

The steel in the center portion of the hub, where the shaft plug goes through, is ¾ inch thick. Trapezoidal threads are turned on the end of the hub to take the thrust collar shown at 12, figure 157.

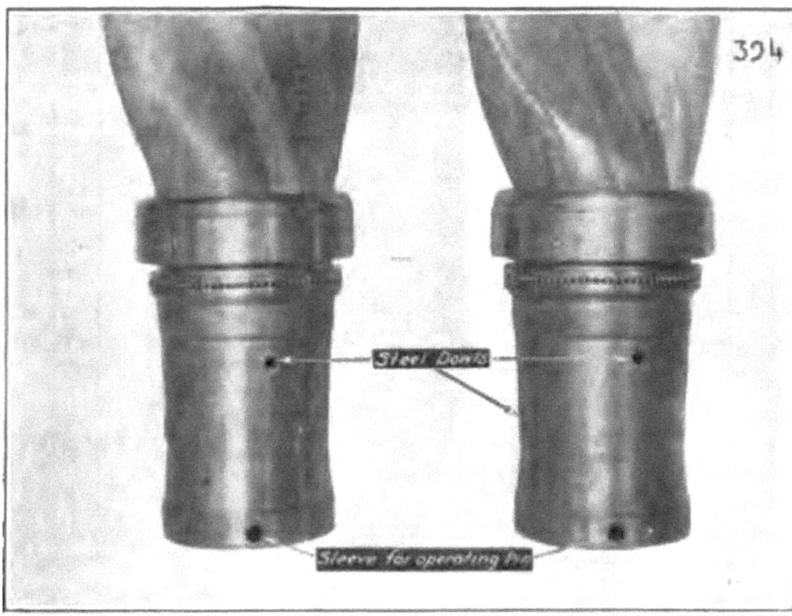

FIG. 159.—BLADE FERRULE, THRUST COLLAR, AND BALL BEARINGS.

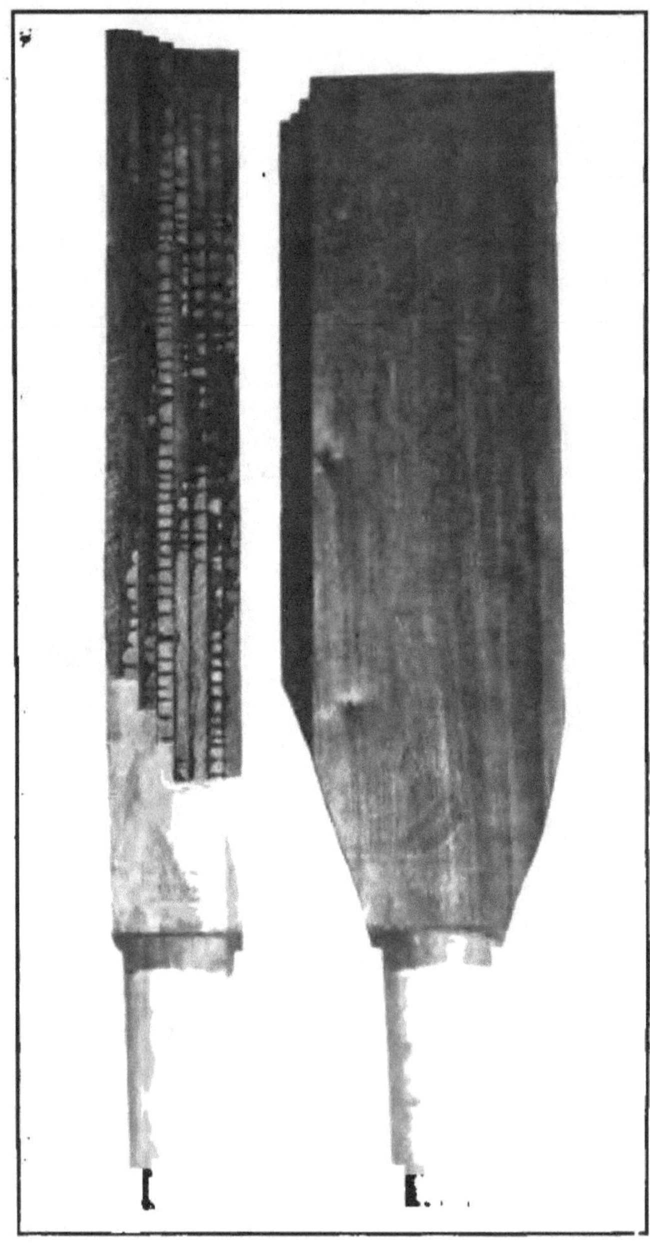

FIG. 160.—GLUED-UP BLOCK TURNED TO RECEIVE FERRULE AND WEDGE.

Two holes (14, fig. 157) ⅝ inch wide and 1¼ inches long are cut through the hub on each side of the shaft plug and directly opposite each other. These are to accommodate the operating pins (shown at 1). The operating pins attach to the sliding ball race at (9). The sliding member is an aluminum casting around which is a ball race containing 12 steel balls $\frac{5}{16}$ inch in diameter, held equally spaced by a spacer ring. The outside ring (shown at 10) of this ball race has two lugs for attaching the control levers from the pilot's cockpit. Each ferrule has two ball race rings which fit between the thrust collar and the ferrule, as shown in figure 159. The wooden blades are glued up in blocks of eight laminations of ⅞ inch thickness and the desired length. These blocks are then put in a lathe and the root of the blade turned to fit the steel ferrule. Eight saw cuts are then made in the wedge end to allow the wood to break uniformly while the wedge is being pressed in. When the blocks are turned to fit the ferrules a stub is left just the same diameter as the inside of the circular steel wedge. This is used as a guide for the wedge while it is being pressed into the blades. These operations are illustrated in figure 160. After the circular steel wedge has been inserted wooden wedges are driven into the saw cuts to expand the wood circumferentially. A $\frac{5}{16}$-inch steel dowel is driven through the wedge and the wood for the purpose of tying them together, so that in case the wood should have a tendency to move outward it would carry the wedge with it. These dowels are hollow and hardened except at the ends, where they are riveted over. They are shown in figure 159. The first ferrules that were made were machined to fit the metal hub before being pressed into the wood. Later in pressing them onto the blades they went out of round to such an extent that after the high spots were taken off they did not fit well. This has been overcome by leaving the ferrules oversize and turning them to fit the hub after the wood has been pressed in. The ferrules are pressed on by a hydraulic press. To do this requires about 10 tons pressure. The wedges are also pressed in with the same press. This requires about 34 tons pressure. The wedge for the Liberty propeller is 4½ inches inside diameter, 2$\frac{3}{16}$ inches long, and ¼ inch thick at one end and tapered on the outside to a thin edge at the other end. They have two slots at the thickest end, through which the sliding-pin sleeve passes after the wedge is pressed into the wood. (See item 7, fig. 157.) These slots are cut out of the wedges so that in case the wood pulls in the ferrule it will not cause any pressure on the sliding-pin sleeve. Any pressure on this sleeve will tend to bend it and thereby bind the sliding pin so as to make it inoperative. After the hole has been drilled for the sliding-pin sleeve the wood is cut out back of the hole so as to eliminate any possibility of pressure on the sleeve. After the ferrules on the wood have been turned to fit the hub, the

FIG. 161.—HART VARIABLE-PITCH PROPELLER INSTALLED ON DH-9A PLANE.

propeller is assembled as shown in figure 158. The tips are then carved to form and finished in the usual manner, as described in Chapter II.

The Liberty propeller when finished was run on the test rig for 10 hours at 600 horsepower. At the end of the run the blades were closely examined for any sign of failure. They seemed to be in perfect condition and had not pulled in the ferrules. The ball races were also examined and found to be in perfect condition. This propeller is now being installed with a suitable control to the pilot's cockpit on a *DH-9A* for flight tests. (See fig. 161.)

While these propellers were being built for the Army, the Navy aeronautical department asked that two adjustable pitch propellers be built for the *F-5-L* flying boat. Work on these propellers was immediately commenced and they were finished and each tested for 10 hours at 600 horsepower on the test rig at McCook Field. The ferrules did not fit well in the hubs on account of having been made to size before the wood was pressed into them. It was therefore decided to make new tips for these hubs and to turn the ferrules to fit the hub after the wood had been pressed in. This was done and a much better fit was obtained. After running each of them on the test rig for 10 hours at 600 horsepower they were found to be in perfect condition. Probably the principal reason for the wood not pulling in these two propellers was that the ferrules used had the straight sides, with the exception of the inside taper at the wedge end. It has been definitely decided that the ferrule with the taper running toward the blade is not as good as the ones without this taper. Flight tests will soon be made with these propellers on the *F-5-L* flying boat.

Another propeller of 9-foot diameter was made for flight tests on the Loening monoplane. A new control to the pilot's cockpit had also to be made for this plane, as there was no way by which the previous controls could be fitted to it. A bracket was designed and cast of aluminum. This carried a shaft on which were two throw levers which connected with the lugs on the sliding ball race. Another lever was attached to the end of this shaft. A rod ran from the lower end of this lever along the side of the fuselage and connected with another lever attached to a shaft which ran through the side of the fuselage to a point just in front of the pilot's seat. On the inside end of this shaft a hand lever was attached which moved in a quadrant. On this hand lever a pointer was mounted which indicated on the quadrant the pitch at which the blades were set. The propeller was given a flight test April 29, 1918, but was in the air only a few minutes. On this test the pilot was unable to change the pitch, due to some hitch occurring in the controls. Upon examination it was found that one of the tapered pins that held one of the

161988--21----18

levers to the shaft had sheared off. Before the proper repairs could be completed the Loening plane was taken out of flying commission for the purpose of mounting guns on it. Further flight tests with this propeller have been impossible as there is no other similar plane available with a Hispano 300-horsepower motor.

*Reversible adjustable pitch propellers.*—In the latter part of September, 1918, the Navy requested that a reversible propeller, along the same lines of the adjustable pitch propeller, be made for trial tests on the Blimp, as it was believed that with such a propeller a Blimp could make a much safer and quicker landing if the pusher propellers then in use could be changed into tractors which would serve as a brake. Also while in flight one propeller could be used as a tractor and the other as a pusher, which would give the ship a much quicker turn.

Some difficulty was experienced in designing such a propeller as the diameter could not exceed 8 feet 6 inches, because the motors on a Blimp are set on both sides of the fuselage with the distance from the fuselage to the center of the crank shaft of only 4 feet 9 inches. Also it became necessary to increase the number of degrees through which the blades could be changed so as to throw the blades enough past center to give a reverse thrust. The shaft plug in the hub was made one-half inch longer in order to give the sliding member more travel, and the operating pins were put in the blades at a greater angle. By these changes it was possible to get a pitch change of 20°. Fifteen degrees was allowed for the blades while they were set as a tractor, which allowed 15° reverse.

The first propeller built under these conditions was designed for a 150-horsepower Hispano-Suiza motor to turn at 1,500 r. p. m. This propeller was put on the test rig at McCook Field and given a whirling test, with the result that it absorbed only 101 horsepower and gave a thrust of only 60 pounds. This was not satisfactory. As the diameter of the propeller could not be increased it was necessary to increase the width of the blades. A second propeller was immediately built and tested, with the result that it absorbed 150 horsepower at 1,500 r. p. m. and gave a reverse thrust of 104 pounds. This was considered satisfactory. Another propeller of the same design was built and the two were then shipped to Akron for installation on a Blimp.

*Reversible adjustable pitch propellers for airplanes.*—There is now being built at McCook Field a reversible adjustable pitch propeller for a Curtiss *JN-4H* plane with a Hispano 150-horsepower motor.

This propeller is to be used to determine in what distance a plane can be stopped in making a landing by using it as a pusher.

Calculations show that a Curtiss *JN-4H* landing at 50 miles per hour can be stopped within 200 feet, and that a Thomas-Morse scout

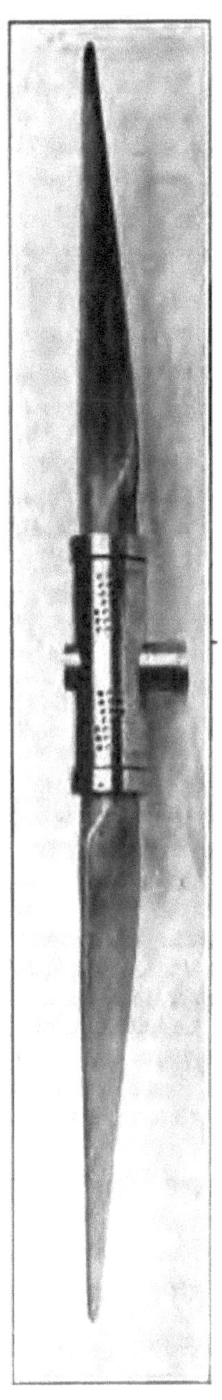

FIG. 162.—MODEL OF HART VARIABLE-PITCH PROPELLER FOR WIND-TUNNEL TESTS.

landing at the same speed can be stopped within 95 feet. This will be an advantage where it is necessary to use small landing fields, such as will have to be used by the aerial mail planes in some places.

This propeller will have a total change of pitch of 29°, 15 of which will be used as a tractor and 14 as a pusher.

A 3-foot model has been made as shown in figure 162. This will be used for wind tunnel tests at Stanford University.

---

## Steel Propellers.

Twenty-nine steel propellers have been presented to the Propeller Section, Engineering Division, Air Service, United States Army, for test and approval.

These propellers may be divided into three general classes: First, steel propellers reinforced with some composition; second, fabricated steel propellers; and third, tubular steel propellers.

*Reinforced steel propellers.*—Two propellers of the first class have been tested. As nearly as could be determined after the failure of the propellers, the first had been made of two sheets of soft cold-rolled steel, cut to shape and formed. These sheets were held together by a lapped seam along the leading and trailing edges. For a distance of 2 feet on both sides of the hub, the two sheets forming the propeller were reinforced by two sheets of the same kind of steel soldered to them. A small hole had been cut at the center and the sheet metal drawn inward to form a bushing for the hub.

This hollow-steel framework was then covered with some black compound. No definite information was available as to the exact nature of this material. The inventor who presented the propeller said that the material was a "sort of vulcanized rubber." The material had been forced into the propeller at the hub hole, and afterward drilled to receive a standard *OX-5* engine hub.

The propeller was run on a destructive whirling test for one-half hour at 1,000 r. p. m. and one-half hour at 1,200 r. p. m. While the speed was being raised to 1,400 r. p. m. the propeller failed. Actual failure occurred at 1,330 r. p. m. The failure seems to have been due to the blades not being strong enough to stand the centrifugal forces set up at high speed.

The second propeller of this type was constructed in exactly the same manner as the first except that the reinforcing sheets of steel had been welded to the inner sheets instead of merely lapped and soldered, as in the first propeller.

This propeller failed at 1,690 r. p. m. after running for one-half hour at 1,000, 1,200, and 1,400 r. p. m. The nature of the failure was exactly the same as that of the first propeller tested. Figure 163 shows the condition of the propeller after failure.

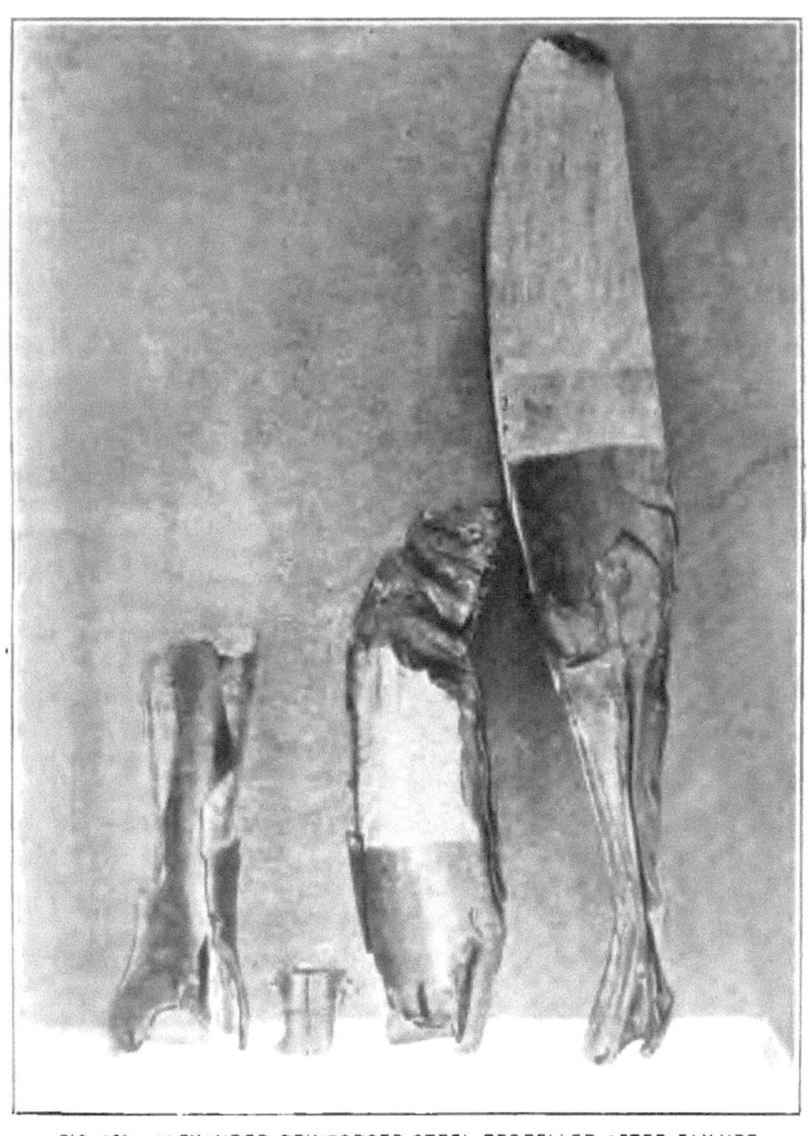

FIG. 163.—ALEXANDER REINFORCED-STEEL PROPELLER AFTER FAILURE.

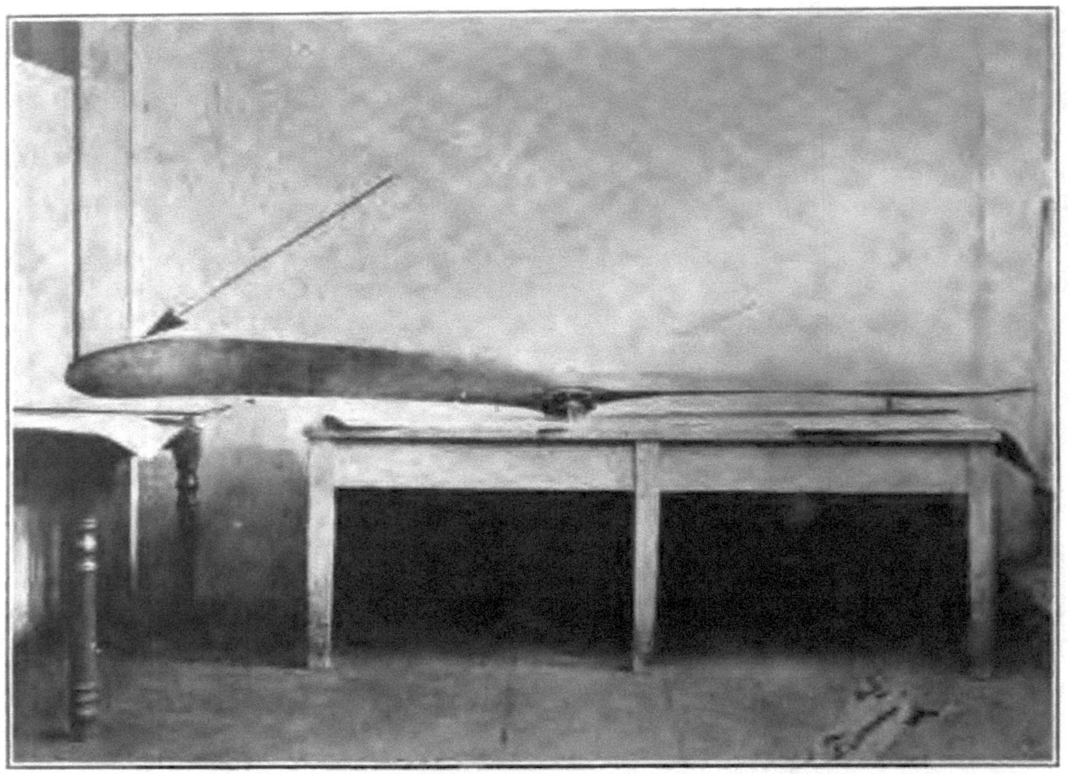

FIG. 164.—LUENSE FABRICATED HOLLOW STEEL PROPELLER AFTER FAILURE.

The failure of these propellers seems to have been due to the fact that too much faith had been put in the strength of the composition and not enough in the strength of the steel. Examination of the break shows that the propeller pulled apart at the hub, due to an insufficient amount of material for the forces involved.

*Fabricated steel propellers.*—This class of propellers may be for convenience subdivided into three groups, the first group including those propellers built of sheet metal with or without some form of internal bracing, the second group including those propellers having the hub and blades built up separately and afterward fastened together, the third group including those propellers having the hub and blades built separately but having the blades forged or machined from solid steel plate or bar stock.

*Group 1.*—Four propellers of this type have been tested, two of which failed, one of which ran for the required length of time but was doubtful, and the fourth, which stood the required test without failure.

The first propeller of this group, known as the Luense hollow steel propeller, was a hollow steel propeller—diameter 8 feet 10 inches, pitch 7 feet—apparently designed for the Curtiss *OX-5* engine. This propeller was built of two sheets of what appeared to be commercial cold-rolled steel. One of these sheets formed the cambered face of the propeller; the other formed the thrust face. Each of these sheets was cut from a suitable pattern and hammered and rolled to give it the proper shape for its half of the propeller. These sheets were then welded together to form the completed propeller. As far as could be determined the sheet metal used in the construction of this propeller was of the same gauge throughout. No extra reinforcement was used at the hub. It should be noted that the method of cutting the sheets for this propeller caused a seam to occur on each side of the propeller at the hub.

This propeller failed after running 24 minutes at 1,404 r. p. m. Previous to failure it had run for one-half hour at 1,008 r. p. m., and one-half hour at 1,215 r. p. m. The failure was due to the welded seam opening for about 15 inches on both the leading and trailing edges of one blade at the tip. There were also two cracks near the hub, one of which was along a seam. The appearance of this propeller after the test is shown in figure 164. It may be noted that an examination of the seam at the point of failure showed that the welding had been very poorly done.

The second propeller of the first group, known as the Kramer steel propeller, was of practically the same construction as the first, the only real difference being that the second propeller was reinforced at the hub by an internal false hub. This consisted of two flanges held apart by one large tube forming the bore, and eight

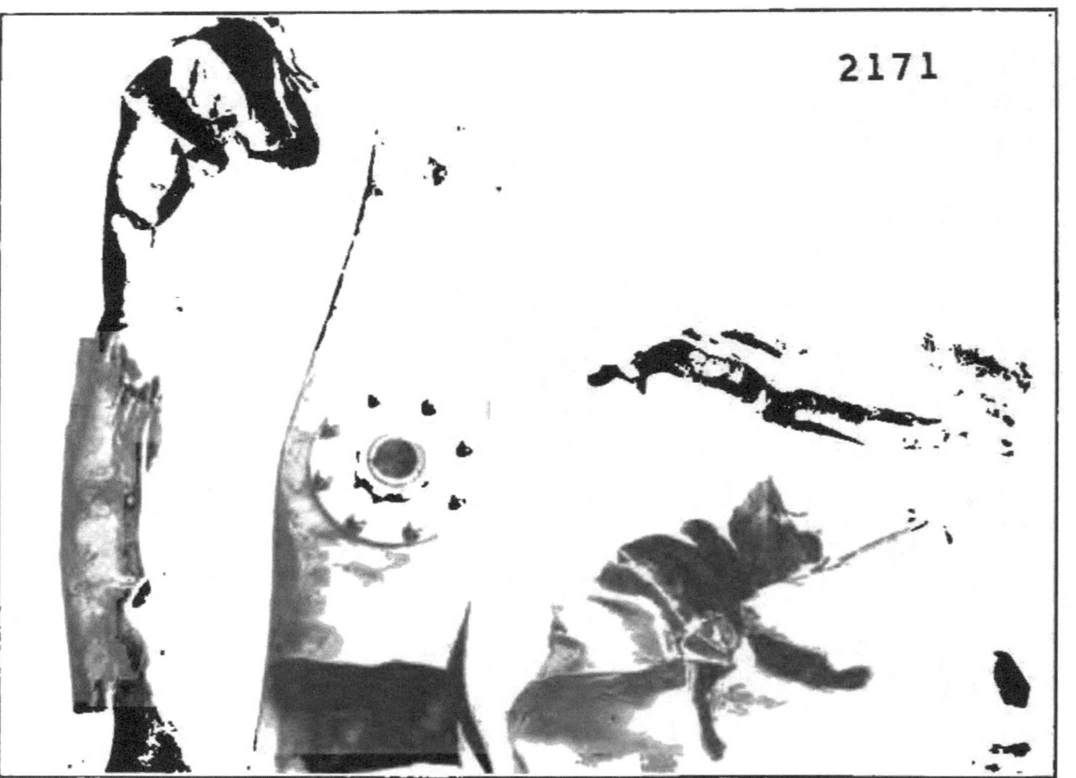


FIG. 165.—KRAMER HOLLOW STEEL PROPELLER AFTER FAILURE.

small tubes forming the holes for the hub bolts. The steel sheet of which this propeller was constructed was $\frac{1}{16}$ inch thick at all points. All the seams in this propeller occurred at the same points as in the first propeller tested.

The diameter of this propeller was 9 feet 2 inches, and judging from its shape it was a copy of propellers built to Curtiss drawing D–5000.

This propeller failed at approximately 1,786 r. p. m. after running on the test rig for two hours. Previous to this test it had been running successively at the following speeds: 1,010, 1,197, 1,391, 1,600, and 1,790 r. p. m. At 1,790 r. p. m. it showed a slight flutter. Figure 165 shows the appearance of this propeller after failure. Its appearance before failure was similar to that shown in figure 164.

The third propeller of the first group, known as the Leitner metal propeller, was of English design and construction. It was 8 feet 8 inches in diameter and had a pitch of 6.34 feet.

This propeller was built of two sheets of steel—thickness unknown—which formed the thrust and cambered faces of the propeller. The arrangement of these sheets was different from the arrangement of the sheets in the propellers described above in that each sheet forms the thrust face of one blade and the cambered face of the other blade. These large pieces were stiffened by other pieces of graduated lengths like a leaf spring placed inside the propeller and riveted together and to the large pieces. There were also two pieces of an elongated diamond shape which form the faces of the hub. A bushing for the engine shaft was inserted in the propeller. The exact method of fastening this bushing could not be determined without damaging the propeller.

A power curve was run on this propeller from 600 to 1,420 r. p. m. in steps of 200 r. p. m. each. It was then run for 10 hours at approximately 1,400 r. p. m. At the end of this test the propeller had a crack 2½-inches long in the leading edge of one blade. This crack had been 1¾-inches long when the test was started.

This propeller was designed for an engine of approximately 150 horsepower, and at 1,400 r. p. m. it absorbed approximately 230 horsepower. This represents an overload of 50 per cent. Since the propeller passed the foregoing test in a comparatively satisfactory manner it would probably be safe for use on an engine of the horsepower for which it was designed. It should be said, however, that the test given this propeller was not nearly as severe as is demanded on steel propellers presented to the Air Service for adoption.

Figure 166 shows this propeller before the test. The arrows 1 and 2 show the position of the crack before the test was started. Arrows 1 and 3 show the increase in length of the crack during the test.

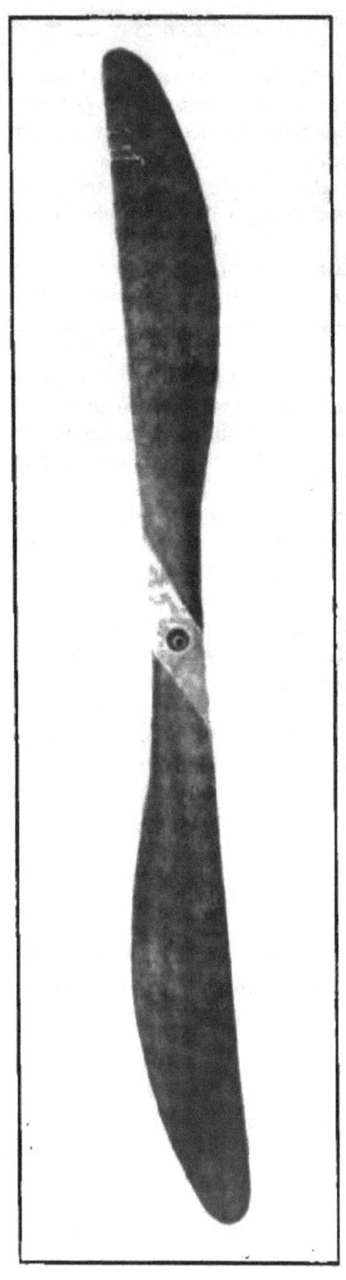

FIG. 166.—LEITNER STEEL PROPELLER AFTER FAILURE.

The fourth propeller of the first group was designed and built by Capt. Vladimir Olhovsky for the Russian Army. It was 9 feet 3 inches in diameter and had a pitch of 7.3 feet.

This propeller was built of two sheets of steel in very much the same manner as the propellers previously described. There was no reinforcing in the propeller except at the hub, at which point a wooden block had been inserted in order to prevent the center of the propeller being mashed by the hub. The workmanship on this propeller was very poor, especially on the seams which were welded.

A power curve was taken on this propeller from 1,000 to 1,657 r. p. m. in steps of approximately 200 r. p. m. each. At 1,657 r. p. m. it absorbed approximately 600 horsepower, and was run at this speed for 10 hours. It may be noted that this propeller ran with practically no fluttering during the test. At the end of the test the propeller appeared to be in as good shape as at the beginning.

Figure 167 shows this propeller at the conclusion of the 10-hour run at 600 horsepower.

*Fabricated steel propeller (Group II).*—The first propeller of this group, known as the Ingells steel propeller No. 3, was 8 feet ½ inch in diameter and had a pitch of 3.75 feet.

It was made of two blades riveted to a cast-steel hub. The blades themselves were made of two sheets of steel, one forming the camber face and the other the thrust face. These sheets were fastened together by lapping the edges of the sheet forming the thrust face around the edges of the other and welding it. The blades were fastened to the hub by means of five rivets. (See fig. 168.) The workmanship on this propeller was poor, the thrust face of one blade being concave and the other being convex.

A power curve was taken on this propeller from 1,000 to 1,800 r. p. m. in steps of approximately 200 r. p. m. each. Failure occurred at 1,800 r. p. m. while the speed was being adjusted. The rivets holding the blades to the hub sheared, causing a total wreck of the propeller. The extent of the failure is shown in figure 168.

The second propeller of this group, known as the Ingells steel propeller No. 4, was 8 feet in diameter; the pitch was unknown.

This propeller, like the preceding one, was built of two blades riveted to a cast-steel hub. The blades themselves were made of two sheets of steel, one forming the thrust face and the other the camber face. In this propeller these sheets were fastened together by lapping the sheet forming the thrust face ⅜ inch over the camber face. This lap was then spot welded and soldered.

Each blade was fastened to its respective arm on the cast-steel hub by four $\frac{9}{16}$-inch elliptical rivets. These rivets were made elliptical in shape in order to increase the bearing area.

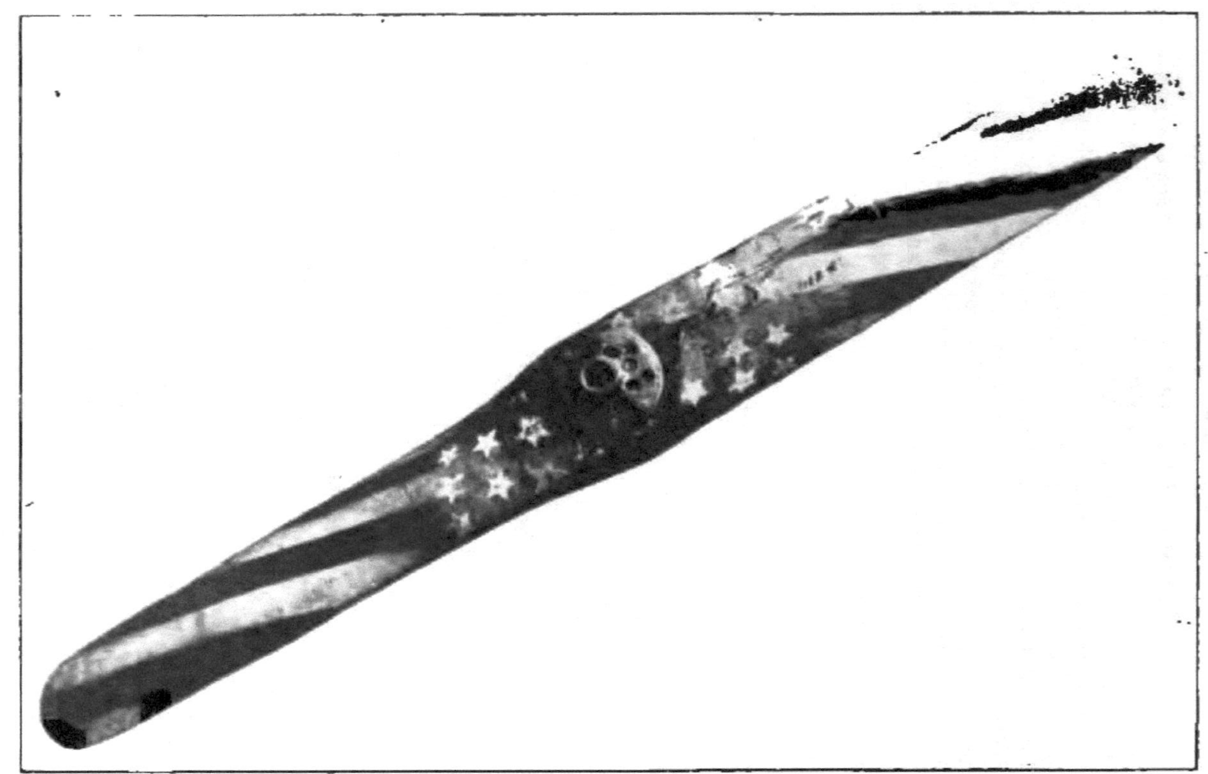

FIG. 167.—VLADIMER OLHOVSKY STEEL PROPELLER AFTER TEST.

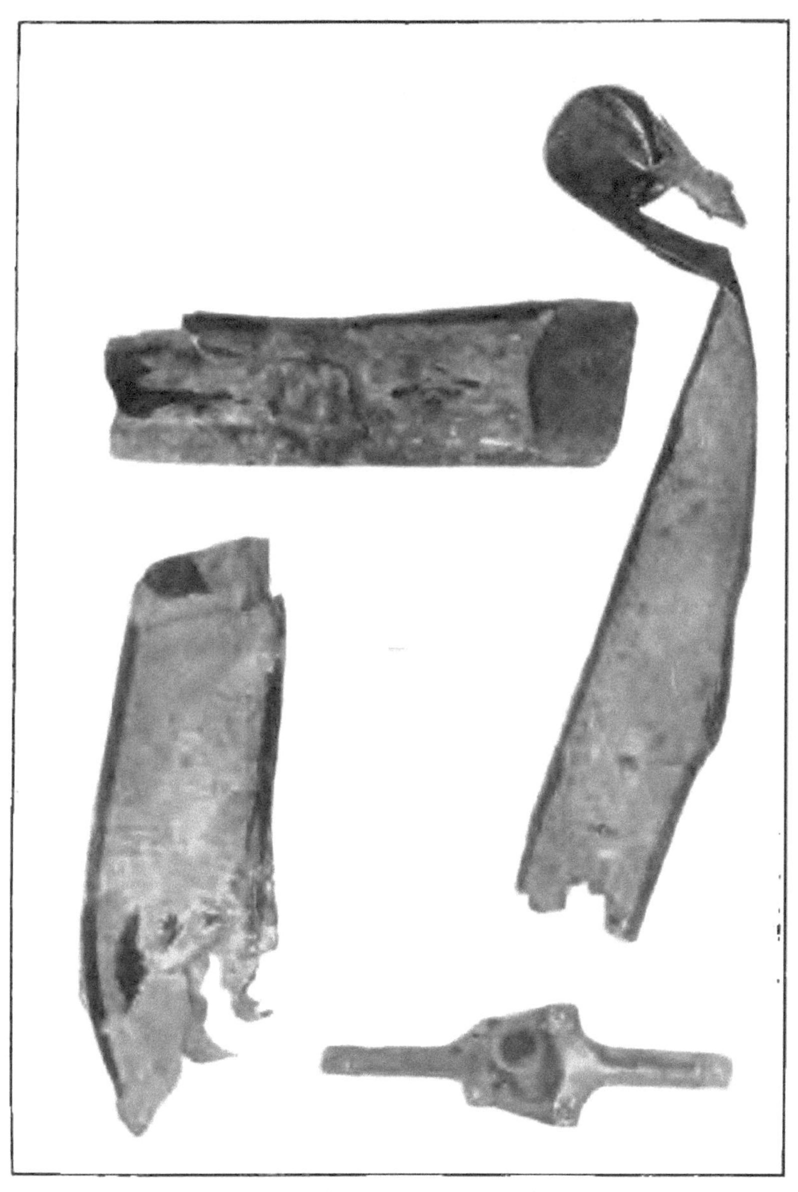

FIG. 168.—INGELLS STEEL PROPELLER NO. 3.

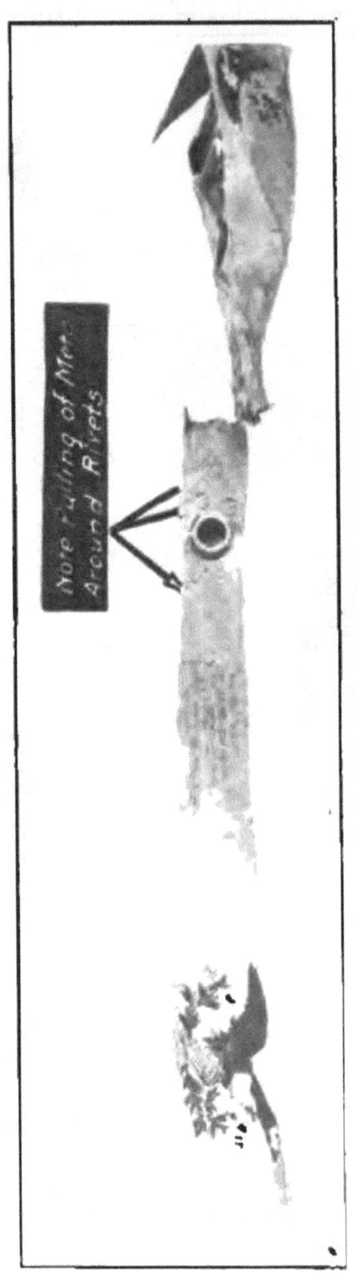


FIG. 169.—INGELLS STEEL PROPELLER NO. 4 AFTER FAILURE.

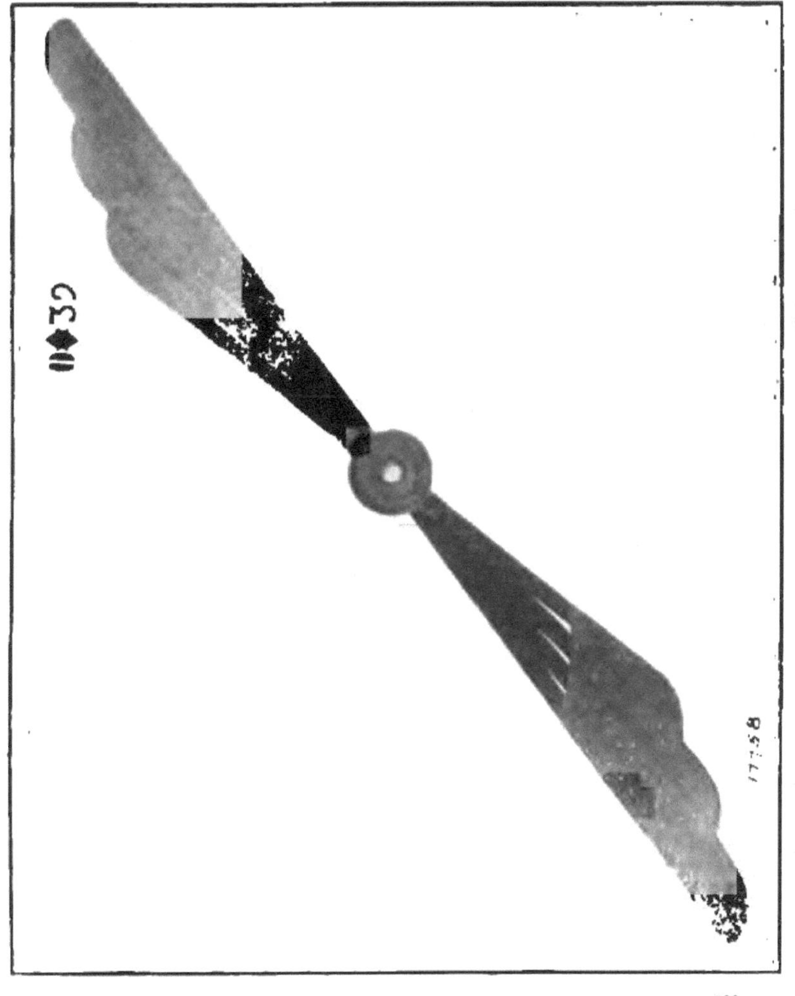

FIG. 170.—FAEHRMANN STEEL PROPELLER BEFORE TEST—THRUST FACE

A power curve was taken on this propeller in steps of 100 r. p. m., from 1,000 up to 1,700 r. p. m., at which speed it failed, due to one blade breaking off at 8½ inches from the hub. An examination of the metal remaining attached to the hub after the failure showed that the metal back of each rivet was failing at the time the blade broke. The propeller is shown in figure 169 after failure. From this figure a very good idea can be obtained of the appearance of the propeller before testing.

The third propeller of this group, known as the Faehrmann steel propeller, was 8 feet 6 inches in diameter and 5.5 feet pitch. It was designed for the *OX-5* engine.

This propeller, shown in figure 170, consisted essentially of two blades and the hub. Each blade was made of four boiler-plate ribs extending from the hub to the tips of the different sections of the blade. On the front face of these ribs, for a distance of 24¾ inches inward from the tips of the blades on the leading edges and 7 11/16 inches on the trailing edges, there were riveted three plates of commercial cold-rolled steel plate approximately 0.03 inch thick.

The ribs which fastened onto the hub were bent through an angle of 90° at their inner ends. At this bend or angle there was a projection ⅜ inch long perpendicular to the flat surface of each rib. This projection on each rib engaged behind a shoulder on the inside of one of the rings forming part of the hub. The ribs were spaced apart at the hub by washers and retaining rings. The attachment of the blades to the hub is illustrated in figure 171. This hub was much heavier than was necessary on account of the fact that it was designed to use a standard *OX-5* engine hub. It could have been built to fit directly on the engine shaft, thereby reducing the weight very materially.

The workmanship on this propeller was of the highest character throughout.

A power curve was run on this propeller from 1,000 to 1,800 r. p. m., in steps of approximately 200 r. p. m. each. At 1,800 r. p. m. it absorbed approximately 230 horsepower, and was run at this speed for 30 hours. When shut down at the end of the test it was found that the steel plating had cracked on both blades along the trailing edges. This failure is not considered serious in view of the severe test applied. None of the essential parts of the propeller showed any signs of strain.

The design of this propeller is not believed to be an efficient one aerodynamically, but it incorporates features of design which would seem to be correct for a steel propeller.

The fourth propeller of this group, known as the Diggle patent all-steel air screw, was 10 feet ⅜ inch in diameter. It had a pitch of

8.12 feet. This propeller was brought to the United States by Diggle, of London, England. Propellers of the same general struction had been previously tested in England. These prope

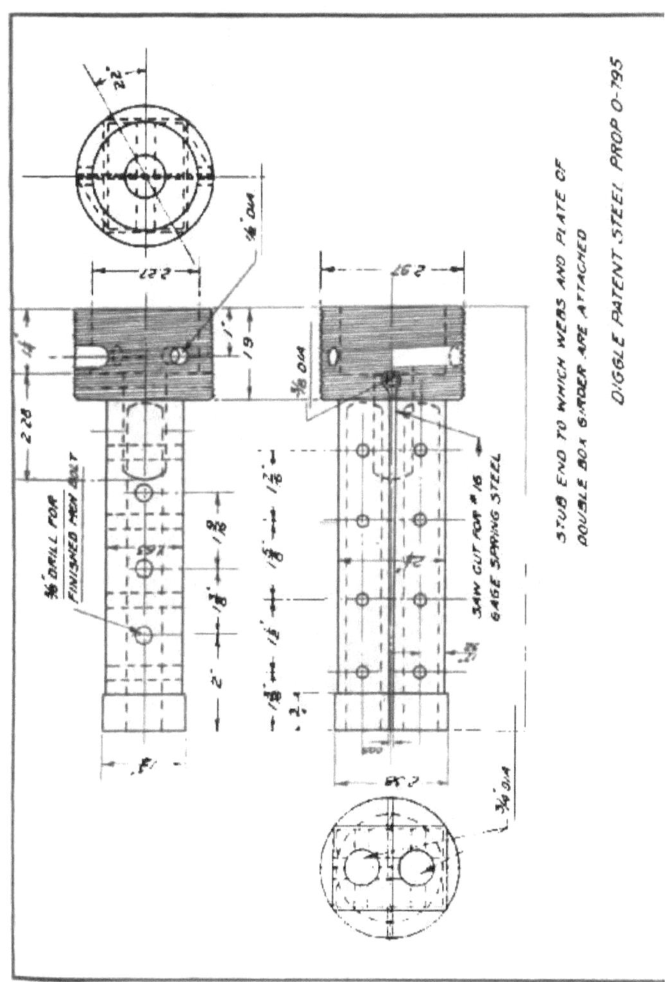


were all for engines of much smaller horsepower than the Liberty for which this one had been designed.

Essentially this propeller consisted of a steel hub and two fa cated steel blades. The hub of this propeller was machined fro

FIG. 171.—FAEHRMANN STEEL PROPELLER—HUB CONSTRUCTION.

solid steel forging. The ends of this hub into which the blades were fastened were threaded to receive the threaded ends of the blades.

The blades themselves consisted of a threaded stub, to which was riveted a double box girder carrying transverse ribs or spacers around and to which the covering of the blades was attached. Each stub was made from a steel forging. This forging was turned and threaded on one end to screw into the threaded part of the hub. The end to which the double box girder was attached was oblong in cross section and was 6.912 inches long. On the outside end of this stub there was a shoulder against which the reinforcing plates in the box girder rested. The skeleton framework of a model blade without the steel covering is shown in figure 172. A detail of the threaded stub is given in figure 173.

The tapered double-box girder was constructed of No. 16 gauge spring steel. Its general construction will be evident from an examination of figure 172. On the inner end of this girder, on the inside of the two outside webs, and on the inside of both plates there were riveted reinforcing plates of No. 16 gauge spring steel, which, when the blade was assembled, rested against the shoulders on the stub referred to in the preceding paragraph.

The transverse ribs serve three purposes, viz, to give the blades the proper section, to give the blades the proper angle, and transmit the stress from the covering to the box girder. These bulkheads were made of No. 16 gauge spring steel. They were slotted so as to slip through notches in the plates of the girder. This was for the purpose of keeping them in place against the action of centrifugal force. All the ribs were tied together around the edges by a spring steel band fitted into slots in the edges. This band was attached to a disk screwed to the stub. The band was riveted and welded to the disk. The photograph of the model blade, figure 172, shows the internal construction of the blades very clearly.

The cambered faces of the blades were made of 0.025-inch commercial spring steel, and the thrust faces of the blades were made of No. 24 gauge (0.022-inch) cold-rolled, close-annealed steel. The thrust face of each blade was lapped over the edge of the cambered face and welded. At the base of the blades the cover was fastened to the disks by eight cap screws.

This propeller was run for two minutes at 615 r. p. m., two minutes at 815 r. p. m., two minutes at 1,003 r. p. m. It failed at approximately 1,100 r. p. m., as the speed was being raised to 1,200 r. p. m. The failure seems to have been due to the transverse bulkheads slipping out of the notches in the plates of the girders, thereby allowing the whole cover to leave the girders. Figure 174 shows the remains of this propeller after failure occurred.

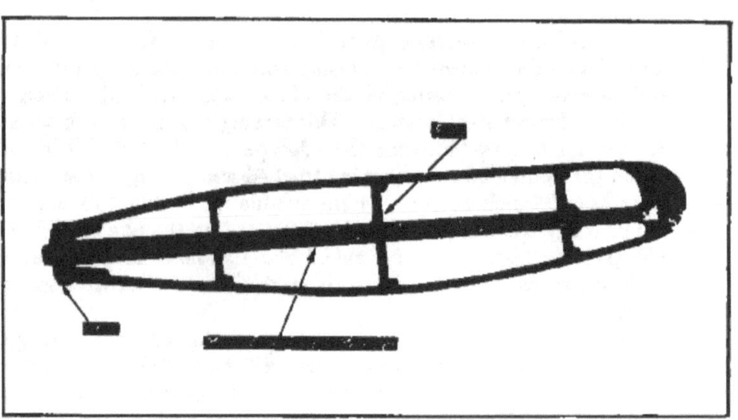

FIG. 172.—MODEL OF SKELETON FRAMEWORK OF DIGGLE STEEL PROPELLER BLADE.

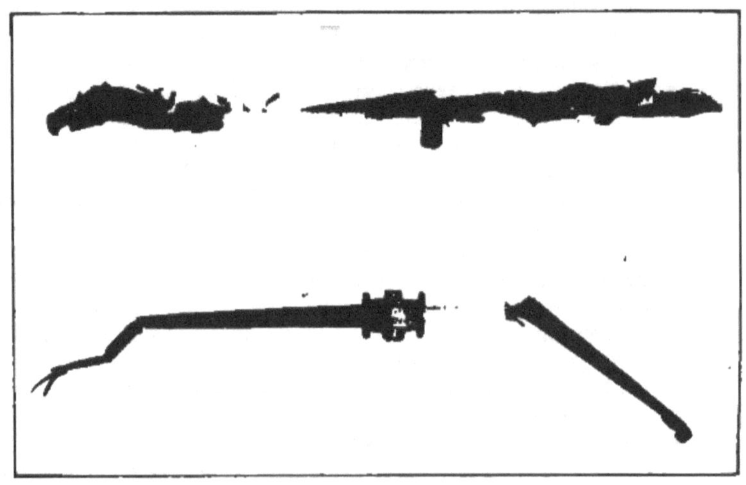

FIG. 174.—DIGGLE PATENT STEEL PROPELLER.

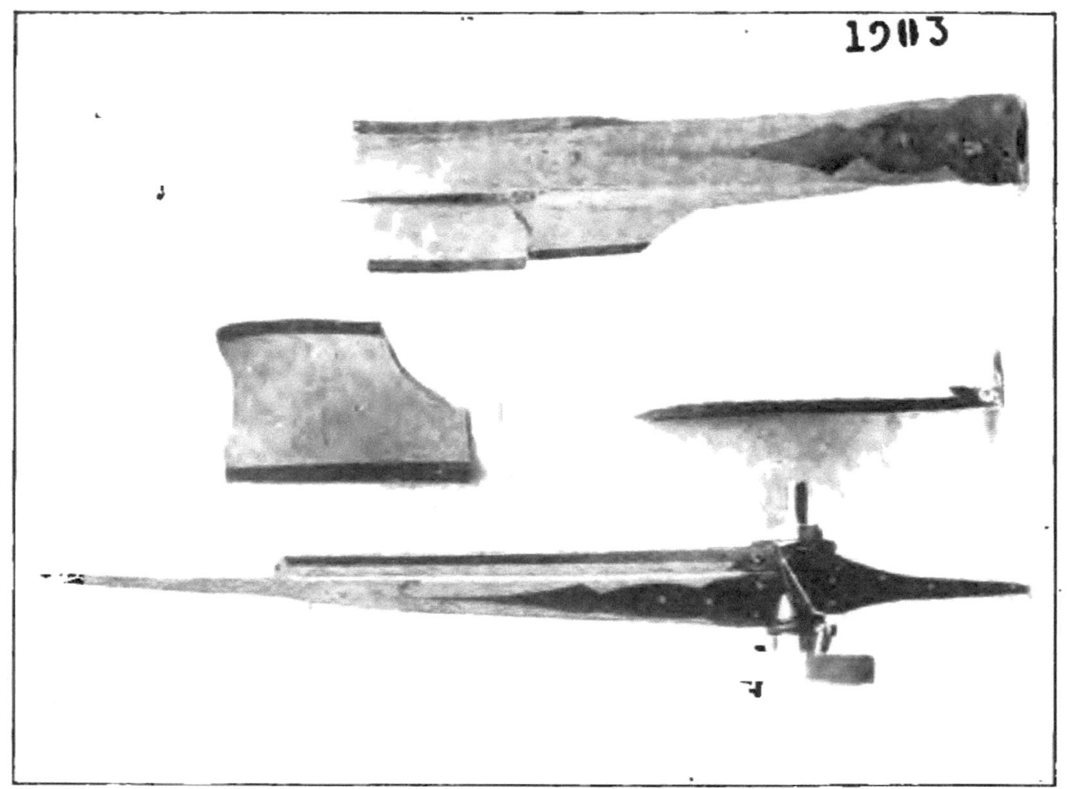


FIG. 175.—INGELLS STEEL PROPELLER NO. 2.

The Diggle steel propeller presents a number of very good features, and it is possible that a satisfactory steel propeller can be constructed along similar lines. It is believed, however, that a considerable amount of work will have to be done in order to develop a method of attaching the covering to the girder and the girder to the stub before the propeller will be satisfactory.

*Fabricated steel propeller (Group III)*.—The first propeller of this group, known as the Ingells Steel Propeller No. 1, was 8 feet in diameter and had a pitch of 2 feet.

The propeller was made of two solid steel blades riveted to a cast-steel hub. The blades themselves were made of heavy steel plate bent into a form somewhat similar to the back side of conventional air foils. Each blade was reinforced by two plates, one on the front and one on the back of the blade. The rivets holding the blades to the hub pass through the front reinforcing plate, through the blade, through the hub casting, and then through the second reinforcing plate on the back of the blade. The rivet nearest the tip was made oval shaped in order to increase its shear without a corresponding decrease in the tension area of the reinforcing plates.

A power curve was taken on this propeller from 1,000 to 1,800 r. p. m. in steps of approximately 200 r. p. m. each. It failed at 1,800 r. p. m., due to one blade pulling away from the hub.

The second propeller of this group, known as the Ingells' Steel Propeller No. 2, was 8 feet in diameter and had a pitch of 2 feet.

The construction of this propeller was exactly the same as that of the first propeller of this group tested except that the reinforcing pieces had been extended in this propeller about 6 inches farther out on the blades.

A power curve was taken on this propeller from 1,000 to 1,520 r. p. m. in steps of approximately 200 r. p. m. each. This propeller failed at 1,520 r. p. m. The failure was caused by the blades shearing the rivets attaching them to the hub. Figure 175 shows the appearance of this propeller after failure occurred.

The third propeller of this group was a propeller built by Mr. Ingells to A. E. D. drawing X-7141. This propeller was designed for the *DH-4* airplane and was a four-blade propeller 11 feet 6 inches in diameter. For test purposes the design was modified and the propeller built as a two-bladed one to absorb 200 horsepower.

The hub of this propeller was of cast steel and was designed for the standard Liberty shaft. The ends of the hub into which the blades were placed were bored with a conical hole. The outside of the ends was threaded with a buttress thread.

The blades themselves were forged from spring steel and afterward tempered to a spring temper. They were held in the hub by collars which screwed on the outside ends of the hub. On the inner

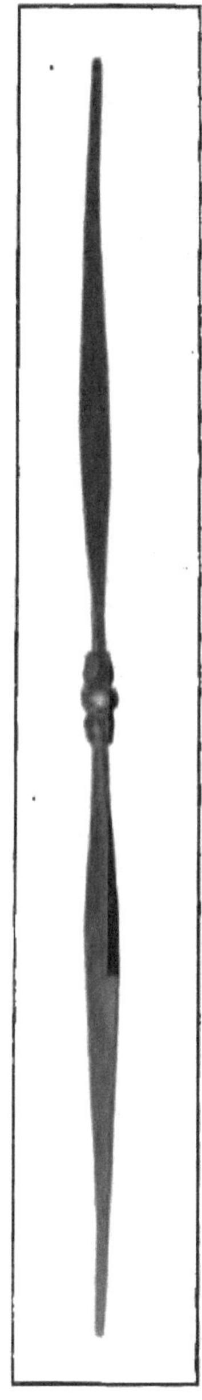

FIG. 176.—A. E. D. SOLID STEEL PROPELLER.

ends of the blades there were shoulders against which the collars pulled in order to press the tapered ends of the blades solidly into the recesses in the ends of the hub. The complete propeller is shown in figure 176.

A power curve was run on this propeller from 600 to 1,800 r. p. m. in steps of approximately 200 r. p. m. each. It was then run at 1,800 r. p. m. for 10 hours. When shut down at the end of 10 hours it was found that a small piece had been thrown off the extreme tip of one blade. This piece was 2¼ inches long and about 1 inch wide. It constituted the tip of the propeller blade which was very narrow at the end.

At the end of this 10-hour run the propeller was speeded up to 2,400 r. p. m., at which speed the cast-steel hub failed. When the hub failed the blades were thrown out against the bombproof and totally destroyed.

An examination of the hub after failure showed that there were a number of flaws in it which undoubtedly contributed to the failure.

*Tubular steel propellers.*—The tubular steel propellers which have been tested may be divided into two classes, viz, first, those which have the blades and hub formed separately, and second, those in which the propeller is formed as a whole from a steel tube.

*Group I.*—All the propellers received for test in this group were designed for the *Liberty 12* engine.

The first propeller of this group, known as Smith Steel Propeller No. 1, was 8 feet 4 inches in diameter and had a pitch of 8 feet 10 inches.

The blades of this propeller were made of 4½-inch steel tubing. The thickness of the wall varied from about $\frac{3}{16}$-inch near the hub to less than $\frac{1}{16}$-inch near the tip. The thinner ends of the tube had been pinched together, deforming the original circular section to approximate air-foil sections. The pinched ends of the tube were welded together to form the tip of the propeller, while the thicker ends had been welded to a steel internal collar and drilled for bolts which connected to a special flanged cylinder. In order to stiffen the walls of the hollow blades, rivets were used. These rivets were welded to the walls of the tube on the thrust and camber faces. There were 41 such rivets in one blade and 43 in the other. The hub end of the blades retained their original circular shape, and the internal collars were bolted to a special flanged cylinder about 8 inches long. There were 14 ⅜-inch bolts holding the blades to this cylinder. Two nuts were used in each bolt. The flanges on the other ends of the cylinders were each drilled for eight ½-inch studs. These studs were not distributed evenly about the bolt circle but were grouped on opposite sides of the axis.

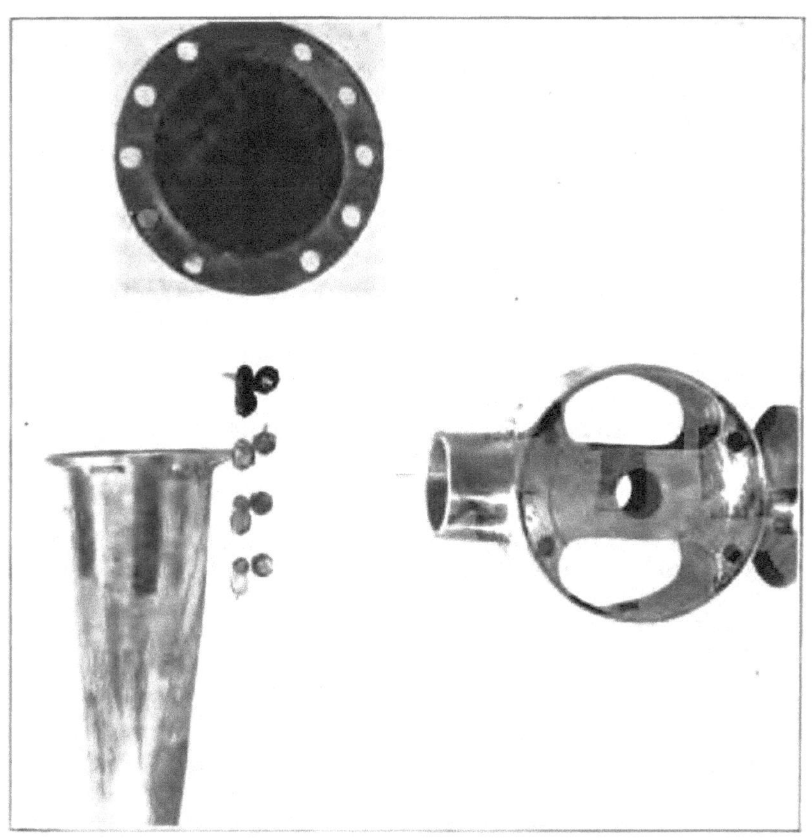

FIG. 177.—COMPONENT PARTS OF SMITH STEEL PROPELLER NO. 2.

The hub was made from a steel forging. It was bored for a standard *Liberty 12* engine shaft. The sides of this hub were faced and finished to take the flange of the 8-inch intermediate cylinder.

This propeller was run successively at 997, 1,220, 1,402, and 1,609 r. p. m. It was shut down after each of the above runs for inspection. Two rivets in each blade were found to be failing after the 30-minute run at 1,402 r. p. m. When the propeller was shut down at the end of a 30-minute run at 1,609 r. p. m. one blade was found to have cracked on the cambered face. This crack extended completely across the face, and had the propeller been run a few minutes longer the whole tip would have left the blade. The crack seems to have been due to the fact that the component of the centrifugal force parallel to the axis on the outer part of the blade was sufficient to overcome the thrust of the propeller and bend the blade downstream a sufficient amount to cause it to fail as noted.

The second propeller of this group, known as the Smith Steel Propeller No. 2, was 9 feet 6 inches in diameter and had a pitch of 7 feet.

This propeller, like the first propeller of this group, consisted of a hub and two tubular blades. In this propeller, however, no intermediate cylinders were used, the blades being fastened directly to the hub. Each blade was formed from a steel tube having walls of varying thickness. In this case the original tube was $5\frac{1}{4}$ inches outside diameter with $\frac{3}{8}$-inch walls. This tube was turned down until the walls were $\frac{5}{32}$-inch at the hub and 0.060-inch at the tip. Flanges were forged and turned on the hub end of each tube for attaching it to the hub. The air-foil shape and the pitch were then formed in a die. Inside the blades, between the faces, spools were placed in order to reinforce the propeller. These spools were held in place by rivets.

The hub was machined from a steel forging. The blades were held together and to the hub by eight bolts. The component parts of this propeller are shown in figure 177.

A power curve was taken on this propeller from 985 to 1,828 r. p. m. in steps of approximately 200 r. p. m. When the propeller was shut down after this curve it was found that three of the eight bolts holding the blades together had broken and had been thrown out. At this time both blades were in good condition except for the fact that the flanges on both blades were slightly bent. A run of a few minutes longer at 1,800 r. p. m. would undoubtedly have totally wrecked the propeller.

The use of bolts to hold the blades of a propeller together, as in the two foregoing propellers, is not believed to be safe on account of the difficulty in tightening the bolts in such a manner as to be sure that each will take its proportionate share of the load.

The third propeller of this group, known as Smith Steel Propeller No. 3, was 9 feet 7½ inches in diameter.

The construction of this propeller was the same as that of the second propeller of this group except that the ends of the blades were threaded instead of being flanged for attaching them to the hub. These threads were V-shaped, ⅛-inch deep, eight threads to the inch. The threads were not standard V threads, being flatter at the top of the threads and between the threads at the bottom.

The blades screwed into threaded recesses in the sides of the hub. There were four and one-half threads holding in this propeller. The blades were held at the proper angle in the hub by $\frac{5}{16}$-inch pins. These pins were placed in holes drilled through the hub and thread ends of the blades and were held in place by cotters.

A power curve was run on this propeller from 1,000 to 1,822 r. p. m. in steps of approximately 200 r. p. m. each. It broke at a speed above 1,822 r. p. m. as the speed was being adjusted to give an input to the propeller of 600 horsepower. The failure was caused by both blades being thrown out of the hub. An examination of the threads on the ends of the blades and the threads in the hub indicated that the failure had been due to the hub expanding and the base of the blades collapsing.

The fourth propeller of this group, known as the Smith Steel Propeller No. 4, was 9 feet 10½ inches in diameter.

The construction of this propeller was the same as that of Smith Steel Propeller No. 3 except that the threads on the ends of the blades and in the hub were buttress threads instead of V threads. There were 8 threads to the inch, and there were 11 threads holding in this propeller instead of 4½, as in the preceding propeller. The metal of the hub outside the threaded portion had been made heavier in this propeller than it was in the preceding one.

This propeller failed after running for five hours at 600 horsepower and approximately 1,900 r. p. m. A piece 15 inches long was thrown off the face of one blade at the tip. The other blade was cracked for about 30 inches inward along the trailing edge. The hub and inner ends of the blades were in good condition after the failure except for the fact that the blades could be shaken slightly in the hub, indicating that the threads were looser than at the beginning of the test. This looseness was probably due to the vibration wearing small burrs off the threads. The failure of this propeller seems to indicate that the welding around the tip of the propeller had been poorly done.

The fifth propeller of this group, known as Smith Steel Propeller No. 5, was 9 feet 2 inches in diameter and had a pitch of 7 feet.

The blades of this propeller were the same shape as those of Smith Steel Propeller No. 4. In this case they were stiffened by the addition of a steel truss on the inside of the blades for a distance of 29 inches inward from the tips. This truss was riveted and welded to the thrust and cambered faces of the blades. There were 10 equally spaced rivets in the leading edge of each blade, beginning at a point 12¼ inches from the tip. These rivets went through the leading edge from the front to the back face. Beginning at a point 18 inches from the tip on the trailing edge, there were eight equally spaced plugs welded in holes drilled in the cambered face but not in the thrust face of the blade. These plugs were put in by simply filling the holes drilled in the cambered face with metal by the use of an acetylene torch. The joint around the whole end of the propeller had been welded.

This propeller failed after running for approximately two and three-quarter hours at 1,914 r. p. m. One blade broke clean across at a point just inside where the trusswork stopped.

The sixth propeller of this group, known as the Smith Steel Propeller No. 6, was 10 feet in diameter and had a pitch of 7.5 feet.

The blades of this propeller were built in exactly the same manner as those of Smith Steel Propeller No. 5, with the exception that the reinforcing trusswork down the backbone of the blade had been extended in toward the hub. Both tips of the blades had been bent farther upstream than they had been in the preceding propellers, with the idea of having the component of the centrifugal force on the tips parallel to the axis balance the thrust. The method of welding the tips was the same in this propeller as in the Smith Steel Propeller No. 5.

A power curve was taken on this propeller from 1,014 to 1,762 r. p. m. At this speed it absorbed approximately 600 horsepower. It was run at this input for 20 hours. During this test the propeller ran very steadily and at no time showed any signs of distress.

The appearance of this propeller after the 20-hour run at 1,762 r. p. m. and 600 horsepower was the same as at the beginning of the test.

After making the foregoing test on this propeller the blade angles were decreased in order to allow the propeller to turn at an increased number of revolutions per minute at an input of 600 horsepower. A power curve was taken at this setting from 1,006 to 2,052 r. p. m., at which speed the propeller was absorbing approximately 600 horsepower. It was run at this speed for five hours, at the end of which time it was noted that the blades appeared to be running out of track. For this reason it was shut down and examined. The cambered face of one blade was cracked completely across at a point 24½ inches from the tip.

It may be noted that at 2,052 r. p. m. the deflection of this propeller was downstream, showing that at this speed the component of the centrifugal force on the tips of the blade parallel to the axis was greater than the thrust. The value of this centrifugal force in excess of the thrust was sufficient to produce the break in the back of the blade previously noted.

The seventh propeller of this group, known as the Smith Steel Propeller No. 7, was 10 feet 2 inches in diameter, and for this test its pitch was set at 4.64 feet. This propeller was built exactly like the Smith Steel Propeller No. 6, except that the thickness of the metal in the blades had been slightly increased throughout the entire length of each blade. The blade width had also been slightly increased. The edges of the blades at the tips had been merely welded with an oxy-acetylene torch instead of having holes drilled in the cambered faces on the edges and later plugged with metal by the use of a blowtorch, as had been the practice in building No. 5 and No. 6.

The hub of this propeller was the same as the hub of propeller No. 6, except that the flange on the back of the hub, which was originally designed with the intention of using a cowling, had been omitted. The blades were held in the hub in exactly the same manner as the blades of the Smith Steel Propeller No. 6; that is, with buttress threads. The blades were set in this test to make the propeller a pusher, giving 12½ threads holding in the hub.

A power curve was taken on this propeller from 1,000 to 2,103 r. p. m., at which speed it absorbed 604 horsepower. It was run at this input for 17 hours and 19 minutes, being shut down at the end of every 5 hours for inspection. When inspected at the end of the first two 5-hour periods no signs of failure were observed. At the end of 15 hours it was observed that one blade was split for a distance of 8 inches along the trailing edge, beginning at a point 14½ inches from the tip, the crack running inward along the trailing edge of the blade toward the hub. The other blade showed no signs of failure. The propeller failed after 17 hours and 19 minutes, due to a piece being thrown off the cambered face of the blade at the tip. This piece was 17¾ inches long and weighed 2 pounds. Figure 178 shows the appearance of the end of the blade of this propeller after failure occurred. It also shows very clearly the truss and the spooled rivets used to stiffen the blades.

The eighth propeller of this group, known as the Smith Steel Propeller No. 8, was 10 feet 2 inches in diameter and had a pitch of 5.04 feet.

The construction of this propeller was the same as the construction of Smith Steel Propeller No. 7 except that the welding on the leading and trailing edges of the blades had been more carefully done. The threads on the ends of the blades had been slightly changed. This change consisted of making the thrust face of the threads 3° less than a right angle instead of a right angle, as they had been in Smith

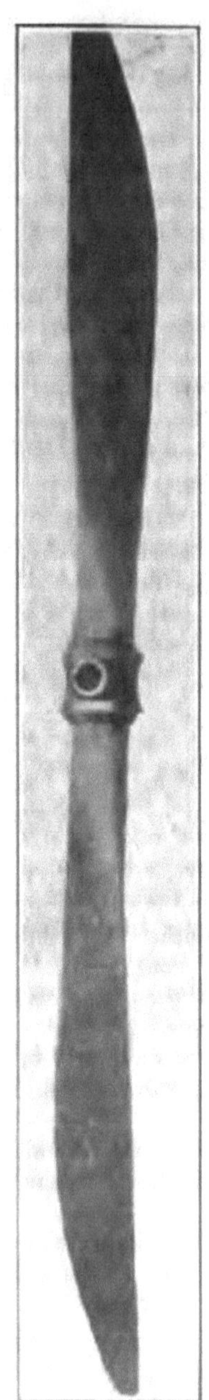

FIG. 179.—SMITH STEEL PROPELLER NO. 8, AFTER A 30-HOUR DESTRUCTIVE WHIRLING TEST.

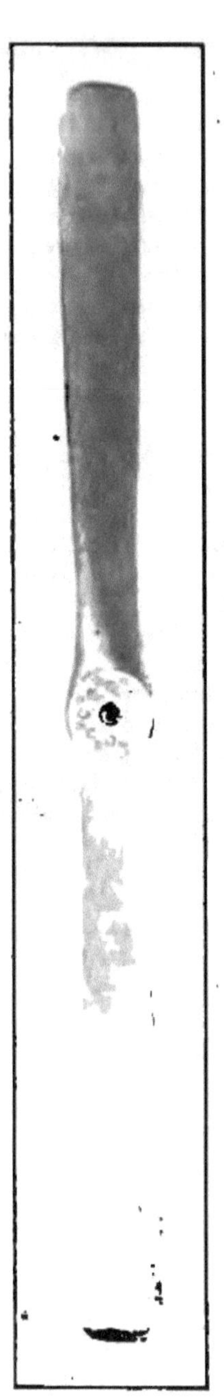

FIG. 180.—DICKS-LUTTRELL STEEL PROPELLER NO. 1.

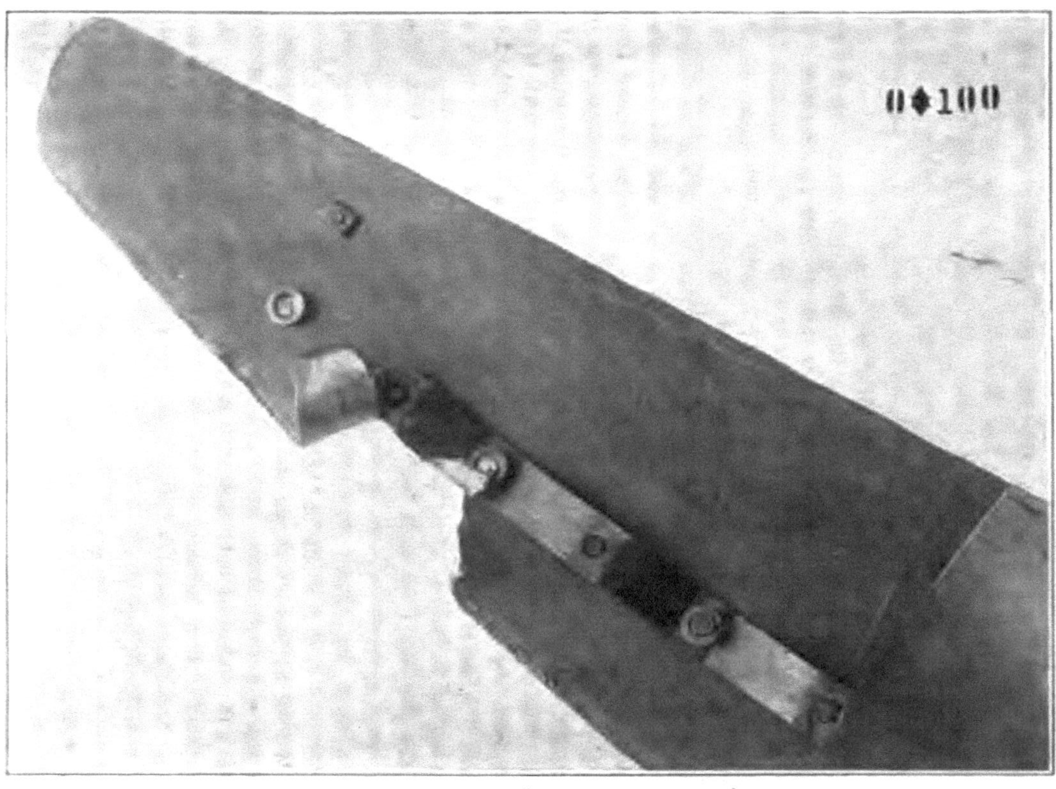

FIG. 178.—FAILURE AT 2,052 REVOLUTIONS PER MINUTE AND 600 HORSEPOWER.

305

Steel Propeller No. 7. The threads in the hub had been changed to correspond to the threads on the blades.

A power curve was taken on this propeller from 992 to 1,925 r. p. m., at which speed it absorbed 600 horsepower. It was run at this input for a total time of 30 hours. At the conclusion of this 30-hour test no signs of failure could be observed except a slight looseness of the blades in the hub. This looseness was probably due to small burrs on the threads being worn off by vibration during the test.

It may be noted that this propeller showed very little deflection at any time during the test and ran with very little flutter.

Figure 179 shows the appearance of this propeller at the conclusion of the 30-hour run at 600 horesepower.

The ninth propeller of this group, known as the Smith Steel Propeller No. 9, was 11 feet 2 inches in diameter, and for the test was set at a pitch of 4.68 feet. The construction of this propeller was exactly the same as the construction of Smith Steel Propeller No. 7, the only difference being the increase in diameter.

A power curve was run on this propeller in steps of approximately 200 r. p. m. from 593 to 1,843 r. p. m., at which speed the propeller absorbed 644 horsepower. It was run at this input for a total time of 31 hours. Periodically during this time the axial movement of the tips of the blades, due to looseness in the fit of the thread in the hub, was measured. During the first 22 hours of this test there seemed to be a gradual increase in this movement, the movement of blade 1 being 0.13 inch at the beginning of the run and 0.31 inch at the end of 22 hours, while the movement of blade 2 at the beginning of the run was 0.16 inch and at the end of 22 hours was 0.29 inch. At the conclusion of the 22-hour run on this propeller it was placed in storage for about one month before the test was completed. When the test was again started the movement of the two blades at the tip was 0.17 inch and 0.14 inch for blades 1 and 2, respectively. At the end of 31 hours the axial movement at the tip of the blades was 0.16 inch and 0.15 inch for blades 1 and 2, respectively. The decrease during storage in the movement of the blade tips, noted from the above, was probably due to a slight rusting of the threads on the blades and in the hub.

Figure 179 is typical of the appearance and construction of all the Smith steel propellers from Nos. 3 to 9, inclusive.

*Group II.*—All the propellers of this group received for test were designed for the *Liberty 12* engine. Four propellers of this group have been tested, all being built according to Signal Corps drawing X-5411.

All these propellers were constructed from a single tube. This tube was turned down on a taper from a point just outside the hub

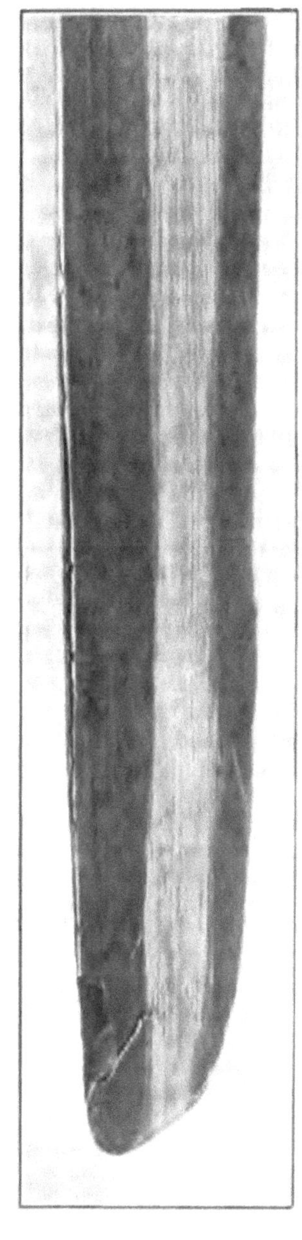

FIG. 181.—DICKS-LUTTRELL STEEL PROPELLER NO. 2 AFTER FAILURE.

161988—21——20

to the tips of the blades. The ends of the tube were then split and cut to the proper shape. The whole tube was then placed in a die and the tube forced against the sides of the die, giving it the proper shape for the completed propeller. The split ends of the tube were then welded and the hub hole drilled. A special hub made by the manufacturer was put in the propeller and it was balanced by means of solder attached to the blades.

The first of these propellers, known as Dicks-Luttrell Steel Propeller No. 1, was constructed as described in the preceding paragraph. A power curve was run on this propeller from 600 to 1,795 r. p. m., at which speed it absorbed 768 horsepower. At the conclusion of this curve the propeller was shut down for examination, and it was found that one blade was cracked along the trailing edge for a distance of 18 inches, beginning at a point 6 inches from the tip. The propeller was started on a 10-hour run at 600 horsepower and failed after running for 45 minutes at this input. Blade No. 2 broke off 31 inches from the center of the hub. The general appearance of this propeller before the test is shown in figure 180.

The second propeller of this group, known as Dicks-Luttrell Steel Propeller No. 2, was constructed exactly as the preceding one. A power curve was run on this propeller from 600 to 1,762 r. p. m. in steps of approximately 200 r. p. m. each. It was then run for 25 minutes at 1,762 r. p. m. and was shut down for inspection. It was found that both blades were cracked near the tips on the leading and trailing edges. The propeller was then started again and ran for 10 minutes, at the end of which time it was seen to be failing and was shut down. The ends of both blades were split open along the weld. It should be noted that both blades of this propeller had been damaged in shipping.

The appearance of the failure is shown in figure 181.

The third propeller of this group, known as Dicks-Luttrell Steel Propeller No. 3, was of the same construction as the other two propellers.

A power curve was run on this propeller from 605 to 1,752 r. p. m. in steps of approximately 200 r. p. m. each. At this speed the propeller absorbed 663 horsepower. It was run at this speed for 27 minutes, at the end of which time it was observed to be failing and was shut down. It was found that one blade was split along the welded seam on the leading edge near the tip.

The manufacturer opened the seams at the tips of both blades and repaired them by brazing the seam instead of welding it. After being repaired the propeller was again tested and split in the same manner as it had in the first test.

It may be noted that all the propellers of this type tested fluttered. This flutter was undoubtedly largely responsible for the splitting of these propellers.

A microscopic examination of the metal in these propellers at the point of failure shows that there was a minute crack along the edges caused by bending the metal around the sharp radii during manufacture. No heat treatment was given these propellers after they were formed. The material used was low carbon steel.

The fourth propeller of this group was constructed in the same dies as were the first three. The results of the tests on the first three propellers enabled the manufacturer to produce a better propeller in his fourth attempt.

An additional feature in the construction of this propeller was the inclusion of spacer spools between the cambered and pressure faces of the blades at points near the tip. Two of these spools were supplied to each blade. They were riveted in place by one rivet through the center of each spool.

The usual power curves were run for this propeller, after which it ran for 10 hours at 1,750 r. p. m. At this speed it absorbed 524 horsepower.

During this run the rivets through the spacer spool reinforcements sheared at one end; the other end held intact, thereby keeping the spool in place against the action of centrifugal force. The failure of the rivets was due to the shearing stress arising out of bending of the blade under thrust.

An additional 10-hour test was given this propeller. At the conclusion of this run the blades were developing cracks along the edges near the tips.

After this test a 30-minute water-spray test was run. Aside from a slight erosion of the steel on the leading edge at the extreme tip, no damage resulted from this test.

At the conclusion of the water-spray test, the propeller was given a power input of 607 horsepower, which caused it to turn 1,850 r. p. m.

It ran for seven hours at this input. At this time a small piece was thrown from one blade. The test was discontinued, as a few minutes more would undoubtedly have resulted in a complete failure.

*Conclusion.*—The foregoing tests, which have been on steel propellers of various designs and constructions, prove conclusively that it is possible to build a steel propeller which will be amply strong for use on any airplane engine in use at present. Of the various propellers tested the fabricated steel propeller, built by Mr. James Ingells, to A. E. D. drawing X–7141 (shown in fig. 176), the Smith Steel Propellers 8 and 9, and the Dicks-Luttrell No. 4, to A. E. D. drawing X–5411, are believed to represent the most promising types.

All these propellers have stood satisfactorily the tests applied to determine their strength with the exception of the last-mentioned one.

Of the above propellers only one, namely, the Smith Steel Propeller No. 8, has been flown on a plane in order to determine its performance. This flight indicated that the design was not as efficient as that of the standard wooden propellers built for the same engine and plane.

A comparative test of the static thrust showed that that of the Smith Steel Propeller No. 8 was 1,695 pounds at 1,590 r. p. m. against 2,290 pounds thrust at 1,585 r. p. m. for a wooden propeller to design No. 8-45. The thrust-torque ratio of the Smith steel propeller at 1,590 r. p. m. was 1.22 against 1.38 for the wooden propeller design No. 8-45.

## CHAPTER VII.

### FLIGHT TESTS OF PROPELLERS.

Flight tests are, of course, the ultimate criterion of the over-all efficiency of the motor, propeller, and airplane combined. The effect of each of these upon the other is such that they must be considered together and not separately if we are to arrive at the best combination.

Comparative flight tests are extremely difficult to carry out even under the most favorable conditions. They should never be undertaken except by the most experienced and skillful pilots, and all of the instruments used must be carefully calibrated and checked from time to time by means of the best controls available.

There are many sources of error in flight tests, one of the most difficult being the presence of vertical air currents, either rising or falling. Horizontal winds, if steady, can be allowed for with some degree of accuracy and only need to be taken into account in the calibration of the air-speed meter, which will be described later. Another source of error is the failure to align all parts of the airplane properly before beginning the flight. The alignment must be carried out as accurately as possible. A standard gross weight for the machine under test must be determined, and the machine should be weighed at the start of each flight to make sure that the load is standard.

Aside from variations in the airplane, any variation in the engine power will affect the performance. Loss of engine power may be due to many causes, and each different type requires expert inspection and overhauling before the test. The following control will afford the test pilot an opportunity to check up the engine power before each flight:

(1) Choose a propeller (preferably one of heavy, thick design, which is less apt to warp than the thinner blades) and check up the angles on the inspection table, figure 51, page 114.

(2) Attach this propeller to an engine which is known to be functioning properly and delivering full power. Find the r. p. m. at which the engine turns the propeller at full throttle with the plane at fixed point.

(3) Before the start of each test attach this standard propeller and check the r. p. m. of the engine. In case the engine fails to reach the specified r. p. m. it must be overhauled until the r. p. m. is brought up to the standard.

(4) The angles of the propeller must be checked from time to time in order to guard against warping.

(5) This test of the engine must be run in a spot where there is no wind, as the presence of wind will affect the r. p. m. of the propeller.

(6) The test should also be run about 100 feet away from the nearest obstruction. The r. p. m. at full throttle should not vary more than about 10 for a given engine.

Having determined that the engine is functioning properly on the ground, it is necessary to guard against loss of power in the air. A few of the causes of this are:

(1) *Overheating or overcooling.*—In the case of water-cooled engines this may be controlled by regulating the radiator shutters so as to maintain a uniform temperature of cooling water.

(2) Oil pressure too low, leading to insufficient oiling and consequent overheating, or oil pressure too high, leading to fouling of spark plugs from excess of oil and consequent loss of power.

(3) *Carburetion.*—In case altitude adjustments are used during the test, they must be used in exactly the same way for each test, as no comparative results can otherwise be obtained.

(4) There are many other sources of loss of engine power, such as partial breakdown of the ignition system, etc., and the pilot must maintain the most constant vigilance and care in order to avoid them.

In analyzing the results of flight tests the following partial check provides a useful control, namely, for a given air density and pressure and a given plane speed the r. p. m. of an identical propeller should always be the same:

Before proceeding with a flight test the propeller should be carefully checked up with the drawing. (See method on p. 115.) The angles should check within two-tenths of 1°, and all other dimensions should check within the tolerance limits. (See Specification No. 29500D, par. 29.) If the propeller does not check within these limits it should be rejected for flight test and another propeller of the same design chosen.

It may be seen from the above that tests must be carried out under the most carefully regulated conditions and by the most skillful pilots, as it is otherwise easy to reach a conclusion that is entirely misleading.

*Instruments.*—Altimeter, aneroid, recording barograph, air-speed meter, strut thermometer, tachometer, recording air-speed meter and tachometer (Toussaint-Le Pere), thrust meter, thrust and torque meter, statoscope, flow meter.

*Air-speed meter.*—Since all air-speed indicators based on pressure differences are of the Pitot or Pitot and Venturi type, a simple explanation of the principles involved will be excluded.

(*a*) *Pitot tube.*—In figure 182 is shown a Pitot tube in diagrammatic form. It consists of two concentric tubes, the inner open to the wind, the outer closed and communicating with the current of air only by a series of fine holes. The tubes are connected to the two arms of a pressure gauge, which measures the difference in pressure between them. The inner tube, open to the wind, brings the air impinging on it to rest, and the pressure on it is therefore a measure of both the static pressure in the stream and of the kinetic energy head of the stream. If $p$ is the static pressure of the stream, $V$ the velocity, the total pressure on the inner tube will be given by

$$p+\frac{\rho V^2}{2g}$$

The outer tube, on the other hand, being closed to the wind, will, if the holes are small enough, read the static pressure of the air flow $p$.

Hence, the differences in pressures read on gauge will be

$$\frac{\rho V^2}{2g}$$

and the gauge reading will be a measure of the velocity.

Pitot tubes with suitable gauges are widely used in laboratory practice, but, owing to the small difference in head $\frac{\rho V^2}{2g}$, the forces acting on the gauge are very small and hard to record.

(*b*) *Combination of Pitot and Venturi.*—To increase the pressure differences, and thus get practicable forces on the gauges, the Venturi tube is coupled with the pressure part of the Pitot. Such a combination is shown diagrammatically in figure 183. Here the velocity at the throat will be considerably greater than that acting on the suction side of a Pitot, and therefore has a considerably greater effect. The mathematical theory of the Venturi is a little more complicated than that of the Pitot, and the theoretical suction heads are not always in accord with practical results. From the simple formulæ of figure 183 it can be seen that the gauge readings will be proportional to $\frac{\rho V^2}{2g}$, hence are a measure of the velocity.

*Typical recording air-speed meter.*—In the Toussaint-Le Pere air-speed meter the dynamic pressure of the wind is measured by a combination of Pitot tube and Venturi meter. This pressure is transmitted to a clockwork recording device by a gauge consisting

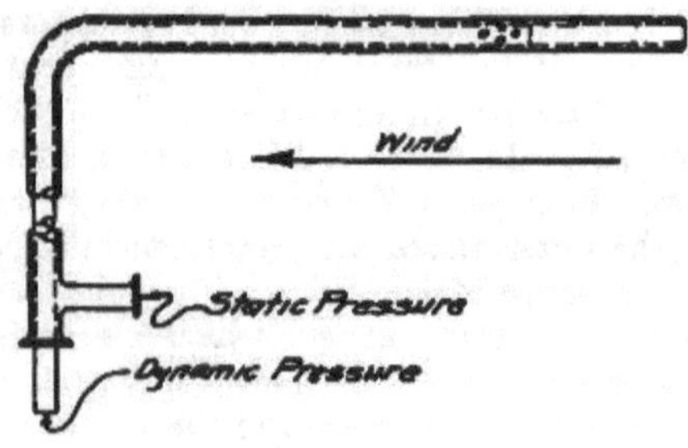


Fig. 182 Standard form of Pitot tube

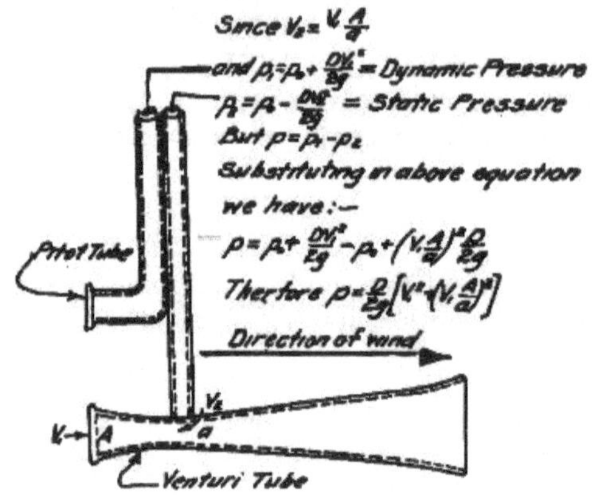


Fig. 183 Combination of Pitot and Venturi tubes

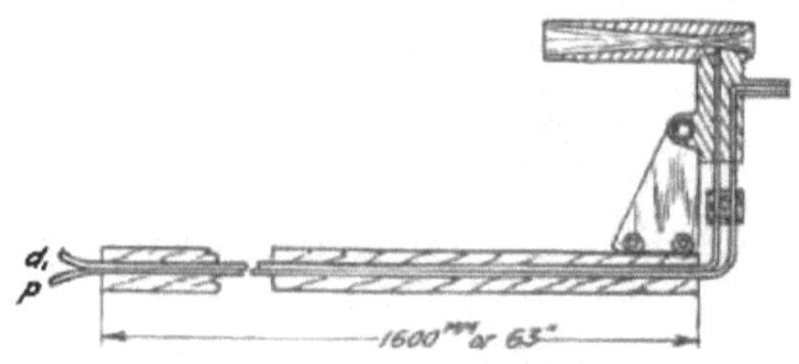


Fig 184 - Antena and support arm

of bellows and a tension spring. The Pitot tube and Venturi meter are combined in a small casting, conveniently called by the French "antenna," and similar to that of many other speed indicators. This antenna is shown in figure 184. The Venturi is carefully proportioned to give the maximum possible suction with a given air speed. The antenna is supported by a long, slender, hollow arm of light wood, which contains the tubes transmitting the pressure to the recording device, as shown in figure 185. It is fastened to this arm by a light, adjustable clip, in order that the antenna may be turned directly into the wind.

The recording device is shown diagrammatically in figure 186. It has the ordinary clockwork drum and pen. These are described elsewhere. The gauge consists of two movable circular plates $S_1$ and $S_2$, rigidly connected by a rod *a b*. The plates form the tops of the bellows $f_1$ and $f_2$. The sides of these bellows are made of thin rubber that is very flexible; the bottoms are formed by the fixed

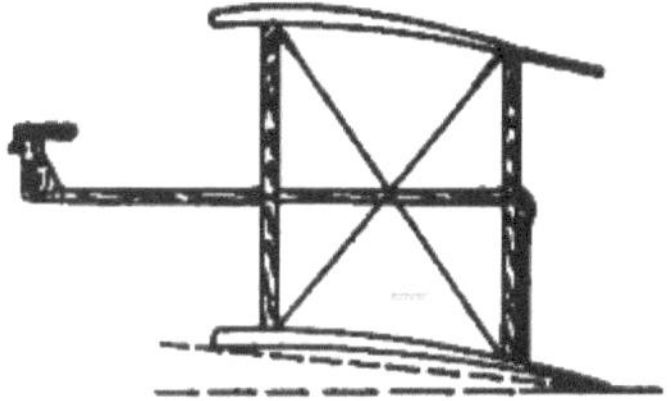

*Fig. 185 Installation of Toussaint-Lepere Air Speed Meter*

plates *m* and *n*. The suction from the Venturi is led to the air-tight chamber *c c*, and so acts on top of the plate $S_1$. The pressure from the Pitot is led to the under side of the plate $S_2$. The top of $S_2$ and the bottom of $S_1$ are open to the air inside of the box. Thus a variation of that pressure causes no motion on the rod *a b*, which is moved only by the difference of the pressure transmitted from the antenna. The rod *a b* is constrained to move vertically by the form bar linkage *a d c b*. The link *b c* carries on one end the marking pen *g;* on the other a counterweight for the movable parts of the instrument. At the end of this link is fastened the spring *R*, whose tension balances the pressure of the pen. This spring is so placed that the displacement of the pen is nearly proportional to the wind speed. The recording apparatus is inclosed in a box about 9 by 6 by 5 inches; total weight, about 4½ pounds. The apparatus slides out of this box to facilitate adjustment of paper on the drum.

The complete speed indicator must be calibrated and a chart or table made for converting the readings on this drum into true wind speeds. This chart of course is only correct for readings in air of standard density.

The recording apparatus is suspended in the airplane by elastic cords or may be held by the passenger in a two seater. The antenna must not be placed near any obstruction or disturbance, including the slip stream, body, etc. The supporting arm is fastened to any convenient part of the airplane, such as a strut. The antenna is then adjusted to point directly into the wind. With this instrument, as with all air-speed meters, a test run in flight must be made over a measured course to determine the effect of interference of the plane upon the air flow to the antenna and to find the correction due to this interference.

*The Foxboro-Zahm direct-reading air-speed meter.*—In figures 187 and 188 are shown views of a very widely used combination of the

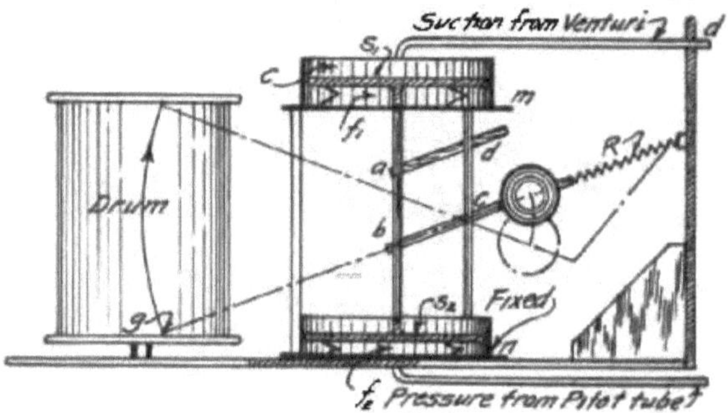


*Fig. 186 Recording Apparatus of Toussaint LePere Air Speed Meter*

Foxboro indicating box and the Zahm-Pitot-Venturi tube (adopted as standard by the Signal Corps). The pressure lead of the Pitot enters the small cylinders located in the indicating case, which in itself is made air-tight by a gasket under the cover. The suction of the Venturi is transmitted to the case itself. When a difference of pressure exists between the inside and outside of the two cylinders, they elongate or contract. The motion is transmitted to the pointer by means of links to a circular rack which engages a pinion on the spindle.

Since the movement of the aneroid boxes is proportional to the square of velocity, the scale on the dial is not uniformly graduated; and were it not for a compensating device, the divisions of the scale for the higher velocities would increase rapidly as the velocity increased. The small springs fastened to the aneroid boxes shown in

FIG. 187. —FOXBORO INDICATOR USED WITH VAHM PITOT-VENTURI TUBE SHOWN IN FIGURE 188.

FIG. 189.—TEL CHROMETRIC TACHOMETERS.

figure 187 restrict the movement of the boxes and shorten the scale divisions.

*Tachometer.*—The only type of tachometer yet developed with sufficient precision and reliability for propeller testing is the chronometric escapement type. The principle of this type of instrument is to actually measure the number of r. p. m. during a given interval of time measured by a chronometer. Two designs widely used are the

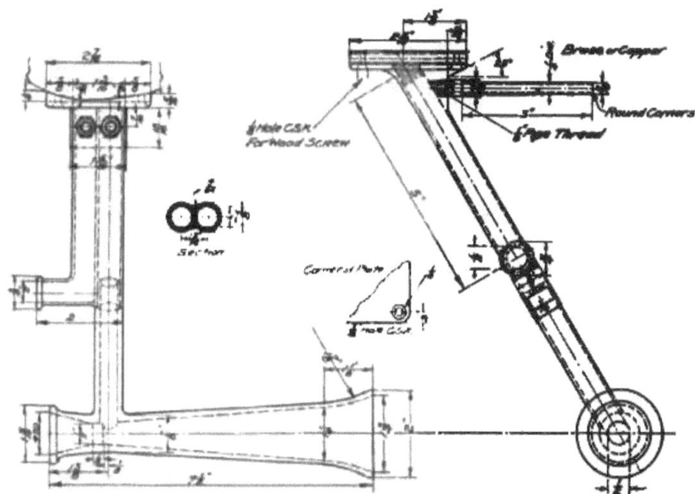

Fig. 188 - Zahm Pitot-Venturi tube adopted as standard by Signal Corps

Jaeger and the Tel. These are about equally good. A photograph of the Tel instrument is shown in figure 189.

A partial description of this instrument is given as follows by Dr. Washburn:

Every chronometric tachometer of whatever make contains:

(1) A fine-toothed pinion or gear, which we will call the drive pinion, that is connected to the main drive shaft, and therefore rotates at a speed proportional to the speed of the engine.

(2) A fine-toothed rack traveling on another guide (Tel), or a fine-toothed gear (Jaeger, Van Sicklen), either of which we will denote by the general term *counter*.

(3) An escapement mechanism. The escapement is similar to those used in clocks and watches, whence the name *chronometric*. Tachometers of this type are recognized easily by the periodic ticking of the escapement.

The drive pinion is thrown into mesh with the counter (Jaeger, Van Sicklen) or the counter with the drive pinion (Tel) and kept there for a definite period of time; for example, one second. This is done, regardless of whether the tachometer is running at high or low speed, by the escapement, and is its most important function.

It follows, therefore, from the above, that the motion of the counter, that is, the distance through which the toothed rack is moved, or the angle through which the gear is rotated while in mesh with the drive pinion, is proportional to the speed of the engine, the same as the distances traveled by two runners, in equal times, are in ratio of their speeds.

For example, if the tachometer were attached successively to two shafts, the second running twice as fast as the first, the motion of the

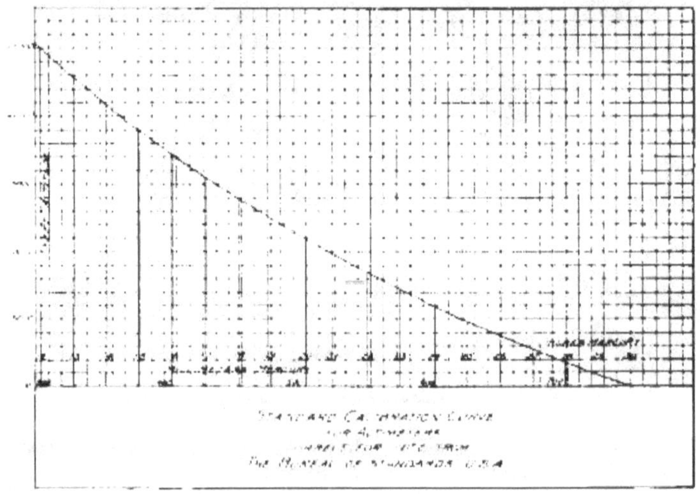

FIG. 190.

counter would be twice as great in the latter case as in the former. In order to make the instrument recording, it remains only to transform the motion of the counter into a proportionate angular rotation of the pointer on the dial. Then the angular deflection of the pointer and counter would be moved the same distance as though the speed of revolution had been constant and equal to the average speed over the whole period and no indication of the assumed variation in speed would be given at all.

A better understanding of the mechanism may be obtained by taking the case of one of the instruments and tracing through the mechanism. All types of tachometers other than the escapement type are at times found to be several hundred revolutions in error and are worthless for test work.

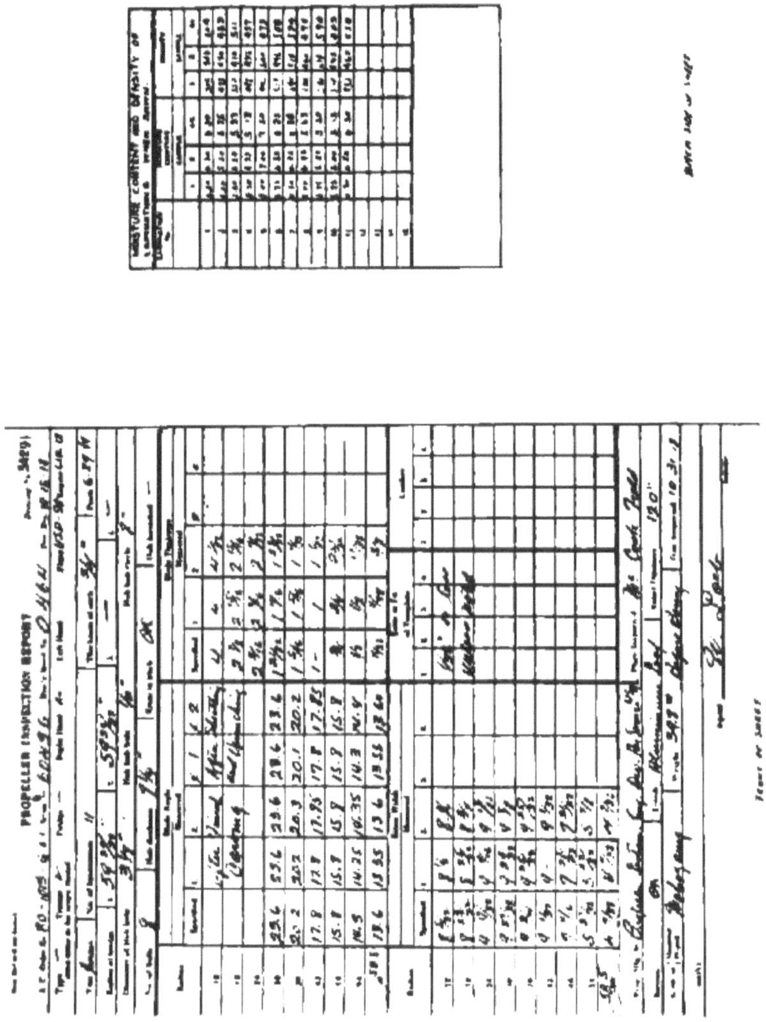

PROPELLER INSPECTION REPORT

FRONT OF SHEET

MOISTURE CONTENT AND DENSITY OF

BACK SIDE OF SHEET

FIG. 191.—TYPICAL PROPELLER INSPECTION REPORT.

*Aneroid barometer.*—This instrument is calibrated on the assumption that the variation of atmospheric pressure with altitude corresponds to the curve shown in figure 190.

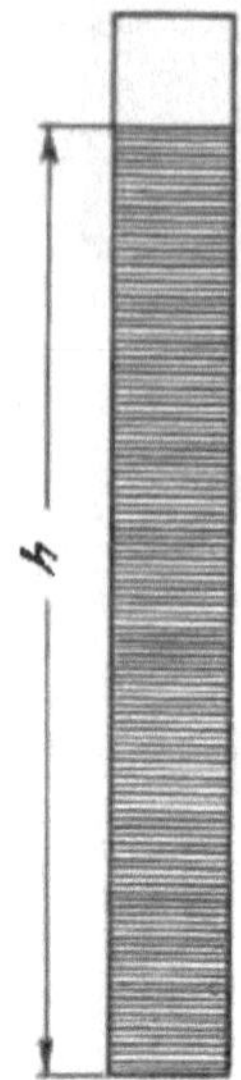


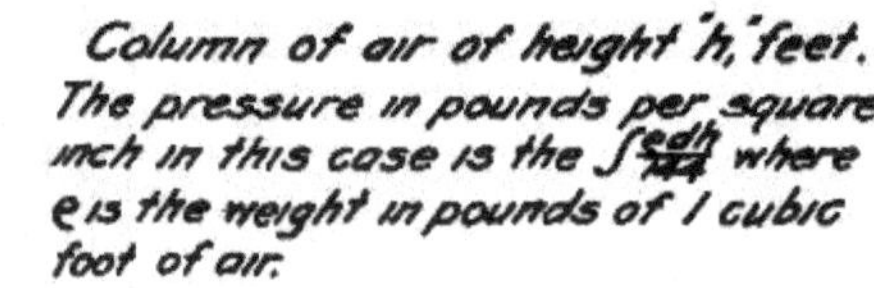


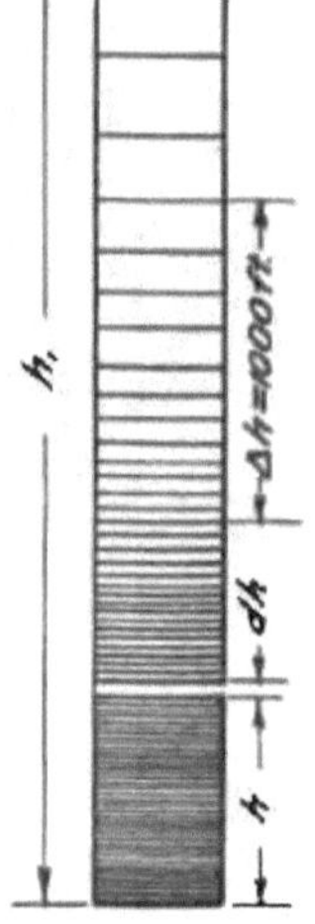


Fig. 192.

This represents what may be termed the standard pressures at various altitudes. The curve is supposed to represent an average condition.

To understand the correction to be applied to this instrument it is necessary to think of what the atmospheric pressure is due to.

It is a familiar fact that the pressure at the bottom of a homogenous, noncompressible fluid is proportional to the product of its height, *h*, times its density. (See fig. 191.) In the case of air, however, we have a column which is continually varying in density from the top to the bottom. This variation does not follow any predetermined law. It is therefore necessary to measure the density at each 1,000 feet and make the approximation of assuming the density correction constant over a height of each 1,000 feet. (See fig. 192.)

In order to measure density we have only to know the pressure and temperature. The familiar expression $PV = RT$ enables us to calculate the density.

The instruments are all calibrated at 10° C. Consequently the divisions of 1,000 feet represent the change in pressure due to a column of air 1,000 feet in height at the corresponding pressure and temperature. At any other temperature $t°$ C., the height of the column will be inversely proportional to the density and will be equal to $\frac{1{,}000 \times 273 + t°}{273 + 10}$, or approximately $964.6 + 3.53\ t°$.

From this formula and the temperature at each 1,000 feet of indicated climb we may compute the true climb, and adding these up for all the 1,000 feet indicated intervals we may find the true altitude.

*Calibration of the air-speed meter.*—The calibration of the air-speed meter is carried out by flying back and forth over a course of known length and taking readings of the air-speed meter and time required to cover the course in each direction. It is also necessary to take readings of the altimeter and strut temperature at various intervals during the flight. If the wind is absolutely steady and along the path of flight it may be allowed for by averaging the velocity in the two directions. If there is a component normal to the path, the wind must be allowed for by the following method.

CASE 1.

$$V_1 = \frac{S}{T_1}. \quad \text{(From fig. 193.)}$$

(1) $$V = \sqrt{(V_1 - V_w \cos \phi)^2 + (V_w \sin \phi)^2}$$

(2) $$V = \sqrt{V^2_1 - 2\ V_1\ V_w \cos \phi + V_w^2}$$

CASE 2.

$$V_2 = \frac{S}{T_2}. \quad \text{(From fig. 194.)}$$

(3) $$V = \sqrt{(V_2 + V_w \cos \phi)^2 + (V_w \sin \phi)^2}$$

(4) $$= (V^2_2 + 2\ V_2 V_w \cos \phi + V^2_w)$$

From cases (1) and (2)—

(5) $$V^2_1 - 2 V_1 V_w \cos \phi = V^2_2 + 2 V_1 V_w \cos \phi$$

161988—21——21

Therefore—

$$V_w \cos\phi = \frac{V_1 - V_2}{2}$$

(6) $$V_x = V_w \sin\phi = (V_w \cos\phi) \times \tan\phi = \left(\frac{V_1 - V_2}{2}\right) \tan\phi$$

(7) $$V^2 = (V_1 - V_w \cos\phi)^2 + V_x^2 = \left(V_1 - \frac{V_1 - V_2}{2}\right)^2 + \left(\frac{V_1 - V_2}{2} \tan\phi\right)^2$$

(8) $$V^2 = \left(\frac{V_1 + V_2}{2}\right) + \left(\frac{(V_1 - V_2)\tan\phi}{2}\right)^2$$

(9) $$V = \sqrt{1/2[(V_1 + V_2) + (V_1 - V_2)\tan\phi]}$$

If the wind is perfectly steady, its angle $\phi$ with the course may be determined and the true velocity found from equation (9). If the wind is not steady, its angle $\phi$ with the course and its magnitude $V_w$ must be found and equations (1) and (3) must be used.

CASE 1.

Wind at an angle $\phi$ to the course - Flying with the wind

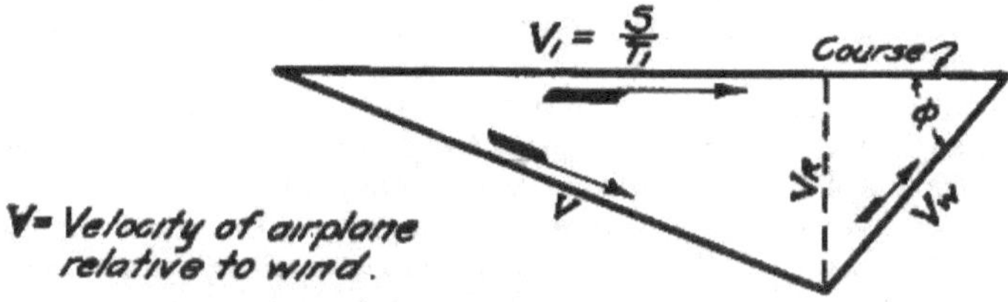


Fig. 198.

If a very quiet period is chosen, it will be sufficiently accurate to assume the wind velocity constant and to use equation (9).

The direction of the wind may be ascertained by means of a pennant at each end of the course provided the trial is carried out at an altitude of less than 100 feet.

For trials at greater heights it may be necessary to have the pilot fire a smoke pistol at each end of the course and observe the direction of the path of the smoke by means of camera obscura or transits located at two distant points. The smoke-puff method has the added advantage of enabling one to detect vertical currents if any are present.

In carrying out the calibration the plane should be flown back and forth at least twice in each direction at the following speeds: (1) Top speed, (2) slow speed, (3) three or four intermediate speeds.

CASE 2.

*Wind at an angle $\phi$ to the course - Flying against the wind*

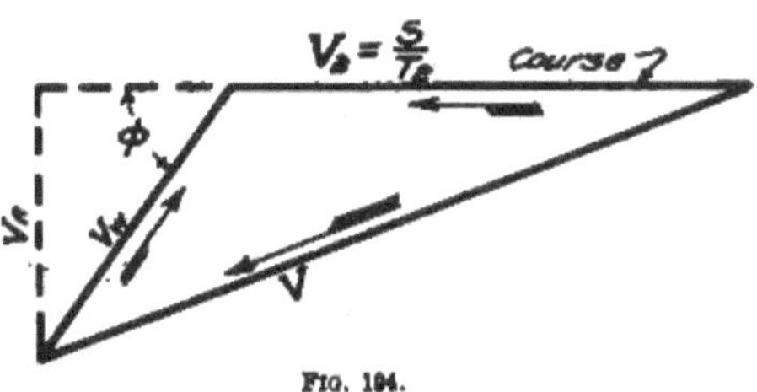


FIG. 194.

Below is tabulated the data from a typical air-speed meter calibration:

LENGTH OF COURSE—OBSERVATIONS.

| Reading of air-speed meter. | Reading of strut temperature. | Corrected temperature, °C. | Reading of altimeter. | Time over course. | Angle of wind with course. | R. p. m. of engine. |
|---|---|---|---|---|---|---|
| | | | | | | |
| | | | | | | |
| | | | | | | |
| | | | | | | |
| | | | | | | |

COMPUTED RESULTS.

| Reading of air-speed meter. | Fraction of standard density = $\Delta$. | True air speed = $V$. | Correct reading of air-speed meter = $V\sqrt{\Delta}$. | Correction for air-speed meter reading. |
|---|---|---|---|---|
| | | | | |
| | | | | |
| | | | | |
| | | | | |
| | | | | |

The correction found from this calibration may be applied directly to the air-speed meter reading. The corrected reading divided by the factor $\sqrt{\Delta}$ will give the true air speed.

The procedure in carrying out a flight test on a propeller is as follows:

(1) Make sure that the plane has been carefully aligned, the propeller checked with the drawing, and all instruments properly calibrated.

(2) Weigh the plane with full load and crew, making sure that this weight corresponds to the standard weight.

(3) Attach the standard test propeller to the engine and try the r. p. m. at full throttle.

(4) In case the r. p. m. of the engine is up to standard the propeller to be tested may be applied and the maximum r. p. m. determined with the plane at fixed point.

(5) Climb the plane to the ceiling, flying at the best climbing speed for the plane on which the propeller is being tested. This best climbing speed may be determined in advance by trial or calculation. In climbing from the ground to the ceiling the reading of the air-speed meter should be maintained constant at all altitudes. The true air speed will then increase inversely as the square root of the density and the flying angle of the plane will be maintained constant.

Readings of all instruments should be taken at each thousand feet as read on the altimeter. The data taken during a typical climbing test is given on page 327.

(6) After reaching the ceiling the plane should be flown level at full throttle and all readings taken. The statoscope is useful in maintaining level flight.

In descending, the plane should be leveled off at full throttle at each interval of 5,000 feet altitude and readings of all instruments taken. This level flight should be maintained for at least four minutes in order to reach equilibrium. Three readings of all instruments should be taken at each height.

An example of the data obtained in these level flight tests is also given on page 327.

This climbing and level flight test should now be followed immediately by a similar test on the propeller with which comparison is to be made.

(7) A climb to 10,000 feet altitude should now be made at full throttle with the plane flying at each of the following air speeds:

The minimum flying speed at each 10-mile increment of speed up to the maximum flying speed. An example of the data obtained from this test is given on pages 328 and 329.

(8) *Flow-meter test.*—If a reliable gasoline flow meter is available, a test should be made as follows:

I. At an altitude of 2,000 feet fly level at full throttle, taking readings of all instruments, including the flow meter. Then fly level with the engine throttled to give level flight speeds in steps of 10 miles per hour between maximum flying speed and minimum flying speed, taking readings of all instruments, including the flow meter.

PERFORMANCE DATA SHEET.

[Pilot, Mr. Hoover. Airplane, Curtiss R-4. Date, June 26, 1918. Observer, Sergt. Smith. Propeller No. X-4875. Test No., P-20.]

(*A*) *Climb* (add height of Dayton, 750 feet above sea level).—Set barograph and aneroid at barometric pressure.

| Height. | Time. | | R. p. m. | Speed. | Strut temperature. | Flow meter. | Remarks. |
|---|---|---|---|---|---|---|---|
| | Minutes. | Seconds. | | | | | |
| 700 (ground). | .... | .... | 1,400 | .... | .... | .... | |
| 1,700 | 2 | .... | 1,450 | 70 | 30 | .... | Aneroid was set at 700 feet on ground to correspond with barometric pressure. |
| 2,700 | 4 | 20 | 1,420 | 62 | 25 | .... | |
| 3,700 | 7 | 10 | 1,540 | 70 | 21 | .... | |
| 4,700 | 9 | 10 | 1,420 | 83 | 20 | .... | |
| 5,700 | 11 | 10 | 1,570 | 84 | 17 | .... | |
| 6,700 | 12 | 50 | 1,570 | 69 | 10 | .... | |
| 7,000 | 15 | .... | 1,570 | 69 | 10 | .... | |
| 8,700 | 16 | 50 | 1,570 | 65 | 10 | .... | |
| 9,700 | 18 | 45 | 1,540 | 70 | 10 | .... | |
| 10,700 | 21 | 20 | 1,550 | 70 | 15 | .... | |
| 11,700 | 24 | 10 | 1,540 | 73 | 12 | .... | |
| 12,700 | 27 | 5 | 1,520 | 69 | 10 | .... | |
| 13,700 | 30 | 40 | 1,575 | 67 | 7 | .... | |
| 14,700 | 35 | 35 | 1,520 | 66 | 5 | .... | |
| 15,700 | 45 | 35 | 1,520 | 65 | 7 | .... | |
| 16,800 | 56 | 30 | 1,520 | 65 | 5 | .... | |
| 17,000 | .... | .... | .... | .... | .... | .... | |
| 18,000 | .... | .... | .... | .... | .... | .... | |
| 19,000 | .... | .... | .... | .... | .... | .... | |
| 20,000 | .... | .... | .... | .... | .... | .... | |

(*B*) *Level* (flights for speed, during descent).

| Height. | R. p. m. | Speed. | Strut temperature. | Remarks. |
|---|---|---|---|---|
| 16,800 | 1,550 | 75 | 5 | |
| 14,700 | 1,550 | 76 | 5 | |
| 12,700 | 1,590 | 80 | 10 | |
| 10,700 | 1,590 | 86 | 15 | |
| 8,700 | 1,600 | 89 | 10 | |
| 5,700 | 1,620 | 94 | 17 | |
| 2,700 | 1,640 | 99 | 25 | |

*Flight test log.*

PROPELLER PERFORMANCE.

[Pilot, Mr. J. D. Hill. Airplane, DH-4, P-34. Propeller, drawing, 8-45. Observer, Sergt. Langham. Date, Oct. 22, 1918. Insp. No. 0-135. Test No. F-122. Signal Corps, No. 60175.]

| | Observed height. | Time. | Strut temperature. Ind. | Strut temperature. Cortd. | Climbing calculations. Air speed. Ind. | Air speed. Cortd. | Engine r. p. m. Ind. | Engine r. p. m. Cortd. | Average temperature. | Altitude cortn. | True altitude. | K. | True air speed. |
|---|---|---|---|---|---|---|---|---|---|---|---|---|---|
| | *Feet.* | | *°F.* | *°C.* | | *M.p.h.* | | | *°C.* | *Feet.* | *Feet.* | | *M. p. h.* |
| F | 500 | ...... | ...... | 20 | ...... | ...... | ...... | ...... | ...... | ...... | 500 | ...... | ...... |
| | 1,000 | ...... | ...... | ¹19 | ...... | ...... | ...... | ...... | 19.5 | 517 | 1,017 | 1.026 | ...... |
| | 2,000 | ...... | ...... | ¹17 | ...... | ...... | ...... | ...... | 18 | 1,028 | 2,045 | 1.042 | ...... |
| | 3,000 | 0.57 | ...... | 15 | 71 | 71 | 1,550 | 1,550 | 16 | 1,021 | 3,066 | 1.058 | 75.2 |
| | 4,000 | 1.59 | ...... | 13 | 71 | 71 | 1,560 | 1,560 | 14 | 1,014 | 4,080 | 1.072 | 76 |
| | 5,000 | 3.02 | ...... | 10 | 70 | 70 | 1,560 | 1,560 | 11.5 | 1,005 | 5,085 | 1.087 | 76 |
| | 6,000 | 4.11 | ...... | 8 | 70 | 70 | 1,560 | 1,560 | 2 | 996 | 6,081 | 1.102 | 77.2 |
| | 7,000 | 5.22 | ...... | 6.5 | 70 | 70 | 1,560 | 1,560 | 7.3 | 990 | 7,071 | 1.118 | 78.3 |
| | 8,000 | 6.40 | ...... | 4.5 | 70 | 70 | 1,560 | 1,560 | 5.5 | 984 | 8,055 | 1.134 | 79.4 |
| | 9,000 | 8.00 | ...... | 2.5 | 69 | 69 | 1,560 | 1,560 | 3.5 | 977 | 9,032 | 1.152 | 79.5 |
| | 10,000 | 9.17 | ...... | 1 | 70 | 70 | 1,550 | 1,550 | 1.7 | 971 | 10,003 | 1.169 | 81.8 |
| G | 500 | ...... | ...... | 20 | ...... | ...... | ...... | ...... | ...... | ...... | 500 | ...... | ...... |
| | 1,000 | ...... | ...... | ¹19 | ...... | ...... | ...... | ...... | 19.5 | 517 | 1,017 | 1.026 | ...... |
| | 2,000 | ...... | ...... | ¹17 | ...... | ...... | ...... | ...... | 18 | 1,028 | 2,045 | 1.042 | ...... |
| | 3,000 | .54 | ...... | 15 | 60 | 60 | 1,520 | 1,520 | 16 | 1,021 | 3,066 | 1.058 | 63.5 |
| | 4,000 | 1.57 | ...... | 13 | 60 | 60 | 1,520 | 1,520 | 14 | 1,014 | 4,080 | 1.072 | 64.2 |
| | 5,000 | 3.01 | ...... | 10 | 60 | 60 | 1,525 | 1,525 | 11.5 | 1,005 | 5,085 | 1.087 | 65.2 |
| | 6,000 | 4.06 | ...... | 8 | 61 | 61 | 1,525 | 1,525 | 9 | 996 | 6,081 | 1.102 | 67.2 |
| | 7,000 | 5.15 | ...... | 6 | 60 | 60 | 1,525 | 1,525 | 7 | 989 | 7,070 | 1.118 | 67.2 |
| | 8,000 | 6.27 | ...... | 4.5 | 61 | 61 | 1,530 | 1,530 | 5.3 | 983 | 8,053 | 1.134 | 69.2 |
| | 9,000 | 7.44 | ...... | 3 | 60 | 60 | 1,530 | 1,530 | 3.8 | 978 | 9,031 | 1.152 | 69.1 |
| | 10,000 | 9.12 | ...... | 1 | 60 | 60 | 1,530 | 1,530 | 2 | 972 | 10,003 | 1.169 | 70.2 |
| C | 500 | ...... | ...... | 20 | ...... | ...... | ...... | ...... | ...... | ...... | 500 | ...... | ...... |
| | 1,000 | ...... | ...... | ¹18.7 | ...... | ...... | ...... | ...... | 19.4 | 517 | 1,017 | 1.025 | ...... |
| | 2,000 | ...... | ...... | ¹16.3 | ...... | ...... | ...... | ...... | 17.5 | 1,026 | 2,043 | 1.04 | ...... |
| | 3,000 | 1.55 | ...... | 14 | 100 | 100 | 1,640 | 1,640 | 15.1 | 1,017 | 3,060 | 1.055 | 105.5 |
| | 4,000 | 4.03 | ...... | 12 | 99 | 99 | 1,640 | 1,640 | 13 | 1,011 | 4,071 | 1.07 | 105.8 |
| | 5,000 | 6.36 | ...... | 10 | 100 | 100 | 1,635 | 1,635 | 11 | 1,004 | 5,075 | 1.086 | 108.6 |
| | 6,000 | 10.19 | ...... | 7 | 100 | 100 | 1,650 | 1,650 | 8.5 | 995 | 6,070 | 1.10 | 110 |
| | 7,000 | 16.30 | ...... | 6 | 100 | 100 | 1,650 | 1,650 | 6.5 | 987 | 7,057 | 1.118 | 111.8 |
| | 7,600 | 24.09 | ...... | 1 | 101 | 101 | 1,650 | 1,650 | 3.5 | 586 | 7,643 | 1.119 | 113.1 |
| D | 500 | ...... | ...... | 20 | ...... | ...... | ...... | ...... | ...... | ...... | 500 | ...... | ...... |
| | 1,000 | ...... | ...... | ¹19 | ...... | ...... | ...... | ...... | 19.5 | 517 | 1,017 | 1.026 | ...... |
| | 2,000 | ...... | ...... | ¹17 | ...... | ...... | ...... | ...... | 18 | 1,028 | 2,045 | 1.042 | ...... |
| | 3,000 | 1.25 | ...... | 15 | 92 | 92 | 1,630 | 1,630 | 16 | 1,021 | 3,066 | 1.058 | 97.5 |
| | 4,000 | 2.43 | ...... | 13 | 92 | 92 | 1,620 | 1,620 | 14 | 1,014 | 4,080 | 1.072 | 98.5 |
| | 5,000 | 4.22 | ...... | 11 | 91 | 91 | 1,620 | 1,620 | 12 | 1,007 | 5,087 | 1.089 | 99.2 |
| | 6,000 | 6.02 | ...... | 8 | 91 | 91 | 1,620 | 1,620 | 9.5 | 998 | 6,085 | 1.102 | 100.2 |
| | 7,000 | 7.51 | ...... | 6 | 91 | 91 | 1,640 | 1,640 | 7 | 989 | 7,074 | 1.118 | 101.7 |
| | 8,000 | 10.01 | ...... | 4 | 90 | 90 | 1,635 | 1,635 | 5 | 983 | 8,057 | 1.134 | 102 |
| | 9,000 | 12.39 | ...... | 3 | 90 | 90 | 1,640 | 1,640 | 3.5 | 977 | 9,034 | 1.152 | 103.6 |
| | 10,000 | 16.12 | ...... | 1 | 91 | 91 | 1,640 | 1,640 | 2 | 972 | 10,006 | 1.169 | 106.3 |
| E | 500 | ...... | ...... | 20 | ...... | ...... | ...... | ...... | ...... | ...... | 500 | ...... | ...... |
| | 1,000 | ...... | ...... | ¹19 | ...... | ...... | ...... | ...... | 19.5 | 517 | 1,017 | 1.026 | ...... |
| | 2,000 | ...... | ...... | ¹17 | ...... | ...... | ...... | ...... | 18 | 1,028 | 2,045 | 1.042 | ...... |
| | 3,000 | 1.02 | ...... | 15 | 82 | 82 | 1,600 | 1,600 | 16 | 1,021 | 3,066 | 1.058 | 86.8 |
| | 4,000 | 2.08 | ...... | 13 | 81 | 81 | 1,600 | 1,600 | 14 | 1,014 | 4,080 | 1.072 | 86.8 |
| | 5,000 | 3.24 | ...... | 10 | 80 | 80 | 1,595 | 1,595 | 11.5 | 1,005 | 5,085 | 1.087 | 87 |
| | 6,000 | 4.35 | ...... | 8 | 80 | 80 | 1,590 | 1,590 | 2 | 996 | 6,081 | 1.102 | 88.2 |
| | 7,000 | 5.53 | ...... | 7 | 81 | 81 | 1,600 | 1,600 | 7.5 | 991 | 7,072 | 1.12 | 90.7 |
| | 8,000 | 7.17 | ...... | 5 | 81 | 81 | 1,600 | 1,600 | 6 | 986 | 8,058 | 1.135 | 92 |
| | 9,000 | 8.49 | ...... | 3 | 80 | 80 | 1,600 | 1,600 | 4 | 979 | 9,037 | 1.152 | 92.2 |
| | 10,000 | 10.38 | ...... | 1 | 80 | 80 | 1,590 | 1,590 | 2 | 972 | 10,009 | 1.169 | 93.5 |

¹ Temperature by interpolation.

Remarks:

Barometer reading on ground—29.3" merc.
Air temperature on ground—20° C.
Weight of machine, including pilot, observer, military load, and ballast—3,700 pounds.
Thrust on ground 1,410 pounds at 1,480 r. p. m.

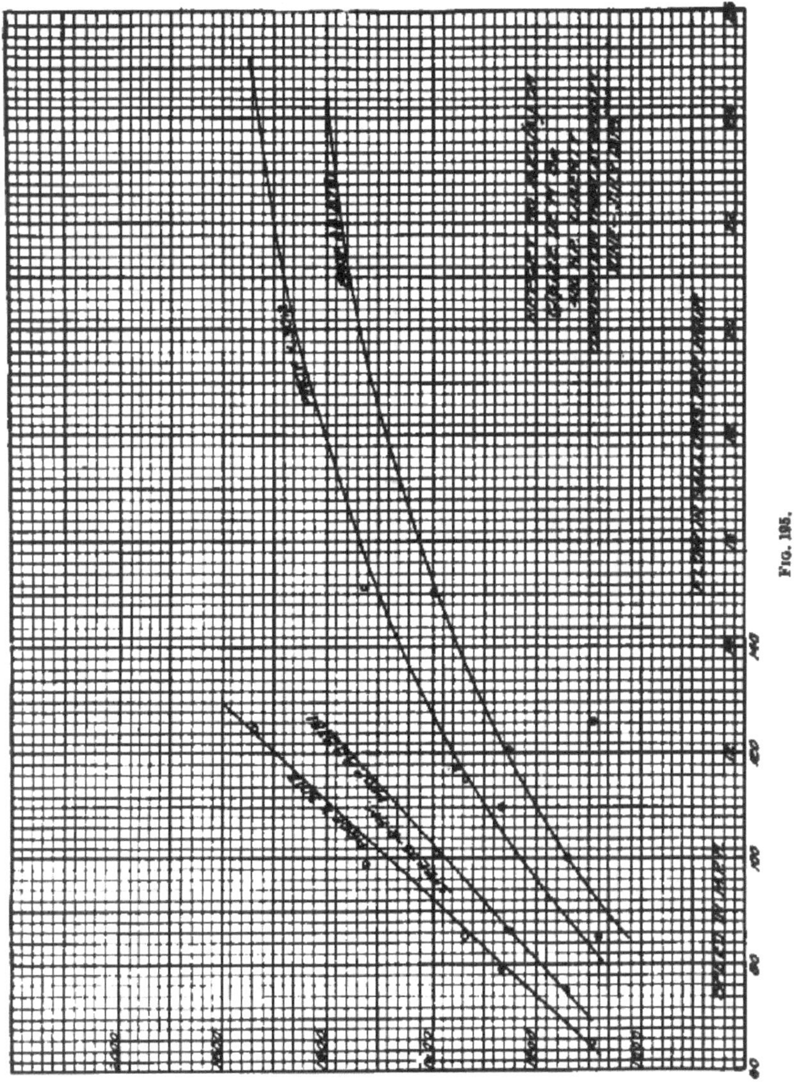

Fig. 195.

II. This test should be repeated at altitude of 5,000, 10,000, and 15,000 feet.

The data from a typical flow-meter test is given on page 326, figure 195.

*Analysis of results.*—After applying correction to all the instrument readings taken from the climbing tests, curves showing the time required to reach various altitudes may be plotted. From these

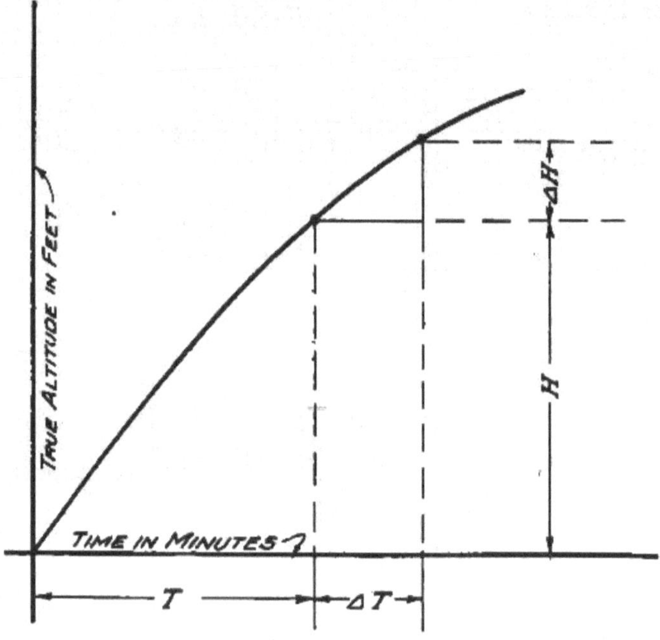


FIG. 196.

curves the rate of climb at various altitudes may be found by drawing a tangent to the curve of time against altitude at various altitudes. (See fig. 196.)

The rate of climb at any altitude then will be $\frac{\Delta H}{\Delta T}$, as shown in figure 196.

To be absolutely comparative, the rate of climb for two propellers at a given air speed must be found not only at the same air density but at the same air pressure as well. This is true because the engine horsepower depends only on the pressure, while the power required to drive the plane depends on the air density, hence both on the

ENGINE R.P.M.
RATE OF CLIMB
LEVEL R.P.M.
CLIMB R.P.M.
LEVEL M.P.H.
CLIMB M.P.H.
CLIMB
TRUE ALTITUDE FEET
TIME MINUTES
AIR SPEED M.P.H.
FEET PER MINUTE
FLIGHT CHARACTERISTICS FOR

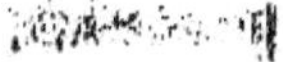

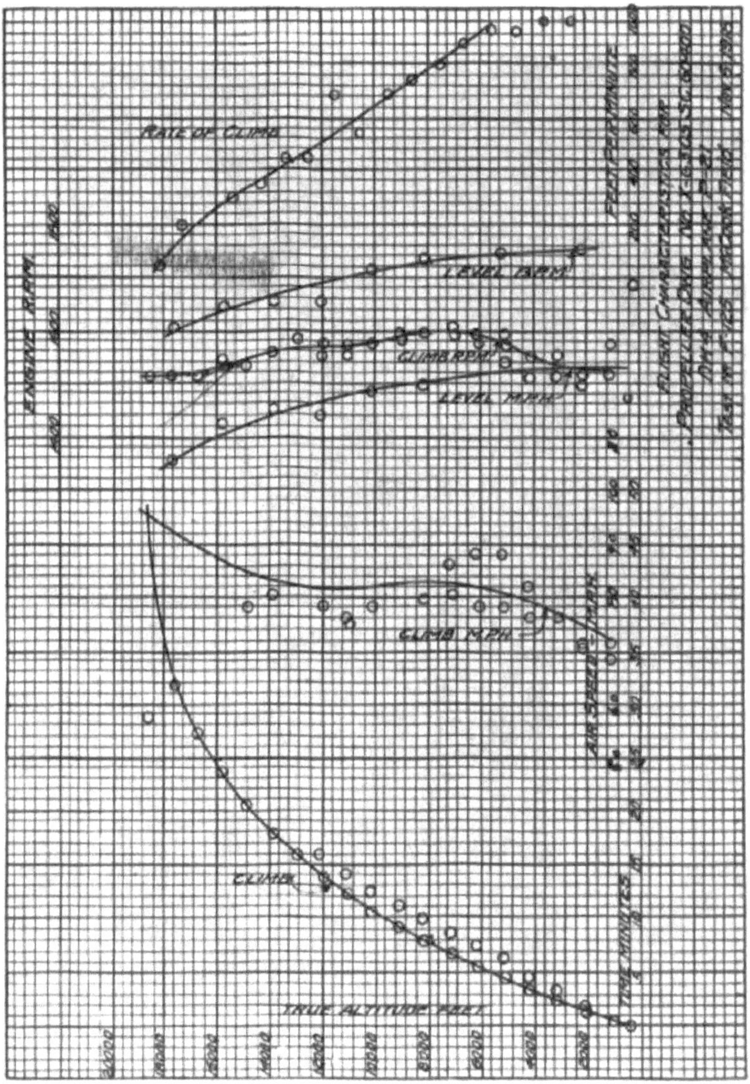

FIG. 197.

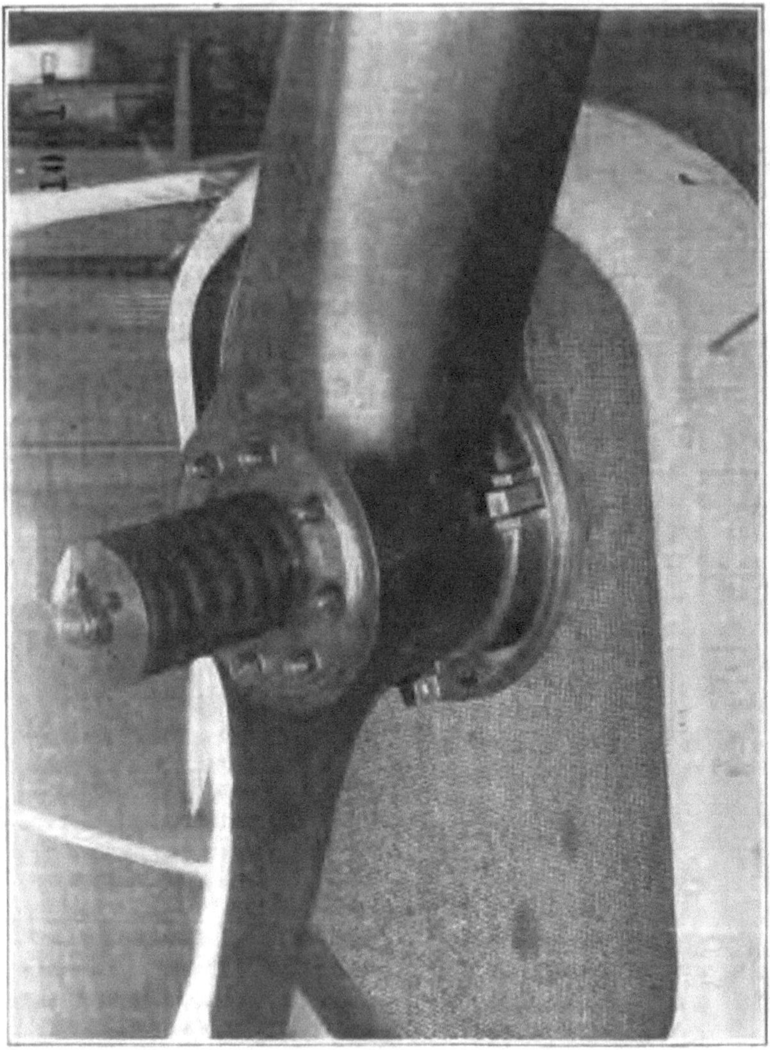

FIG. 198.

FIG. 199.

pressure of the air and its temperature. For this reason it is apparent that it is necessary to run comparative tests on the same day and as nearly as possible at the same time of day.

The results obtained from a typical test are shown in the charts given in figure 197.

*Thrust meter.*—In order to check up the calculations for both propeller thrust and resistance of the airplane, it is desirable to actually measure the thrust of the propeller while the airplane is in flight.

In order to accomplish this purpose, the thrust meter shown on page 332, figure 198, was designed at McCook Field.

By examining the photograph it can be seen that a large helical spring absorbs the thrust of the propeller, the amount of movement of the spring being proportional to the propeller thrust. Anyone interested in the details of this instrument and test may obtain same

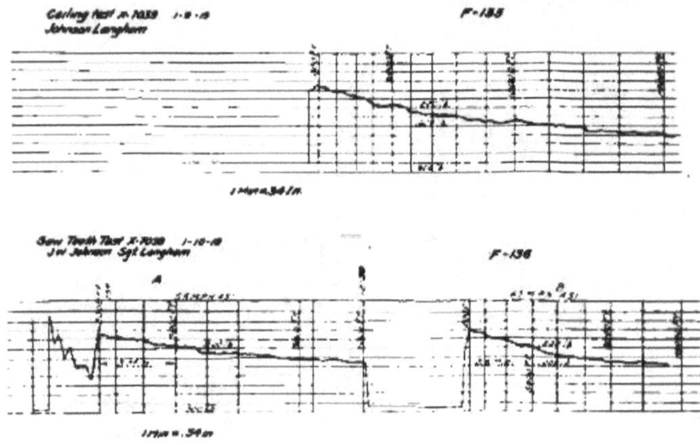

Fig. 200.

from McCook Field Report, Serial No. 133, and McCook Field Report Serial No. 880.

The amount of movement of the spring is recorded on a recording drum shown on page 333, figure 199.

Two sample thrust records are shown in figure 200, page 334.

The result of the series of tests on the propeller made from drawing 37593 is shown on page 335, figure 201.

A complete explanation of the method of obtaining these curves would go beyond the scope of this book.

Figure 202, page 336, shows the efficiency of the propeller as obtained from the thrust meter compared to the computed efficiency figured out from wind tunnel tests. The actual thrust-meter efficiency is a little higher than the computed efficiency, owing to the interference of the airplane fuselage, which increases the propeller efficiency slightly.

Thrust = T ρ/g V² D²
ρ/g = Density of air in mass units
V = Velocity of plane
D = Diameter of Propeller
T = Non-dimensional coefficient
Thrustmeter Tests on P-31
V/ND

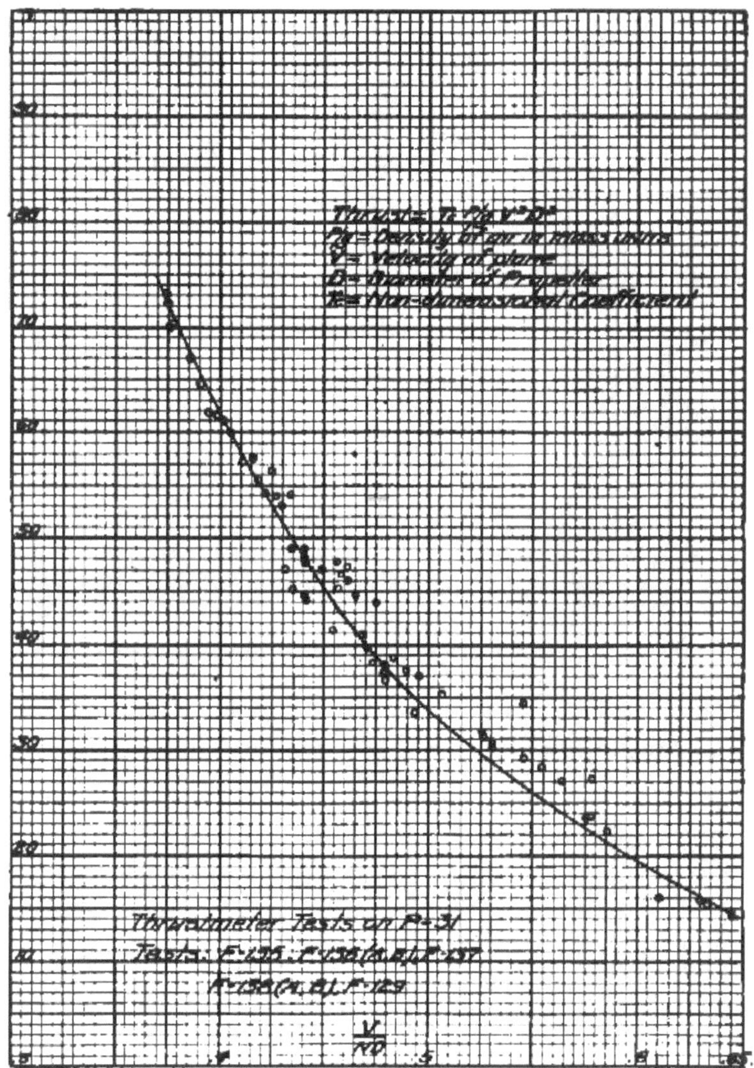

FIG. 201.

80
70
60
50
40
30
20
10
Percent Efficiency
EFFICIENCY CONSTANTS FROM THRUST METER TESTS
COMPUTED EFFICIENCY
V/ND
.35
.40
.45
.50
.55
.60
.65
.70

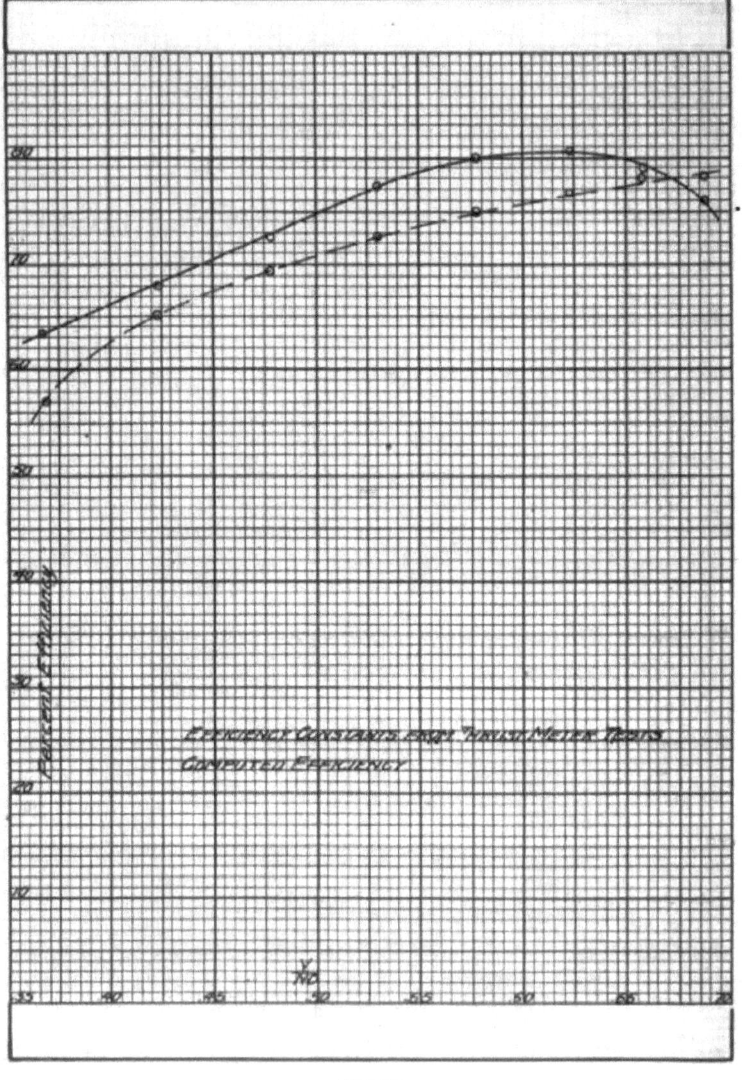

FIG. 202.

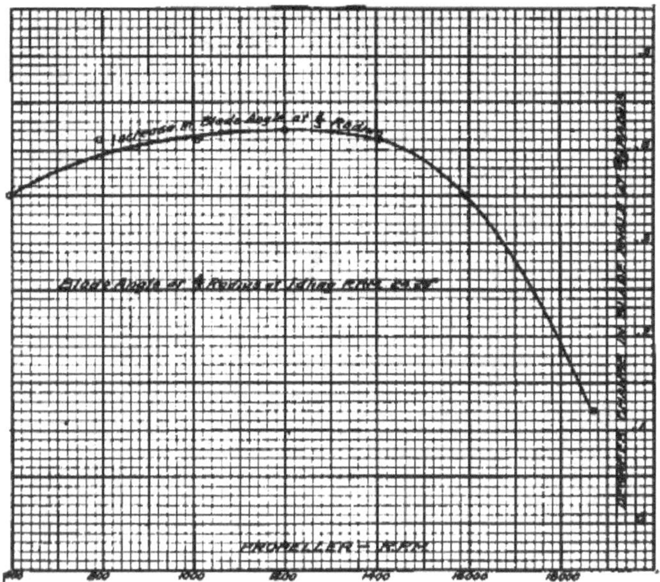


Characteristics of A. E. D. propeller dwg. No. X-13855 insp. No. O-843, S. C. No. 105228 from ground tests. McCook field April 28, 1919. Destructive whirling test No. 213.

FIG. 203.

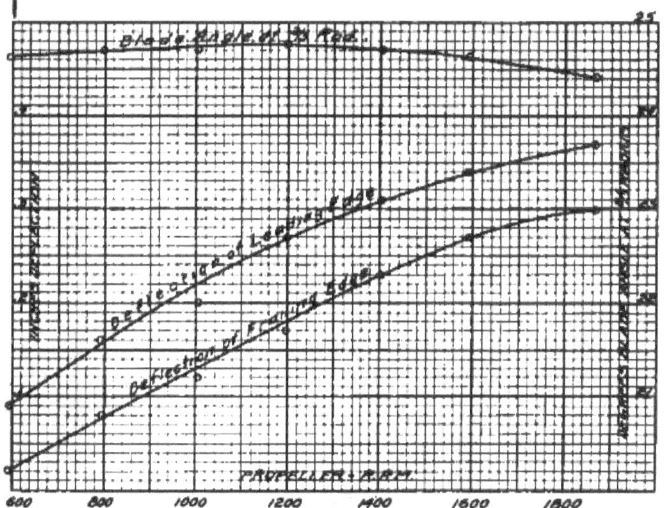


Characteristics of A. E. D. propeller dwg. No. X-13855 insp. No. O-843, S. C. No. 105228 from ground tests. McCook field April 28, 1919. Destructive whirling test No. 213.

FIG. 204.

Zeitfracht Medien GmbH
Ferdinand-Jühlke-Straße 7
99095 Erfurt, Deutschland
produktsicherheit@kolibri360.de